Inequalities

Zdravko Cvetkovski

Inequalities

Theorems, Techniques and Selected Problems

Dipl. Math. Zdravko Cvetkovski
Informatics Department
European University-Republic of Macedonia
Skopje, Macedonia
zdrcvet@gmail.com

ISBN 978-3-642-23791-1 e-ISBN 978-3-642-23792-8
DOI 10.1007/978-3-642-23792-8
Springer Heidelberg Dordrecht London New York

Library of Congress Control Number: 2011942926

Mathematics Subject Classification (2010): 26D20, 97U40, 97Axx

Printed on acid-free paper

Springer is part of Springer Science+Business Media (www.springer.com)

Dedicated with great respect
to the memory of Prof. Ilija Janev

Preface

This book has resulted from my extensive work with talented students in Macedonia, as well as my engagement in the preparation of Macedonian national teams for international competitions. The book is designed and intended for all students who wish to expand their knowledge related to the theory of inequalities and those fascinated by this field. The book could be of great benefit to all regular high school teachers and trainers involved in preparing students for national and international mathematical competitions as well. But first and foremost it is written for students—participants of all kinds of mathematical contests.

The material is written in such a way that it starts from elementary and basic inequalities through their application, up to mathematical inequalities requiring much more sophisticated knowledge. The book deals with almost all the important inequalities used as apparatus for proving more complicated inequalities, as well as several methods and techniques that are part of the apparatus for proving inequalities most commonly encountered in international mathematics competitions of higher rank. Most of the theorems and corollaries are proved, but some of them are not proved since they are easy and they are left to the reader, or they are too complicated for high school students.

As an integral part of the book, following the development of the theory in each section, solved examples have been included—a total of 175 in number—all intended for the student to acquire skills for practical application of previously adopted theory. Also should emphasize that as a final part of the book an extensive collection of 310 "high quality" solved problems has been included, in which various types of inequalities are developed. Some of them are mine, while the others represent inequalities assigned as tasks in national competitions and national olympiads as well as problems given in team selection tests for international competitions from different countries.

I have made every effort to acknowledge the authors of certain problems; therefore at the end of the book an index of the authors of some problems has been included, and I sincerely apologize to anyone who is missing from the list, since any omission is unintentional.

My great honour and duty is to express my deep gratitude to my colleagues Mirko Petrushevski and Đorđe Baralić for proofreading and checking the manuscript, so that with their remarks and suggestions, the book is in its present form. Also I want to thank my wife Maja and my lovely son Gjorgji for all their love, encouragement and support during the writing of this book.

There are many great books about inequalities. But I truly hope and believe that this book will contribute to the development of our talented students—future national team members of our countries at international competitions in mathematics, as well as to upgrade their knowledge.

Despite my efforts there may remain some errors and mistakes for which I take full responsibility. There is always the possibility for improvement in the presentation of the material and removing flaws that surely exist. Therefore I should be grateful for any well-intentioned remarks and criticisms in order to improve this book.

Skopje Zdravko Cvetkovski

Contents

Chapter 1
Basic (Elementary) Inequalities and Their Application

There are many trivial facts which are the basis for proving inequalities. Some of them are as follows:

1. If $x \geq y$ and $y \geq z$ then $x \geq z$, for any $x, y, z \in \mathbb{R}$.
2. If $x \geq y$ and $a \geq b$ then $x + a \geq y + b$, for any $x, y, a, b \in \mathbb{R}$.
3. If $x \geq y$ then $x + z \geq y + z$, for any $x, y, z \in \mathbb{R}$.
4. If $x \geq y$ and $a \geq b$ then $xa \geq yb$, for any $x, y \in \mathbb{R}^+$ or $a, b \in \mathbb{R}^+$.
5. If $x \in \mathbb{R}$ then $x^2 \geq 0$, with equality if and only if $x = 0$. More generally, for $A_i \in \mathbb{R}^+$ and $x_i \in \mathbb{R}, i = 1, 2, \ldots, n$ holds $A_1x_1^2 + A_2x_2^2 + \cdots + A_nx_n^2 \geq 0$, with equality if and only if $x_1 = x_2 = \cdots = x_n = 0$.

These properties are obvious and simple, but are a powerful tool in proving inequalities, particularly *Property 5*, which can be used in many cases.

We'll give a few examples that will illustrate the strength of *Property 5*.

Firstly we'll prove few "elementary" inequalities that are necessary for a complete and thorough upgrade of each student who is interested in this area.

To prove these inequalities it is sufficient to know elementary inequalities that can be used in a certain part of the proof of a given inequality, but in the early stages, just basic operations are used.

The following examples, although very simple, are the basis for what follows later. Therefore I recommend the reader pay particular attention to these examples, which are necessary for further upgrading.

Exercise 1.1 Prove that for any real number $x > 0$, the following inequality holds

$$x + \frac{1}{x} \geq 2.$$

Solution From the obvious inequality $(x - 1)^2 \geq 0$ we have

$$x^2 - 2x + 1 \geq 0 \quad \Leftrightarrow \quad x^2 + 1 \geq 2x,$$

and since $x > 0$ if we divide by x we get the desired inequality. Equality occurs if and only if $x - 1 = 0$, i.e. $x = 1$.

Z. Cvetkovski, *Inequalities*,
DOI 10.1007/978-3-642-23792-8_1, © Springer-Verlag Berlin Heidelberg 2012

Exercise 1.2 Let $a, b \in \mathbb{R}^+$. Prove the inequality

$$\frac{a}{b} + \frac{b}{a} \geq 2.$$

Solution From the obvious inequality $(a-b)^2 \geq 0$ we have

$$a^2 - 2ab + b^2 \geq 0 \quad \Leftrightarrow \quad a^2 + b^2 \geq 2ab \quad \Leftrightarrow \quad \frac{a^2+b^2}{ab} \geq 2 \quad \Leftrightarrow \quad \frac{a}{b} + \frac{b}{a} \geq 2.$$

Equality occurs if and only if $a - b = 0$, i.e. $a = b$.

Exercise 1.3 (Nesbitt's inequality) Let a, b, c be positive real numbers. Prove the inequality

$$\frac{a}{b+c} + \frac{b}{c+a} + \frac{c}{a+b} \geq \frac{3}{2}.$$

Solution According to Exercise 1.2 it is clear that

$$\frac{a+b}{b+c} + \frac{b+c}{a+b} + \frac{a+c}{c+b} + \frac{c+b}{a+c} + \frac{b+a}{a+c} + \frac{a+c}{b+a} \geq 2+2+2=6. \qquad (1.1)$$

Let us rewrite inequality (1.1) as follows

$$\left(\frac{a+b}{b+c} + \frac{a+c}{c+b}\right) + \left(\frac{c+b}{a+c} + \frac{b+a}{a+c}\right) + \left(\frac{b+c}{a+b} + \frac{a+c}{b+a}\right) \geq 6,$$

i.e.

$$\frac{2a}{b+c} + 1 + \frac{2b}{c+a} + 1 + \frac{2c}{a+b} + 1 \geq 6$$

or

$$\frac{a}{b+c} + \frac{b}{c+a} + \frac{c}{a+b} \geq \frac{3}{2},$$

a s required.

Equality occurs if and only if $\frac{a+b}{b+c} = \frac{b+c}{a+b}$, $\frac{a+c}{c+b} = \frac{c+b}{a+c}$, $\frac{b+a}{a+c} = \frac{a+c}{b+a}$, from where easily we deduce $a = b = c$.

The following inequality is very simple but it has a very important role, as we will see later.

Exercise 1.4 Let $a, b, c \in \mathbb{R}$. Prove the inequality

$$a^2 + b^2 + c^2 \geq ab + bc + ca.$$

Solution Since $(a-b)^2 + (b-c)^2 + (c-a)^2 \geq 0$ we deduce

$$2(a^2 + b^2 + c^2) \geq 2(ab + bc + ca) \quad \Leftrightarrow \quad a^2 + b^2 + c^2 \geq ab + bc + ca.$$

Equality occurs if and only if $a = b = c$.

As a consequence of the previous inequality we get following problem.

Exercise 1.5 Let $a, b, c \in \mathbb{R}$. Prove the inequalities

$$3(ab+bc+ca) \le (a+b+c)^2 \le 3(a^2+b^2+c^2).$$

Solution We have

$$\begin{aligned}3(ab+bc+ca) &= ab+bc+ca+2(ab+bc+ca)\\ &\le a^2+b^2+c^2+2(ab+bc+ca) = (a+b+c)^2\\ &= a^2+b^2+c^2+2(ab+bc+ca)\\ &\le a^2+b^2+c^2+2(a^2+b^2+c^2) = 3(a^2+b^2+c^2).\end{aligned}$$

Equality occurs if and only if $a = b = c$.

Exercise 1.6 Let $x, y, z > 0$ be real numbers such that $x + y + z = 1$. Prove that

$$\sqrt{6x+1}+\sqrt{6y+1}+\sqrt{6z+1} \le 3\sqrt{3}.$$

Solution Let $\sqrt{6x+1} = a$, $\sqrt{6y+1} = b$, $\sqrt{6z+1} = c$.
Then

$$a^2+b^2+c^2 = 6(x+y+z)+3 = 9.$$

Therefore

$$(a+b+c)^2 \le 3(a^2+b^2+c^2) = 27, \quad \text{i.e.} \quad a+b+c \le 3\sqrt{3}.$$

Exercise 1.7 Let $a, b, c \in \mathbb{R}$. Prove the inequality

$$a^4+b^4+c^4 \ge abc(a+b+c).$$

Solution By Exercise 1.4 we have that: If $x, y, z \in \mathbb{R}$ then

$$x^2+y^2+z^2 \ge xy+yz+zx.$$

Therefore

$$\begin{aligned}a^4+b^4+c^4 &\ge a^2b^2+b^2c^2+c^2a^2 = (ab)^2+(bc)^2+(ca)^2\\ &\ge (ab)(bc)+(bc)(ca)+(ca)(ab) = abc(a+b+c).\end{aligned}$$

Exercise 1.8 Let $a, b, c \in \mathbb{R}$ such that $a+b+c \ge abc$. Prove the inequality

$$a^2+b^2+c^2 \ge \sqrt{3}abc.$$

Solution We have

$$(a^2+b^2+c^2)^2 = a^4+b^4+c^4+2a^2b^2+2b^2c^2+2c^2a^2$$
$$= a^4+b^4+c^4+a^2(b^2+c^2)+b^2(c^2+a^2)+c^2(a^2+b^2). \quad (1.2)$$

By Exercise 1.7, it follows that

$$a^4+b^4+c^4 \geq abc(a+b+c). \quad (1.3)$$

Also

$$b^2+c^2 \geq 2bc, \qquad c^2+a^2 \geq 2ca, \qquad a^2+b^2 \geq 2ab. \quad (1.4)$$

Now by (1.2), (1.3) and (1.4) we deduce

$$(a^2+b^2+c^2)^2 \geq abc(a+b+c)+2a^2bc+2b^2ac+2c^2ab$$
$$= abc(a+b+c)+2abc(a+b+c) = 3abc(a+b+c). \quad (1.5)$$

Since $a+b+c \geq abc$ in (1.5) we have

$$(a^2+b^2+c^2)^2 \geq 3abc(a+b+c) \geq 3(abc)^2,$$

i.e.

$$a^2+b^2+c^2 \geq \sqrt{3}abc.$$

Equality occurs if and only if $a=b=c=\sqrt{3}$.

Exercise 1.9 Let $a, b, c > 1$ be real numbers. Prove the inequality

$$abc+\frac{1}{a}+\frac{1}{b}+\frac{1}{c} > a+b+c+\frac{1}{abc}.$$

Solution Since $a, b, c > 1$ we have $a > \frac{1}{b}, b > \frac{1}{c}, c > \frac{1}{a}$, i.e.

$$\left(a-\frac{1}{b}\right)\left(b-\frac{1}{c}\right)\left(c-\frac{1}{a}\right) > 0.$$

After multiplying we get the required inequality.

Exercise 1.10 Let a, b, c, d be real numbers such that $a^4+b^4+c^4+d^4=16$. Prove the inequality

$$a^5+b^5+c^5+d^5 \leq 32.$$

Solution We have $a^4 \leq a^4+b^4+c^4+d^4=16$, i.e. $a \leq 2$ from which it follows that $a^4(a-2) \leq 0$, i.e. $a^5 \leq 2a^4$.

Similarly we obtain $b^5 \leq 2b^4$, $c^5 \leq 2c^4$ and $d^5 \leq 2d^4$.

Hence

$$a^5 + b^5 + c^5 + d^5 \leq 2(a^4 + b^4 + c^4 + d^4) = 32.$$

Equality occurs iff $a = 2, b = c = d = 0$ (up to permutation).

Exercise 1.11 Prove that for any real number x the following inequality holds

$$x^{12} - x^9 + x^4 - x + 1 > 0.$$

Solution We consider two cases: $x < 1$ and $x \geq 1$.

(1) Let $x < 1$. We have

$$x^{12} - x^9 + x^4 - x + 1 = x^{12} + (x^4 - x^9) + (1 - x).$$

Since $x < 1$ we have $1 - x > 0$ and $x^4 > x^9$, i.e. $x^4 - x^9 > 0$, so in this case

$$x^{12} - x^9 + x^4 - x + 1 > 0,$$

i.e. the desired inequality holds.

(2) For $x \geq 1$ we have

$$\begin{aligned} x^{12} - x^9 + x^4 - x + 1 &= x^8(x^4 - x) + (x^4 - x) + 1 \\ &= (x^4 - x)(x^8 + 1) + 1 = x(x^3 - 1)(x^8 + 1) + 1. \end{aligned}$$

Since $x \geq 1$ we have $x^3 \geq 1$, i.e. $x^3 - 1 \geq 0$.

Therefore

$$x^{12} - x^9 + x^4 - x + 1 > 0,$$

and the problem is solved.

Exercise 1.12 Prove that for any real number x the following inequality holds

$$2x^4 + 1 \geq 2x^3 + x^2.$$

Solution We have

$$\begin{aligned} 2x^4 + 1 - 2x^3 - x^2 &= 1 - x^2 - 2x^3(1 - x) = (1 - x)(1 + x) - 2x^3(1 - x) \\ &= (1 - x)(x + 1 - 2x^3) = (1 - x)(x(1 - x^2) + 1 - x^3) \\ &= (1 - x)\Big(x(1 - x)(1 + x) + (1 - x)(1 + x + x^2)\Big) \\ &= (1 - x)\Big((1 - x)(x(1 + x) + 1 + x + x^2)\Big) \\ &= (1 - x)^2((x + 1)^2 + x^2) \geq 0. \end{aligned}$$

Equality occurs if and only if $x = 1$.

Exercise 1.13 Let $x, y \in \mathbb{R}$. Prove the inequality

$$x^4 + y^4 + 4xy + 2 \geq 0.$$

Solution We have

$$\begin{aligned} x^4 + y^4 + 4xy + 2 &= (x^4 - 2x^2y^2 + y^4) + (2x^2y^2 + 4xy + 2) \\ &= (x^2 - y^2)^2 + 2(xy + 1)^2 \geq 0, \end{aligned}$$

as desired.

Equality occurs if and only if $x = 1, y = -1$ or $x = -1, y = 1$.

Exercise 1.14 Prove that for any real numbers x, y, z the following inequality holds

$$x^4 + y^4 + z^2 + 1 \geq 2x(xy^2 - x + z + 1).$$

Solution We have

$$\begin{aligned} &x^4 + y^4 + z^2 + 1 - 2x(xy^2 - x + z + 1) \\ &\quad = (x^4 - 2x^2y^2 + x^4) + (z^2 - 2xz + x^2) + (x^2 - 2x + 1) \\ &\quad = (x^2 - y^2)^2 + (x - z)^2 + (x - 1)^2 \geq 0, \end{aligned}$$

from which we get the desired inequality.

Equality occurs if and only if $x = y = z = 1$ or $x = z = 1, y = -1$.

Exercise 1.15 Let x, y, z be positive real numbers such that $x + y + z = 1$. Prove the inequality

$$xy + yz + 2zx \leq \frac{1}{2}.$$

Solution We will prove that

$$2xy + 2yz + 4zx \leq (x + y + z)^2,$$

from which, since $x + y + z = 1$ we'll obtain the required inequality.

The last inequality is equivalent to

$$x^2 + y^2 + z^2 - 2zx \geq 0, \quad \text{i.e.} \quad (x - z)^2 + y^2 \geq 0,$$

which is true.

Equality occurs if and only if $x = z$ and $y = 0$, i.e. $x = z = \frac{1}{2}, y = 0$.

Exercise 1.16 Let $a, b \in \mathbb{R}^+$. Prove the inequality

$$a^2 + b^2 + 1 > a\sqrt{b^2 + 1} + b\sqrt{a^2 + 1}.$$

Solution From the obvious inequality

$$(a - \sqrt{b^2+1})^2 + (b - \sqrt{a^2+1})^2 \geq 0, \tag{1.6}$$

we get the desired result.

Equality occurs if and only if

$$a = \sqrt{b^2+1} \quad \text{and} \quad b = \sqrt{a^2+1}, \quad \text{i.e.} \quad a^2 = b^2+1 \quad \text{and} \quad b^2 = a^2+1,$$

which is impossible, so in (1.6) we have strictly inequality.

Exercise 1.17 Let $x, y, z \in \mathbb{R}^+$ such that $x+y+z=3$. Prove the inequality

$$\sqrt{x} + \sqrt{y} + \sqrt{z} \geq xy + yz + zx.$$

Solution We have

$$3(x+y+z) = (x+y+z)^2 = x^2+y^2+z^2+2(xy+yz+zx).$$

Hence it follows that

$$xy+yz+zx = \frac{1}{2}(3x - x^2 + 3y - y^2 + 3z - z^2).$$

Then

$$\begin{aligned}
&\sqrt{x} + \sqrt{y} + \sqrt{z} - (xy+yz+zx) \\
&\quad = \sqrt{x} + \sqrt{y} + \sqrt{z} + \frac{1}{2}(x^2 - 3x + y^2 - 3y + z^2 - 3z) \\
&\quad = \frac{1}{2}((x^2 - 3x + 2\sqrt{x}) + (y^2 - 3y + 2\sqrt{y}) + (z^2 - 3z + 2\sqrt{z})) \\
&\quad = \frac{1}{2}(\sqrt{x}(\sqrt{x}-1)^2(\sqrt{x}+2) + \sqrt{y}(\sqrt{y}-1)^2(\sqrt{y}+2) \\
&\qquad + \sqrt{z}(\sqrt{z}-1)^2(\sqrt{z}+2)) \geq 0,
\end{aligned}$$

i.e.

$$\sqrt{x} + \sqrt{y} + \sqrt{z} \geq xy + yz + zx.$$

Chapter 2
Inequalities Between Means (with Two and Three Variables)

In this section, we'll first mention and give a proof of *inequalities between means*, which are of particular importance for a full upgrade of the student in solving tasks in this area. It ought to be mentioned that in this section we will discuss the case that treats two or three variables, while the general case will be considered later in Chap. 5.

Theorem 2.1 *Let $a, b \in \mathbb{R}^+$, and let us denote*

$$QM = \sqrt{\frac{a^2+b^2}{2}}, \qquad AM = \frac{a+b}{2}, \qquad GM = \sqrt{ab} \quad \text{and} \quad HM = \frac{2}{\frac{1}{a}+\frac{1}{b}}.$$

Then

$$QM \geq AM \geq GM \geq HM. \tag{2.1}$$

Equalities occur if and only if $a = b$.

Proof Firstly we'll show that $QM \geq AM$.

For $a, b \in \mathbb{R}^+$ we have

$$\begin{aligned}
&(a-b)^2 \geq 0 \\
&\Leftrightarrow \quad a^2+b^2 \geq 2ab \quad \Leftrightarrow \quad 2(a^2+b^2) \geq a^2+b^2+2ab \\
&\Leftrightarrow \quad 2(a^2+b^2) \geq (a+b)^2 \quad \Leftrightarrow \quad \frac{a^2+b^2}{2} \geq \left(\frac{a+b}{2}\right)^2 \\
&\Leftrightarrow \quad \sqrt{\frac{a^2+b^2}{2}} \geq \frac{a+b}{2}.
\end{aligned}$$

Equality holds if and only if $a - b = 0$, i.e. $a = b$.

Z. Cvetkovski, *Inequalities*,
DOI 10.1007/978-3-642-23792-8_2, © Springer-Verlag Berlin Heidelberg 2012

Furthermore, for $a, b \in \mathbb{R}^+$ we have

$$(\sqrt{a}-\sqrt{b})^2 \geq 0 \quad \Leftrightarrow \quad a+b-2\sqrt{ab} \geq 0 \quad \Leftrightarrow \quad \frac{a+b}{2} \geq \sqrt{ab}.$$

So $AM \geq GM$, with equality if and only if

$$\sqrt{a}-\sqrt{b}=0, \quad \text{i.e.} \quad a=b.$$

Finally we'll show that

$$GM \geq HM, \quad \text{i.e.} \quad \sqrt{ab} \geq \frac{2}{\frac{1}{a}+\frac{1}{b}}.$$

We have

$$(\sqrt{a}-\sqrt{b})^2 \geq 0 \quad \Leftrightarrow \quad a+b \geq 2\sqrt{ab} \quad \Leftrightarrow \quad 1 \geq \frac{2\sqrt{ab}}{a+b} \quad \Leftrightarrow \quad \sqrt{ab} \geq \frac{2ab}{a+b}$$

$$\Leftrightarrow \quad \sqrt{ab} \geq \frac{2}{\frac{1}{a}+\frac{1}{b}}.$$

Equality holds if and only if $\sqrt{a}-\sqrt{b}=0$, i.e. $a=b$. □

Remark The numbers QM, AM, GM and HM are called the quadratic, arithmetic, geometric and harmonic mean for the numbers a and b, respectively; the inequalities (2.1) are called *mean inequalities.*

These inequalities usually well be use in the case when $a, b \in \mathbb{R}^+$.

Also similarly we can define the quadratic, arithmetic, geometric and harmonic mean for three variables as follows:

$$QM=\sqrt{\frac{a^2+b^2+c^2}{3}}, \qquad AM=\frac{a+b+c}{3}, \qquad GM=\sqrt[3]{abc} \quad \text{and}$$

$$HM=\frac{3}{\frac{1}{a}+\frac{1}{b}+\frac{1}{c}}.$$

Analogous to Theorem 2.1, with three variables we have the following theorem.

Theorem 2.2 *Let $a, b, c \in \mathbb{R}^+$, and let us denote*

$$QM=\sqrt{\frac{a^2+b^2+c^2}{3}}, \qquad AM=\frac{a+b+c}{3}, \qquad GM=\sqrt[3]{abc} \quad \textit{and}$$

$$HM=\frac{3}{\frac{1}{a}+\frac{1}{b}+\frac{1}{c}}.$$

Then

$$QM \geq AM \geq GM \geq HM.$$

Equalities occur if and only if $a = b = c$.

Over the next few exercises we will see how these inequalities can be put in use.

Exercise 2.1 Let $x, y, z \in \mathbb{R}^+$ such that $x + y + z = 1$. Prove the inequality

$$\frac{xy}{z} + \frac{yz}{x} + \frac{zx}{y} \geq 1.$$

When does equality occur?

Solution We have

$$\frac{xy}{z} + \frac{yz}{x} + \frac{zx}{y} = \frac{1}{2}\left(\frac{xy}{z} + \frac{yz}{x}\right) + \frac{1}{2}\left(\frac{yz}{x} + \frac{zx}{y}\right) + \frac{1}{2}\left(\frac{zx}{y} + \frac{xy}{z}\right). \tag{2.2}$$

Since $AM \geq GM$ we have

$$\frac{1}{2}\left(\frac{xy}{z} + \frac{yz}{x}\right) \geq \sqrt{\frac{xy}{z}\frac{yz}{x}} = y.$$

Analogously we get

$$\frac{1}{2}\left(\frac{yz}{x} + \frac{zx}{y}\right) \geq z \quad \text{and} \quad \frac{1}{2}\left(\frac{zx}{y} + \frac{xy}{z}\right) \geq x.$$

Adding these three inequalities we obtain

$$\frac{xy}{z} + \frac{yz}{x} + \frac{zx}{y} \geq x + y + z = 1.$$

Equality holds if and only if $\frac{xy}{z} = \frac{yz}{x} = \frac{zx}{y}$, i.e. $x = y = z$. Since $x + y + z = 1$ we get that equality holds iff $x = y = z = 1/3$.

Exercise 2.2 Let $x, y, z > 0$ be real numbers. Prove the inequality

$$\frac{x^2 - z^2}{y + z} + \frac{y^2 - x^2}{z + x} + \frac{z^2 - y^2}{x + y} \geq 0.$$

When does equality occur?

Solution Let $a = x + y, b = y + z, c = z + x$.

Then clearly $a, b, c > 0$, and it follows that

$$\frac{x^2 - z^2}{y + z} + \frac{y^2 - x^2}{z + x} + \frac{z^2 - y^2}{x + y} = \frac{(a - b)c}{b} + \frac{(b - c)a}{c} + \frac{(c - a)b}{a}$$
$$= \frac{ac}{b} + \frac{ba}{c} + \frac{cb}{a} - (a + b + c). \tag{2.3}$$

Similarly as in Exercise 2.1, we can prove that for any $a, b, c > 0$

$$\frac{ac}{b} + \frac{ba}{c} + \frac{cb}{a} \geq a + b + c. \tag{2.4}$$

By (2.3) and (2.4) we get

$$\frac{x^2 - z^2}{y + z} + \frac{y^2 - x^2}{z + x} + \frac{z^2 - y^2}{x + y}$$
$$= \frac{ac}{b} + \frac{ba}{c} + \frac{cb}{a} - (a + b + c) \geq (a + b + c) - (a + b + c) = 0.$$

Equality occurs iff we have equality in (2.4), i.e. $a = b = c$, from which we deduce that $x = y = z$.

Exercise 2.3 Let $a, b, c \in \mathbb{R}^+$. Prove the inequality

$$\left(a + \frac{1}{b}\right)\left(b + \frac{1}{c}\right)\left(c + \frac{1}{a}\right) \geq 8.$$

When does equality occur?

Solution Applying $AM \geq GM$ we get

$$a + \frac{1}{b} \geq 2\sqrt{\frac{a}{b}}, \qquad b + \frac{1}{c} \geq 2\sqrt{\frac{b}{c}}, \qquad c + \frac{1}{a} \geq 2\sqrt{\frac{c}{a}}.$$

Therefore

$$\left(a + \frac{1}{b}\right)\left(b + \frac{1}{c}\right)\left(c + \frac{1}{a}\right) \geq 8\sqrt{\frac{a}{b}} \cdot \sqrt{\frac{b}{c}} \cdot \sqrt{\frac{c}{a}} = 8.$$

Equality occurs if and only if $a = \frac{1}{b}, b = \frac{1}{c}, c = \frac{1}{a}$ i.e. $a = \frac{1}{b} = c = \frac{1}{a}$, from which we deduce that $a = b = c = 1$.

Exercise 2.4 Let a, b, c be positive real numbers. Prove the inequality

$$\frac{ab}{a + b + 2c} + \frac{bc}{b + c + 2a} + \frac{ca}{c + a + 2b} \leq \frac{a + b + c}{4}.$$

Solution Since $AM \geq HM$ we have

$$\frac{ab}{a+b+2c} = \frac{ab}{(a+c)+(b+c)} \leq \frac{ab}{4}\left(\frac{1}{a+c} + \frac{1}{b+c}\right).$$

Similarly we get

$$\frac{bc}{b+c+2a} \leq \frac{bc}{4}\left(\frac{1}{a+b} + \frac{1}{a+c}\right) \quad \text{and} \quad \frac{ca}{c+a+2b} \leq \frac{ca}{4}\left(\frac{1}{a+b} + \frac{1}{b+c}\right).$$

By adding these three inequalities we obtain the required inequality.

Exercise 2.5 Let x, y, z be positive real numbers such that $x+y+z=1$. Prove the inequality

$$xy + yz + zx \geq 9xyz.$$

Solution Applying $AM \geq GM$ we get

$$xy + yz + zx = (xy + yz + zx)(x + y + z) \geq 3\sqrt[3]{(xy)(yz)(zx)} \cdot 3\sqrt[3]{xyz} = 9xyz.$$

Equality occur if and only if $x = y = z = \frac{1}{3}$.

Exercise 2.6 Let $a, b, c \in \mathbb{R}^+$ such that $a^2 + b^2 + c^2 = 3$. Prove the inequality

$$\frac{1}{1+ab} + \frac{1}{1+bc} + \frac{1}{1+ca} \geq \frac{3}{2}.$$

Solution Applying $AM \geq HM$ and the inequality $a^2 + b^2 + c^2 \geq ab + bc + ca$, we get

$$\frac{1}{1+ab} + \frac{1}{1+bc} + \frac{1}{1+ca} \geq \frac{9}{3+ab+bc+ca} \geq \frac{9}{3+a^2+b^2+c^2} = \frac{3}{2}.$$

Exercise 2.7 Let a, b, c be positive real numbers. Prove the inequality

$$\sqrt{\frac{a+b}{c}} + \sqrt{\frac{b+c}{a}} + \sqrt{\frac{c+a}{b}} \geq 3\sqrt{2}.$$

Solution We have

$$\begin{aligned}
\sqrt{\frac{a+b}{c}} + \sqrt{\frac{b+c}{a}} + \sqrt{\frac{c+a}{b}} &\overset{A \geq G}{\geq} 3\sqrt[3]{\sqrt{\left(\frac{a+b}{c}\right)\left(\frac{b+c}{a}\right)\left(\frac{c+a}{b}\right)}} \\
&= 3\sqrt[6]{\frac{(a+b)(b+c)(c+a)}{abc}} \\
&\overset{A \geq G}{\geq} 3\sqrt[6]{\frac{2^3\sqrt{ab} \cdot \sqrt{bc} \cdot \sqrt{ca}}{abc}} = 3\sqrt{2}.
\end{aligned}$$

Equality occurs if and only if $a = b = c$.

Exercise 2.8 Let x, y, z be positive real numbers such that $\frac{1}{x} + \frac{1}{y} + \frac{1}{z} = 1$. Prove the inequality

$$(x-1)(y-1)(z-1) \geq 8.$$

Solution The given inequality is equivalent to

$$\left(\frac{x-1}{x}\right)\left(\frac{y-1}{y}\right)\left(\frac{z-1}{z}\right) \geq \frac{8}{xyz}$$

or

$$\left(1-\frac{1}{x}\right)\left(1-\frac{1}{y}\right)\left(1-\frac{1}{z}\right) \geq \frac{8}{xyz}. \tag{2.5}$$

From the initial condition and $AM \geq GM$ we have

$$1-\frac{1}{x} = \frac{1}{y} + \frac{1}{z} \geq 2\sqrt{\frac{1}{y}\frac{1}{z}} = \frac{2}{\sqrt{yz}}.$$

Analogously we obtain $1 - \frac{1}{y} \geq \frac{2}{\sqrt{zx}}$ and $1 - \frac{1}{z} \geq \frac{2}{\sqrt{xy}}$.

If we multiply the last three inequalities we get inequality (2.5), as required. Equality holds if and only if $x = y = z = 3$.

Exercise 2.9 Let $x, y, z \in \mathbb{R}^+$ such that $x + y + z = 1$. Prove the inequality

$$\frac{x^2+y^2}{z} + \frac{y^2+z^2}{x} + \frac{z^2+x^2}{y} \geq 2.$$

Solution We have

$$\begin{aligned}
&\frac{x^2+y^2}{z} + \frac{y^2+z^2}{x} + \frac{z^2+x^2}{y} \\
&\geq 2\frac{xy}{z} + 2\frac{yz}{x} + 2\frac{zx}{y} = 2\left(\frac{xy}{z} + \frac{yz}{x} + \frac{zx}{y}\right) \\
&= 2\left(\frac{1}{2}\left(\frac{xy}{z} + \frac{yz}{x}\right) + \frac{1}{2}\left(\frac{xy}{z} + \frac{zx}{y}\right) + \frac{1}{2}\left(\frac{yz}{x} + \frac{zx}{y}\right)\right) \\
&\geq 2\left(\sqrt{y^2} + \sqrt{x^2} + \sqrt{z^2}\right) = 2(x+y+z) = 2.
\end{aligned}$$

Exercise 2.10 Let $x, y, z \in \mathbb{R}^+$ such that $xyz = 1$. Prove the inequality

$$\frac{x^2+y^2+z^2+xy+yz+zx}{\sqrt{x}+\sqrt{y}+\sqrt{z}} \geq 2.$$

Solution We have

$$\begin{aligned}\frac{x^2+y^2+z^2+xy+yz+zx}{\sqrt{x}+\sqrt{y}+\sqrt{z}} &= \frac{x^2+yz+y^2+zx+z^2+xy}{\sqrt{x}+\sqrt{y}+\sqrt{z}}\\ &\geq \frac{2\sqrt{x^2yz}+2\sqrt{xy^2z}+2\sqrt{xyz^2}}{\sqrt{x}+\sqrt{y}+\sqrt{z}}\\ &= \frac{2(\sqrt{x}+\sqrt{y}+\sqrt{z})}{\sqrt{x}+\sqrt{y}+\sqrt{z}} = 2.\end{aligned}$$

Equality occurs if and only if $x = y = z = 1$.

Exercise 2.11 Let $a, b, c \in \mathbb{R}^+$. Prove the inequalities

$$\frac{9abc}{2(a+b+c)} \leq \frac{ab^2}{a+b}+\frac{bc^2}{b+c}+\frac{ca^2}{c+a} \leq \frac{a^2+b^2+c^2}{2}.$$

Solution Since $AM \geq HM$ and from the well-known inequality

$$ab+bc+ca \leq a^2+b^2+c^2,$$

we get

$$\begin{aligned}\frac{ab^2}{a+b}+\frac{bc^2}{b+c}+\frac{ca^2}{c+a} &= \frac{1}{1/b^2+1/ab}+\frac{1}{1/c^2+1/bc}+\frac{1}{1/a^2+1/ca}\\ &\leq \frac{b^2+ab}{4}+\frac{c^2+bc}{4}+\frac{a^2+ca}{4}\\ &= \frac{a^2+b^2+c^2+ab+bc+ca}{4}\\ &\leq \frac{2(a^2+b^2+c^2)}{4} = \frac{a^2+b^2+c^2}{2}.\end{aligned}$$

It remains to show the left inequality.

Since $AM \geq GM$ we have

$$\frac{ab^2}{a+b}+\frac{bc^2}{b+c}+\frac{ca^2}{c+a} \geq \frac{3abc}{\sqrt[3]{(a+b)(b+c)(c+a)}}.$$

Therefore it suffices to show that

$$\frac{3abc}{\sqrt[3]{(a+b)(b+c)(c+a)}} \geq \frac{9abc}{2(a+b+c)},$$

i.e.

$$2(a+b+c) \geq 3\sqrt[3]{(a+b)(b+c)(c+a)},$$

which is true, since

$$2(a+b+c)=(a+b)+(b+c)+(c+a)\geq 3\sqrt[3]{(a+b)(b+c)(c+a)}.$$

The following exercises shows how we can use *mean inequalities* in a different, non-trivial way.

Exercise 2.12 Prove that for every positive real number a, b, c we have

$$\frac{a^2}{b}+\frac{b^2}{c}+\frac{c^2}{a}\geq a+b+c.$$

Solution 1 From $AM \geq GM$ we have

$$\frac{a^2}{b}+b\geq 2\sqrt{\frac{a^2}{b}\cdot b}=2a.$$

Analogously we get

$$\frac{b^2}{c}+c\geq 2b \quad \text{and} \quad \frac{c^2}{a}+a\geq 2c.$$

After adding these three inequalities we obtain

$$\frac{a^2}{b}+\frac{b^2}{c}+\frac{c^2}{a}+(a+b+c)\geq 2(a+b+c),$$

i.e.

$$\frac{a^2}{b}+\frac{b^2}{c}+\frac{c^2}{a}\geq a+b+c.$$

Equality occurs if and only if $a=b=c$.

Solution 2 Observe that

$$\frac{a^2}{b}+\frac{b^2}{c}+\frac{c^2}{a}=\frac{a^2-ab+b^2}{b}+\frac{b^2-bc+c^2}{c}+\frac{c^2-ca+a^2}{a}. \tag{2.6}$$

Since for any $x, y \in \mathbb{R}$, we have $x^2-xy+y^2\geq xy$, by (2.6) we get

$$\frac{a^2}{b}+\frac{b^2}{c}+\frac{c^2}{a}\geq \frac{ab}{b}+\frac{bc}{c}+\frac{ca}{a}=a+b+c.$$

Exercise 2.13 Let x, y, z be positive real numbers. Prove the inequality

$$\frac{x^3}{yz}+\frac{y^3}{zx}+\frac{z^3}{xy}\geq x+y+z.$$

Solution Since $AM \geq GM$ we have

$$\frac{x^3}{yz} + y + z \geq 3\sqrt[3]{\frac{x^3}{yz} \cdot y \cdot z} = 3x.$$

Similarly we have

$$\frac{y^3}{zx} + z + x \geq 3y \quad \text{and} \quad \frac{z^3}{xy} + x + y \geq 3z.$$

After adding these inequalities we get the required result.

Equality holds if and only if $x = y = z$.

Exercise 2.14 Let $a, b, c \in \mathbb{R}^+$. Prove the inequality

$$\frac{abc}{(1+a)(a+b)(b+c)(c+16)} \leq \frac{1}{81}.$$

Solution We have

$$\begin{aligned}
&(1+a)(a+b)(b+c)(c+16)\\
&\quad = \left(1+\frac{a}{2}+\frac{a}{2}\right)\left(a+\frac{b}{2}+\frac{b}{2}\right)\left(b+\frac{c}{2}+\frac{c}{2}\right)(c+8+8)\\
&\quad \geq 3\sqrt[3]{\frac{a^2}{4}} \cdot 3\sqrt[3]{\frac{ab^2}{4}} \cdot 3\sqrt[3]{\frac{bc^2}{4}} \cdot 3\sqrt[3]{\frac{64c}{4}} \geq 81abc.
\end{aligned}$$

Thus

$$\frac{abc}{(1+a)(a+b)(b+c)(c+16)} \leq \frac{1}{81}.$$

Exercise 2.15 Let $x, y \in \mathbb{R}^+$ such that $x + y = 2$. Prove the inequality

$$x^3y^3(x^3 + y^3) \leq 2.$$

Solution Since $AM \geq GM$ we have $\sqrt{xy} \leq \frac{x+y}{2} = 1$, i.e. $xy \leq 1$.

Hence $0 \leq xy \leq 1$.

Furthermore

$$\begin{aligned}
x^3y^3(x^3+y^3) &= (xy)^3(x+y)(x^2-xy+y^2) = 2(xy)^3((x+y)^2-3xy)\\
&= 2(xy)^3(4-3xy).
\end{aligned}$$

It's enough to show that

$$(xy)^3(4-3xy) \leq 1.$$

Let $xy = z$ then $0 \leq z \leq 1$ and clearly $4 - 3z > 0$.

Then using $AM \geq GM$ we obtain

$$z^3(4-3z) = z \cdot z \cdot z(4-3z) \leq \left(\frac{z+z+z+4-3z}{4}\right)^4 = 1,$$

as required.

Equality occurs if and only if $z = 4 - 3z$, i.e. $z = 1$, i.e. $x = y = 1$. (Why?)

Exercise 2.16 Let a, b, c, d be positive real numbers such that $a + b + c + d = 4$. Prove the inequality

$$\frac{1}{a^2+1} + \frac{1}{b^2+1} + \frac{1}{c^2+1} + \frac{1}{d^2+1} \geq 2.$$

Solution We have

$$\frac{1}{a^2+1} = 1 - \frac{a^2}{a^2+1} \geq 1 - \frac{a^2}{2a} = 1 - \frac{a}{2}.$$

Similarly we get

$$\frac{1}{b^2+1} \geq 1 - \frac{b}{2}, \frac{1}{c^2+1} \geq 1 - \frac{c}{2} \quad \text{and} \quad \frac{1}{d^2+1} \geq 1 - \frac{d}{2}.$$

After adding these inequalities we obtain

$$\frac{1}{a^2+1} + \frac{1}{b^2+1} + \frac{1}{c^2+1} + \frac{1}{d^2+1} \geq 4 - \frac{a+b+c+d}{2} = 4 - 2 = 2.$$

Equality occurs if and only if $a = b = c = 1$.

Chapter 3
Geometric (Triangle) Inequalities

These inequalities in most cases have as variables the lengths of the sides of a given triangle; there are also inequalities in which appear other elements of the triangle, such as lengths of heights, lengths of medians, lengths of the bisectors, angles, etc.

First we will introduce some standard notation which will be used in this section:

- h_a, h_b, h_c—lengths of the altitudes drawn to the sides a, b, c, respectively.
- t_a, t_b, t_c—lengths of the medians drawn to the sides a, b, c, respectively.
- $l_\alpha, l_\beta, l_\gamma$—lengths of the bisectors of the angles α, β, γ, respectively.
- P—area, s—semi-perimeter, R—circumradius, r—inradius.

Furthermore we will give relations between the lengths of medians and lengths of the bisectors of the angles with the sides of a given triangle.

Namely we have

$$t_a^2 = \frac{b^2+c^2}{2} - \frac{a^2}{4}, \qquad t_b^2 = \frac{a^2+c^2}{2} - \frac{b^2}{4}, \qquad t_c^2 = \frac{a^2+b^2}{2} - \frac{c^2}{4}$$

and

$$l_\alpha^2 = bc\frac{((b+c)^2-a^2)}{(b+c)^2}, \qquad l_\beta^2 = ac\frac{((a+c)^2-b^2)}{(a+c)^2},$$

$$l_\gamma^2 = ab\frac{((a+b)^2-c^2)}{(a+b)^2}.$$

We can rewrite the last three identities in the following form

$$l_\alpha^2 = 4bc\frac{s(s-a)}{(b+c)^2}, \qquad l_\beta^2 = 4ac\frac{s(s-b)}{(a+c)^2}, \qquad l_\gamma^2 = 4ab\frac{s(s-c)}{(a+b)^2}.$$

Also we note that the following properties are true, and we'll present them without proof. (The first inequality follows by using geometric formulas and *mean inequalities*, and the second inequality immediately follows, for instance, according to *Leibniz's theorem*.)

Z. Cvetkovski, *Inequalities*,
DOI 10.1007/978-3-642-23792-8_3,

Proposition 3.1 *For an arbitrary triangle the following inequalities hold*

$$R \geq 2r \quad \text{and} \quad a^2 + b^2 + c^2 \leq 9R^2.$$

Basic inequalities which concern the lengths of the sides of a given triangle are well-known inequalities: $a + b > c, a + c > b, b + c > a$.

But also useful and frequent substitutions are:

$$a = x + y, \quad b = y + z, \quad c = z + x, \quad \text{where } x, y, z > 0. \tag{3.1}$$

The question is whether there are always positive real numbers x, y, z, such that the above identities (3.1) hold and a, b, c are the sides of the triangle.

The answer is positive.

Namely x, y, z are tangent segments dropped from the vertices to the inscribed circle of the given triangle.

From (3.1) we easily get that

$$x = \frac{a + c - b}{2}, \qquad y = \frac{a + b - c}{2}, \qquad z = \frac{c + b - a}{2},$$

and then clearly $x, y, z > 0$.

Remark The substitutions (3.1) are called *Ravi's substitutions.*

Exercise 3.1 Let a, b, c be the lengths of the sides of given triangle. Prove the inequalities

$$\frac{3}{2} \leq \frac{a}{b + c} + \frac{b}{c + a} + \frac{c}{a + b} < 2.$$

Solution Let's prove the right-hand inequality.

Since $a + b > c$ we have $2(a + b) > a + b + c$, i.e. $a + b > s$.

Similarly we get $b + c > s$ and $a + c > s$.

Therefore

$$\frac{a}{b + c} + \frac{b}{a + c} + \frac{c}{b + a} < \frac{a}{s} + \frac{b}{s} + \frac{c}{s} = 2.$$

Let's consider the left-hand inequality.

If we denote $b + c = x, a + c = y, a + b = z$ then we have

$$a = \frac{z + y - x}{2}, \qquad b = \frac{z + x - y}{2}, \qquad c = \frac{x + y - z}{2}.$$

Hence

$$\frac{a}{b + c} + \frac{b}{a + c} + \frac{c}{b + a} = \frac{z + y - x}{2x} + \frac{z + x - y}{2y} + \frac{x + y - z}{2z},$$

i.e.

$$\frac{a}{b+c}+\frac{b}{a+c}+\frac{c}{b+a}=\frac{1}{2}\left(\frac{z}{x}+\frac{y}{x}+\frac{z}{y}+\frac{x}{y}+\frac{x}{z}+\frac{y}{x}-3\right)\geq\frac{1}{2}(2+2+2-3)=\frac{3}{2},$$

as required.

Remark The left-hand inequality is known as *Nesbitt's inequality*, and is true for any positive real numbers a, b and c (Exercise 1.3).

Exercise 3.2 Let a, b, c be the side lengths of a given triangle. Prove the inequality

$$\frac{1}{s-a}+\frac{1}{s-b}+\frac{1}{s-c}\geq\frac{9}{s}.$$

Solution Since $AM \geq HM$ we have

$$\frac{1}{s-a}+\frac{1}{s-b}+\frac{1}{s-c}\geq\frac{9}{(s-a)+(s-b)+(s-c)}=\frac{9}{s}.$$

Equality occurs if and only if $a=b=c$.

Exercise 3.3 Let s and r be the semi-perimeter and inradius, respectively, in an arbitrary triangle. Prove the inequality

$$s\geq 3r\sqrt{3}.$$

Solution 1 We have

$$2s=a+b+c\geq 3\sqrt[3]{abc}=3\sqrt[3]{4PR}=3\sqrt[3]{4srR}\geq 3\sqrt[3]{8sr^2},$$

i.e.

$$s\geq 3\sqrt[3]{sr^2}$$

or

$$s\geq 3r\sqrt{3}.$$

Equality occurs if and only if $a=b=c$.

Solution 2 We have

$$\frac{s}{3}=\frac{(s-a)+(s-b)+(s-c)}{3}\overset{AM\geq GM}{\geq}\sqrt[3]{(s-a)(s-b)(s-c)}. \tag{3.2}$$

Also

$$(s-a)(s-b)(s-c)=\frac{P^2}{s}=\frac{s^2r^2}{s}=sr^2. \tag{3.3}$$

By (3.2) and (3.3) we obtain

$$s \geq 3\sqrt[3]{sr^2}, \quad \text{i.e.} \quad s \geq 3\sqrt{3}r.$$

Equality occurs if and only if $a = b = c$.

Exercise 3.4 Let a, b, c be the side lengths of a given triangle. Prove the inequality

$$(a+b-c)(b+c-a)(c+a-b) \leq abc.$$

Solution 1 We have

$$a^2 \geq a^2 - (b-c)^2 = (a+b-c)(a+c-b).$$

Analogously

$$b^2 \geq (b+a-c)(b+c-a) \quad \text{and} \quad c^2 \geq (c+a-b)(c+b-a).$$

If we multiply these inequalities we obtain

$$\begin{aligned} a^2b^2c^2 &\geq (a+b-c)^2(b+c-a)^2(c+a-b)^2 \\ \Leftrightarrow \quad abc &\geq (a+b-c)(b+c-a)(c+a-b). \end{aligned}$$

Equality holds if and only if $a = b = c$, i.e. the triangle is equilateral.

Solution 2 After setting $a = x + y, b = y + z, c = z + x$, where $x, y, z > 0$, the given inequality becomes

$$(x+y)(y+z)(z+x) \geq 8xyz.$$

Since $AM \geq GM$ we have

$$(x+y)(y+z)(z+x) \geq 2\sqrt{xy} \cdot 2\sqrt{yz} \cdot 2\sqrt{zx} = 8xyz,$$

as required. Equality occurs if and only if $x = y = z$ i.e. $a = b = c$.

Remark This inequality holds for any $a, b, c \in \mathbb{R}^+$ (Problem 47).

Exercise 3.5 Let a, b, c be the side lengths of a given triangle. Prove the inequality

$$a^2 + b^2 + c^2 < 2(ab + bc + ca).$$

Solution Let $a = x + y, b = y + z, c = z + x, x, y, z > 0$.

Then we have

$$\begin{aligned} &(x+y)^2 + (y+z)^2 + (z+x)^2 \\ &\quad < 2((x+y)(y+z) + (y+z)(z+x) + (z+x)(x+y)) \end{aligned}$$

or

$$xy + yz + zx > 0,$$

which is clearly true.

Exercise 3.6 Let a, b, c be the side lengths of a given triangle. Prove the inequality

$$8(a+b-c)(b+c-a)(c+a-b) \leq (a+b)(b+c)(c+a).$$

Solution Since $AM \geq GM$ we have

$$(a+b)(b+c)(c+a) \geq 2\sqrt{ab}2\sqrt{bc}2\sqrt{ca} = 8abc.$$

So, it suffices to show that

$$8abc \geq 8(a+b-c)(b+c-a)(c+a-b),$$

i.e.

$$abc \geq (a+b-c)(b+c-a)(c+a-b),$$

which is true by Exercise 3.4.

Equality occurs if and only if $a = b = c$.

Exercise 3.7 Let a, b, c be the lengths of the sides of a triangle. Prove the inequality

$$\frac{1}{a} + \frac{1}{b} + \frac{1}{c} \leq \frac{1}{a+b-c} + \frac{1}{b+c-a} + \frac{1}{c+a-b}.$$

Solution Since $AM \geq HM$ we have

$$\frac{1}{2}\left(\frac{1}{a+b-c} + \frac{1}{b+c-a}\right) \geq \frac{2}{a+b-c+b+c-a} = \frac{1}{b}.$$

Similarly we deduce

$$\frac{1}{2}\left(\frac{1}{a+b-c} + \frac{1}{c+a-b}\right) \geq \frac{1}{a} \quad \text{and} \quad \frac{1}{2}\left(\frac{1}{b+c-a} + \frac{1}{c+a-b}\right) \geq \frac{1}{c}.$$

Adding these inequalities we get the required inequality.

Equality occurs if and only if $a = b = c$.

Exercise 3.8 Let ABC be a triangle with side lengths a, b, c and $\triangle A_1B_1C_1$ with side lengths $a + \frac{b}{2}, b + \frac{c}{2}, c + \frac{a}{2}$. Prove that $P_1 \geq \frac{9}{4}P$, where P is the area of $\triangle ABC$, and P_1 is the area of $\triangle A_1B_1C_1$.

Solution By *Heron's formula* for $\triangle ABC$ and $\triangle A_1B_1C_1$ we have

$$16P^2 = (a+b+c)(a+b-c)(b+c-a)(a+c-b)$$

and

$$16P_1^2=\frac{3}{16}(a+b+c)(-a+b+3c)(-b+c+3a)(-c+a+3b).$$

Since a,b and c are the side lengths of triangle there exist positive real numbers p,q,r such that $a=q+r, b=r+p, c=p+q$.

Now we easily get that

$$\frac{P^2}{P_1^2}=\frac{16pqr}{3(2p+q)(2q+r)(2r+p)}. \tag{3.4}$$

So it suffices to show that

$$(2p+q)(2q+r)(2r+p)\geq 27pqr.$$

Applying $AM\geq QM$ we obtain

$$\begin{aligned}(2p+q)(2q+r)(2r+p)&=(p+p+q)(q+q+r)(r+r+p)\\&\geq 3\sqrt[3]{p^2q}\cdot 3\sqrt[3]{q^2r}\cdot 3\sqrt[3]{r^2p}=27pqr.\end{aligned} \tag{3.5}$$

By (3.4) and (3.5) we get the desired result.

Exercise 3.9 Let a,b,c be the lengths of the sides of a triangle. Prove that: if $2(ab^2+bc^2+ca^2)=a^2b+b^2c+c^2a+3abc$ then the triangle is equilateral.

Solution We'll show that

$$a^2b+b^2c+c^2a+3abc\geq 2(ab^2+bc^2+ca^2),$$

with equality if and only if $a=b=c$, i.e. the triangle is equilateral.

Let us use *Ravi's substitutions*, i.e. $a=x+y, b=y+z, c=z+x$. Then the given inequality becomes

$$x^3+y^3+z^3+x^2y+y^2z+z^2x\geq 2(x^2z+y^2x+z^2y).$$

Since $AM\geq GM$ we have

$$x^3+z^2x\geq 2x^2z,\ y^3+x^2y\geq 2y^2x,\ z^3+y^2z\geq 2z^2y.$$

After adding these inequalities we obtain

$$x^3+y^3+z^3+x^2y+y^2z+z^2x\geq 2(x^2z+y^2x+z^2y).$$

Equality holds if and only if $x=y=z$, i.e. $a=b=c$, as required.

Exercise 3.10 Let a, b, c be the side lengths, and α, β, γ be the respective angles (in radians) of a given triangle. Prove the inequalities

$$\frac{\pi}{3} \leq \frac{a\alpha + b\beta + c\gamma}{a+b+c} < \frac{\pi}{2}.$$

Solution First let's prove the left inequality.

We can assume that $a \geq b \geq c$ and then clearly $\alpha \geq \beta \geq \gamma$.

So we have

$$\begin{aligned}&(a-b)(\alpha-\beta)+(b-c)(\beta-\gamma)+(c-a)(\gamma-\alpha) \geq 0\\ &\quad\Leftrightarrow\quad 2(a\alpha+b\beta+c\gamma) \geq (b+c)\alpha+(c+a)\beta+(a+b)\gamma,\end{aligned}$$

i.e.

$$3(a\alpha+b\beta+c\gamma) \geq (a+b+c)(\alpha+\beta+\gamma).$$

Hence

$$\frac{a\alpha+b\beta+c\gamma}{a+b+c} \geq \frac{\alpha+\beta+\gamma}{3} = \frac{\pi}{3}.$$

Equality occurs if and only if $a = b = c$.

Let's consider the right inequality.

Since a, b and c are side lengths of a triangle we have $a+b+c > 2a, a+b+c > 2b$ and $a+b+c > 2c$.

If we multiply these inequalities by α, β and γ, respectively, we obtain

$$(a+b+c)(\alpha+\beta+\gamma) > 2(a\alpha+b\beta+c\gamma),$$

i.e.

$$\frac{a\alpha+b\beta+c\gamma}{a+b+c} < \frac{\alpha+\beta+\gamma}{2} = \frac{\pi}{2}.$$

Chapter 4
Bernoulli's Inequality, the Cauchy–Schwarz Inequality, Chebishev's Inequality, Surányi's Inequality

These inequalities fill that part of the knowledge of students necessary for proving more complicated, characteristic inequalities such as mathematical inequalities containing more variables, and inequalities which are difficult to prove with already adopted elementary inequalities. These inequalities are often used for proving different inequalities for mathematical competitions.

Theorem 4.1 (Bernoulli's inequality) *Let $x_i, i = 1, 2, \ldots, n$, be real numbers with the same sign, greater then -1. Then we have*

$$(1+x_1)(1+x_2)\cdots(1+x_n) \geq 1+x_1+x_2+\cdots+x_n. \tag{4.1}$$

Proof We'll prove the given inequality by induction.

For $n = 1$ we have $1 + x_1 \geq 1 + x_1$.

Suppose that for $n = k$, and arbitrary real numbers $x_i > -1, i = 1, 2, \ldots, k$, with the same signs, inequality (4.1) holds i.e.

$$(1+x_1)(1+x_2)\cdots(1+x_k) \geq 1+x_1+x_2+\cdots+x_k. \tag{4.2}$$

Let $n = k + 1$, and $x_i > -1, i = 1, 2, \ldots, k + 1$, be arbitrary real numbers with the same signs.

Then, since $x_1, x_2, \ldots, x_{k+1}$ have the same signs, we have

$$(x_1+x_2+\cdots+x_k)x_{k+1} \geq 0. \tag{4.3}$$

Hence

$$\begin{aligned}
&(1+x_1)(1+x_2)\cdots(1+x_{k+1}) \\
&\overset{(4.2)}{\geq} (1+x_1+x_2+\cdots+x_k)(1+x_{k+1}) = 1+x_1+x_2+\cdots+x_k+x_{k+1} \\
&\quad + (x_1+x_2+\cdots+x_k)x_{k+1} \overset{(4.3)}{\geq} 1+x_1+x_2+\cdots+x_{k+1},
\end{aligned}$$

i.e. inequality (4.1) holds for $n = k + 1$, and we are done. □

Z. Cvetkovski, *Inequalities*,
DOI 10.1007/978-3-642-23792-8_4, © Springer-Verlag Berlin Heidelberg 2012

Corollary 4.1 (Bernoulli's inequality) *Let* $n \in \mathbb{N}$ *and* $x > -1$. *Then* $(1+x)^n \geq 1+nx$.

Proof According to *Theorem 4.1*, for $x_1 = x_2 = \cdots = x_n = x$, we obtain the required result. □

Definition 4.1 We'll say that the function $f(x_1, x_2, \ldots, x_n)$ is *homogenous* with *coefficient of homogeneity k*, if for arbitrary $t \in \mathbb{R}, t \neq 1$, we have

$$f(tx_1, tx_2, \ldots, tx_n) = t^k f(x_1, x_2, \ldots, x_n).$$

Example 4.1 The function $f(x, y) = \frac{x^2+y^2}{2x+y}$ is homogenous with coefficient 1, since

$$f(tx, ty) = \frac{t^2x^2 + t^2y^2}{2tx + ty} = t\frac{x^2 + y^2}{2x + y} = t \cdot f(x, y).$$

The function $f(x, y, z) = x^2 + xy + 3z$ is not homogenous.

If we consider the inequality $f(x_1, x_2, \ldots, x_n) \geq g(x_1, x_2, \ldots, x_n)$ then for this inequality we'll say that it is homogenous if the function

$$h(x_1, x_2, \ldots, x_n) = f(x_1, x_2, \ldots, x_n) - g(x_1, x_2, \ldots, x_n) \text{ is homogenous.}$$

In other words, a given inequality is homogenous if all its summands have equal degree.

Example 4.2 The inequality $x^2 + y^2 + 2xy \geq z^2 + yz$ is homogenous, since all monomials have degree 2.

The inequality $a^2b + b^2a \leq a^3 + b^3$ is also homogenous, but the inequality $a^5 + b^5 + 1 \geq 5ab(1 - ab)$ is not homogenous.

In the case of a homogenous inequality, without loss of generality we may assume additional conditions, which can reduce the given inequality to a much simpler form. In this way we can always reduce the number of variables of the given inequality. This procedure of assigning additional conditions is called *normalization*. An inequality with variables a, b, c can be normalized in many different ways; for example we can assume $a+b+c = 1$, or $abc = 1$ or $ab+bc+ca = 1$, etc. The choice of normalization depends on the problem and the available substitutions.

Example 4.3 Let us consider the homogenous inequality $a^2 + b^2 + c^2 \geq ab + bc + ca$. We may use the additional condition $abc = 1$. The reason is explained below.

Suppose that $abc = k^3$.

Let $a = kx$, $b = ky$ and $c = kz$; then clearly $xyz = 1$ and the given inequality becomes $x^2 + y^2 + z^2 \geq xy + yz + zx$, which is the same as before. Therefore the restriction $xyz = 1$ doesn't change anything in the inequality.

Alternatively, we can assume $a+b+c=1$ or we can assume $ab+bc+ac=1$, etc.

In general if we have a homogenous inequality then without loss of generality we may assign an additional condition such as: $abc, a+b+c, ab+bc+ca$, etc. to be whatever non-zero constant (not necessarily 1) that we choose.

In the case of a conditional inequality, there is a procedure somewhat opposite to normalization. With this procedure (known as *homogenization*) the given condition can be used to homogenize the whole inequality. After that, the newly acquired homogenous inequality can be normalized with some additional condition. For successful homogenization many obvious substitutions can be helpful.

For example, if we have $abc=1$ then we can take $a=\frac{x}{y}, b=\frac{y}{z}, c=\frac{z}{x}$, if we have $a+b+c=1$ then we can take $a=\frac{x}{x+y+z}, b=\frac{y}{x+y+z}, c=\frac{z}{x+y+z}$ and if $a^2+b^2+c^2=1$ we can take $a=\frac{x}{\sqrt{x^2+y^2+z^2}}, b=\frac{y}{\sqrt{x^2+y^2+z^2}}, c=\frac{z}{\sqrt{x^2+y^2+z^2}}$, etc.

Example 4.4 Consider the following conditional inequality

$$xy+yz+zx\geq 9xyz, \quad \text{when } x+y+z=1.$$

Obviously, the given inequality is not homogenous.

We can homogenize it as follows: since $x+y+z=1$ by taking

$$x=\frac{a}{a+b+c}, \qquad y=\frac{b}{a+b+c}, \qquad z=\frac{c}{a+b+c},$$

the inequality becomes

$$\frac{ab}{(a+b+c)^2}+\frac{bc}{(a+b+c)^2}+\frac{ca}{(a+b+c)^2}\geq\frac{9abc}{(a+b+c)^3},$$

i.e.

$$(a+b+c)(ab+bc+ca)\geq 9abc.$$

Now it is homogenous and can be further normalized with $abc=1$, which reduces it to the inequality

$$(ab+bc+ca)(a+b+c)\geq 9.$$

The last inequality is true since

$$\begin{aligned}(ab+bc+ca)(a+b+c)&=a^2b+a^2c+b^2a+b^2c+c^2b+c^2a+3abc\\&=\frac{a}{c}+\frac{a}{b}+\frac{b}{c}+\frac{b}{a}+\frac{c}{a}+\frac{c}{b}+3\\&=\frac{a}{c}+\frac{c}{a}+\frac{a}{b}+\frac{b}{a}+\frac{b}{c}+\frac{c}{b}+3\\&\geq 2+2+2+3=9.\end{aligned}$$

Theorem 4.2 (Cauchy–Schwarz inequality) *Let $a_1, a_2, \ldots, a_n$ and $b_1, b_2, \ldots, b_n$ be real numbers. Then we have*

$$\left(\sum_{i=1}^n a_i^2\right)\left(\sum_{i=1}^n b_i^2\right)\geq\left(\sum_{i=1}^n a_ib_i\right)^2,$$

i.e.

$$(a_1^2+a_2^2+\cdots+a_n^2)(b_1^2+b_2^2+\cdots+b_n^2) \geq (a_1b_1+a_2b_2+\cdots+a_nb_n)^2.$$

Equality occurs if and only if the sequences $(a_1, a_2, \ldots, a_n)$ *and* $(b_1, b_2, \ldots, b_n)$ *are proportional, i.e.* $\frac{a_1}{b_1} = \frac{a_2}{b_2} = \cdots = \frac{a_n}{b_n}$.

Proof 1 The given inequality is equivalent to

$$\sqrt{a_1^2+a_2^2+\cdots+a_n^2} \cdot \sqrt{b_1^2+b_2^2+\cdots+b_n^2} \geq |a_1b_1+a_2b_2+\cdots+a_nb_n|. \quad (4.4)$$

Let $A = \sqrt{a_1^2+a_2^2+\cdots+a_n^2}$, $B = \sqrt{b_1^2+b_2^2+\cdots+b_n^2}$.
If $A = 0$ then clearly $a_1 = a_2 = \cdots = a_n = 0$, and inequality (4.4) is true.
So let us assume that $A, B > 0$.
Inequality (4.4) is homogenous, so we may normalize with

$$a_1^2+a_2^2+\cdots+a_n^2 = 1 = b_1^2+b_2^2+\cdots+b_n^2, \quad (4.5)$$

i.e. we need to prove that

$$|a_1b_1+a_2b_2+\cdots+a_nb_n| \leq 1, \text{ with conditions (4.5).}$$

Since $QM \geq GM$ we have

$$\begin{aligned}|a_1b_1+a_2b_2+\cdots+a_nb_n| &\leq |a_1b_1|+|a_2b_2|+\cdots+|a_nb_n| \\ &\leq \frac{a_1^2+b_1^2}{2}+\frac{a_2^2+b_2^2}{2}+\cdots+\frac{a_n^2+b_n^2}{2} \\ &= \frac{(a_1^2+a_2^2+\cdots+a_n^2)+(b_1^2+b_2^2+\cdots+b_n^2)}{2} = 1,\end{aligned}$$

as required.
Equality occurs if and only if $\frac{a_1}{b_1} = \frac{a_2}{b_2} = \cdots = \frac{a_n}{b_n}$. (Why?) □

Proof 2. Consider the quadratic trinomial

$$\sum_{i=1}^{n}(a_ix-b_i)^2 = \sum_{i=1}^{n}(a_i^2x^2-2a_ib_ix+b_i^2) = x^2\sum_{i=1}^{n}a_i^2 - 2x\sum_{i=1}^{n}a_ib_i + \sum_{i=1}^{n}b_i^2.$$

This trinomial is non-negative for all $x \in \mathbb{R}$, so its discriminant is not positive, i.e.

$$\begin{aligned}&4\left(\sum_{i=1}^{n}a_ib_i\right)^2 - 4\left(\sum_{i=1}^{n}a_i^2\right)\left(\sum_{i=1}^{n}b_i^2\right) \leq 0 \\ \Leftrightarrow\ &\left(\sum_{i=1}^{n}a_ib_i\right)^2 \leq \left(\sum_{i=1}^{n}a_i^2\right)\left(\sum_{i=1}^{n}b_i^2\right),\end{aligned}$$

as required.

Equality holds if and only if $a_i x - b_i = 0, i = 1, 2, \ldots, n$, i.e. $\frac{a_1}{b_1} = \frac{a_2}{b_2} = \cdots = \frac{a_n}{b_n}$. □

Now we'll give several consequences of the *Cauchy–Schwarz inequality* which have broad use in proving other inequalities.

Corollary 4.2 *Let a, b, x, y be real numbers and $x, y > 0$. Then we have*

$$(1)\ \frac{a^2}{x} + \frac{b^2}{y} \geq \frac{(a+b)^2}{x+y}, \qquad (2)\ \frac{a^2}{x} + \frac{b^2}{y} + \frac{c^2}{z} \geq \frac{(a+b+c)^2}{x+y+z}.$$

Proof (1) The given inequality is equivalent to

$$y(x+y)a^2 + x(x+y)b^2 \geq xy(a+b)^2, \quad \text{i.e.} \quad (ay - bx)^2 \geq 0,$$

which is clearly true.

Equality occurs iff $ay = bx$ i.e. $\frac{a}{x} = \frac{b}{y}$.

(2) If we apply inequality from the first part twice, we get

$$\frac{a^2}{x} + \frac{b^2}{y} + \frac{c^2}{z} \geq \frac{(a+b)^2}{x+y} + \frac{c^2}{z} \geq \frac{(a+b+c)^2}{x+y+z}.$$

Equality occurs iff $\frac{a}{x} = \frac{b}{y} = \frac{c}{z}$. □

Also as you can imagine there must be some generalization of the previous corollaries. Namely the following result is true.

Corollary 4.3 *Let $a_1, a_2, \ldots, a_n; b_1, b_2, \ldots, b_n$ be real numbers such that $b_1, b_2, \ldots, b_n > 0$. Then*

$$\frac{a_1^2}{b_1} + \frac{a_2^2}{b_2} + \cdots + \frac{a_n^2}{b_n} \geq \frac{(a_1 + a_2 + \cdots + a_n)^2}{b_1 + b_2 + \cdots + b_n},$$

with equality if and only if $\frac{a_1}{b_1} = \frac{a_2}{b_2} = \cdots = \frac{a_n}{b_n}$.

Proof The proof is a direct consequence of the *Cauchy–Schwarz inequality*. □

Corollary 4.4 *Let $a_1, a_2, \ldots, a_n; b_1, b_2, \ldots, b_n$ be real numbers. Then*

$$\sqrt{a_1^2 + b_1^2} + \sqrt{a_2^2 + b_2^2} + \cdots + \sqrt{a_n^2 + b_n^2}$$
$$\geq \sqrt{(a_1 + a_2 + \cdots + a_n)^2 + (b_1 + b_2 + \cdots + b_n)^2}.$$

Proof By induction by n.

For $n=1$ we have equality.

For $n=2$ we have

$$\sqrt{a_1^2+b_1^2}+\sqrt{a_2^2+b_2^2}\geq\sqrt{(a_1+a_2)^2+(b_1+b_2)^2}$$

$$\Leftrightarrow \quad \sqrt{a_1^2+b_1^2}\cdot\sqrt{a_2^2+b_2^2}\geq(a_1a_2+b_1b_2)$$

$$\Leftrightarrow \quad (a_1^2+b_1^2)\cdot(a_2^2+b_2^2)\geq(a_1a_2+b_1b_2)^2,$$

which is the *Cauchy–Schwarz inequality*.

For $n=k$, let the given inequality hold, i.e.

$$\sqrt{a_1^2+b_1^2}+\sqrt{a_2^2+b_2^2}+\cdots+\sqrt{a_k^2+b_k^2}$$
$$\geq\sqrt{(a_1+a_2+\cdots+a_k)^2+(b_1+b_2+\cdots+b_k)^2}.$$

For $n=k+1$ we have

$$\sqrt{a_1^2+b_1^2}+\sqrt{a_2^2+b_2^2}+\cdots+\sqrt{a_{k+1}^2+b_{k+1}^2}$$
$$=\sqrt{a_1^2+b_1^2}+\sqrt{a_2^2+b_2^2}+\cdots+\sqrt{a_k^2+b_k^2}+\sqrt{a_{k+1}^2+b_{k+1}^2}$$
$$\geq\sqrt{(a_1+a_2+\cdots+a_k)^2+(b_1+b_2+\cdots+b_k)^2}+\sqrt{a_{k+1}^2+b_{k+1}^2}$$
$$\geq\sqrt{(a_1+a_2+\cdots+a_{k+1})^2+(b_1+b_2+\cdots+b_{k+1})^2}.$$

So the given inequality holds for every positive integer n. □

The next result is due to *Walter Janous*, and is considered by the author to be a very important result, which has broad use in proving inequalities.

Corollary 4.5 *Let a,b,c and x,y,z be positive real numbers. Then*

$$\frac{x}{y+z}(b+c)+\frac{y}{z+x}(c+a)+\frac{z}{x+y}(a+b)\geq\sqrt{3(ab+bc+ca)}.$$

Proof The given inequality is homogenous, in the variables a,b and c, so we can normalize with $a+b+c=1$.

And we can rewrite the inequality as

$$\frac{x}{y+z}(1-a)+\frac{y}{z+x}(1-b)+\frac{z}{x+y}(1-c)\geq\sqrt{3(ab+bc+ca)}.$$

Hence

$$\frac{x}{y+z}+\frac{y}{z+x}+\frac{z}{x+y}\geq\sqrt{3(ab+bc+ca)}+\frac{ax}{y+z}+\frac{by}{z+x}+\frac{cz}{x+y}. \quad (4.6)$$

By the *Cauchy–Schwarz inequality* we have

$$\frac{ax}{y+z}+\frac{by}{z+x}+\frac{cz}{x+y}+\sqrt{3(ab+bc+ca)}$$
$$\leq\sqrt{\left(\frac{x}{y+z}\right)^2+\left(\frac{y}{z+x}\right)^2+\left(\frac{z}{x+y}\right)^2}\cdot\sqrt{a^2+b^2+c^2}$$
$$+\sqrt{\frac{3}{4}}\sqrt{ab+bc+ca}+\sqrt{\frac{3}{4}}\sqrt{ab+bc+ca},$$

and after one more usage of the *Cauchy–Schwarz inequality* we get

$$\sqrt{\left(\frac{x}{y+z}\right)^2+\left(\frac{y}{z+x}\right)^2+\left(\frac{z}{x+y}\right)^2}\cdot\sqrt{a^2+b^2+c^2}$$
$$+\sqrt{\frac{3}{4}}\sqrt{ab+bc+ca}+\sqrt{\frac{3}{4}}\sqrt{ab+bc+ca}$$
$$\leq\sqrt{\left(\frac{x}{y+z}\right)^2+\left(\frac{y}{z+x}\right)^2+\left(\frac{z}{x+y}\right)^2+\frac{3}{2}}$$
$$\times\sqrt{a^2+b^2+c^2+2(ab+bc+ac)}$$
$$=\sqrt{\left(\frac{x}{y+z}\right)^2+\left(\frac{y}{z+x}\right)^2+\left(\frac{z}{x+y}\right)^2+\frac{3}{2}}.$$

So we have

$$\frac{ax}{y+z}+\frac{by}{z+x}+\frac{cz}{x+y}+\sqrt{3(ab+bc+ca)}$$
$$\leq\sqrt{\left(\frac{x}{y+z}\right)^2+\left(\frac{y}{z+x}\right)^2+\left(\frac{z}{x+y}\right)^2+\frac{3}{2}}.$$

It suffices to show that

$$\left(\frac{x}{y+z}\right)^2+\left(\frac{y}{z+x}\right)^2+\left(\frac{z}{x+y}\right)^2+\frac{3}{2}\leq\left(\frac{x}{y+z}+\frac{y}{z+x}+\frac{z}{x+y}\right)^2,$$

which is equivalent to

$$\frac{yz}{(x+y)(x+z)}+\frac{xz}{(y+x)(y+z)}+\frac{xy}{(z+x)(z+y)}\geq\frac{3}{4}. \tag{4.7}$$

After clearing the denominators inequality (4.7) becomes

$$x^2y+y^2x+y^2z+z^2y+z^2x+x^2z\geq 6xyz,$$

which is a direct consequence of $AM\geq GM$. □

Theorem 4.3 (Chebishev's inequality) *Let $a_1 \le a_2 \le \cdots \le a_n$ and $b_1 \le b_2 \le \cdots \le b_n$ be real numbers. Then we have*

$$\left(\sum_{i=1}^{n} a_i\right)\left(\sum_{i=1}^{n} b_i\right) \le n \sum_{i=1}^{n} a_i b_i,$$

i.e.

$$(a_1 + a_2 + \cdots + a_n)(b_1 + b_2 + \cdots + b_n) \le n(a_1 b_1 + a_2 b_2 + \cdots + a_n b_n).$$

Equality occurs if and only if $a_1 = a_2 = \cdots = a_n$ or $b_1 = b_2 = \cdots = b_n$.

Proof For all $i, j \in \{1, 2, \ldots, n\}$ we have

$$(a_i - a_j)(b_i - b_j) \ge 0, \tag{4.8}$$

i.e.

$$a_i b_i + a_j b_j \ge a_i b_j + a_j b_i. \tag{4.9}$$

By (4.9) we get

$$\begin{aligned}
\left(\sum_{i=1}^{n} a_i\right)\left(\sum_{i=1}^{n} b_i\right) &= a_1 b_1 + a_1 b_2 + a_1 b_3 + \cdots + a_1 b_n \\
&\quad + a_2 b_1 + a_2 b_2 + a_2 b_3 + \cdots + a_2 b_n \\
&\quad + a_3 b_1 + a_3 b_2 + a_3 b_3 + \cdots + a_3 b_n \\
&\quad \cdots \quad \cdots \quad \cdots \quad \cdots \quad \cdots \\
&\quad + a_n b_1 + a_n b_2 + a_n b_3 + \cdots + a_n b_n \\
&\le a_1 b_1 \\
&\quad + a_1 b_1 + a_2 b_2 + a_2 b_2 \\
&\quad + a_1 b_1 + a_3 b_3 + a_2 b_2 + a_3 b_3 + a_3 b_3 \\
&\quad \cdots \quad \cdots \quad \cdots \quad \cdots \quad \cdots \\
&\quad + a_1 b_1 + a_n b_n + a_2 b_2 + a_n b_n + \cdots + a_n b_n = n \sum_{i=1}^{n} a_i b_i.
\end{aligned}$$

Equality holds iff we have equality in (4.8), i.e. $a_1 = a_2 = \cdots = a_n$ or $b_1 = b_2 = \cdots = b_n$. □

Note *Chebishev's inequality* is also true in the case when $a_1 \ge a_2 \ge \cdots \ge a_n$ and $b_1 \ge b_2 \ge \cdots \ge b_n$. But if $a_1 \le a_2 \le \cdots \le a_n, b_1 \ge b_2 \ge \cdots \ge b_n$ (or the reverse) then we have

$$\left(\sum_{i=1}^{n} a_i\right)\left(\sum_{i=1}^{n} b_i\right) \ge n \sum_{i=1}^{n} a_i b_i.$$

Let us note that the inequality from Corollary 4.1 is true not just in case when $n \in \mathbb{N}$, but it is also true in the cases $n > 1, n \in \mathbb{Q}$ and $n \in [1, \infty), n \in \mathbb{R}$.

We prove this statement bellow in the case when $n \geq 1, n \in \mathbb{Q}$, and the second case will be left to the reader.

Corollary 4.6 *Let $x > -1$ and $r \geq 1, r \in \mathbb{Q}$. Then*

$$(1+x)^r \geq 1+rx.$$

Proof Let $r = \frac{p}{q}$, $Gcd(p,q) = 1$. Then clearly $p > q$.

Let $a_1 = a_2 = \cdots = a_q = 1 + rx$ and $a_{q+1} = a_{q+2} = \cdots = a_p = 1$.

If $1 + rx \leq 0$, then we are done.

So let us suppose that $1 + rx > 0$.

Since $AM \geq GM$ we have

$$\begin{aligned} 1+x &= \frac{px+p}{p} = \frac{q+rqx+p-q}{p} = \frac{q(1+rx)+p-q}{p} \\ &= \frac{a_1+a_2+\cdots+a_q+a_{q+1}+\cdots+a_p}{p} \geq \sqrt[p]{a_1a_2\cdots a_p} \\ &= \sqrt[p]{(1+rx)^q} = (1+rx)^{\frac{q}{p}} = (1+rx)^{\frac{1}{r}}, \end{aligned}$$

and we easily obtain $(1+x)^r \geq 1+rx$. □

Corollary 4.7 *Let $x > -1$ and $\alpha \in [1, \infty), \alpha \in \mathbb{R}$. Then*

$$(1+x)^\alpha \geq 1+\alpha x.$$

Theorem 4.4 (Surányi's inequality) *Let $a_1, a_2, \ldots, a_n$ be non-negative real numbers, and let n be a positive integer. Then*

$$\begin{aligned} &(n-1)(a_1^n + a_2^n + \cdots + a_n^n) + na_1a_2\cdots a_n \\ &\quad \geq (a_1+a_2+\cdots+a_n)(a_1^{n-1} + a_2^{n-1} + \cdots + a_n^{n-1}). \end{aligned}$$

Proof We will use induction.

Due to the symmetry and homogeneity of the inequality we may assume that

$$a_1 \geq a_2 \geq \cdots \geq a_{n+1} \quad \text{and} \quad a_1 + a_2 + \cdots + a_n = 1.$$

For $n = 1$ equality occurs.

Let us assume that for $n = k$ the inequality holds, i.e.

$$(k-1)(a_1^k + a_2^k + \cdots + a_k^k) + ka_1a_2\cdots a_k \geq a_1^{k-1} + a_2^{k-1} + \cdots + a_k^{k-1}.$$

We need to prove that:

$$k\sum_{i=1}^{k}a_i^{k+1}+ka_{k+1}^{k+1}+ka_{k+1}\prod_{i=1}^{k}a_i+a_{k+1}\prod_{i=1}^{k}a_i-(1+a_{k+1})\left(\sum_{i=1}^{k}a_i^k+a_{k+1}^k\right)\geq 0.$$

But from the inductive hypothesis we have

$$(k-1)(a_1^k+a_2^k+\cdots+a_k^k)+ka_1a_2\cdots a_k\geq a_1^{k-1}+a_2^{k-1}+\cdots+a_k^{k-1}.$$

Hence

$$ka_{k+1}\prod_{i=1}^{k}a_i\geq a_{k+1}\sum_{i=1}^{k}a_i^{k-1}-(k-1)a_{k+1}\sum_{i=1}^{k}a_i^k.$$

Using this last inequality, it remains to prove that:

$$\left(k\sum_{i=1}^{k}a_i^{k+1}-\sum_{i=1}^{k}a_i^k\right)-a_{k+1}\left(k\sum_{i=1}^{k}a_i^k-\sum_{i=1}^{k}a_i^{k-1}\right)$$
$$+a_{k+1}\left(\prod_{i=1}^{k}a_i+(k-1)a_{k+1}^k-a_{k+1}^{k-1}\right)\geq 0.$$

We prove that

$$a_{k+1}\left(\prod_{i=1}^{k}a_i+(k-1)a_{k+1}^k-a_{k+1}^{k-1}\right)\geq 0,$$

and

$$\left(k\sum_{i=1}^{k}a_i^{k+1}-\sum_{i=1}^{k}a_i^k\right)-a_{k+1}\left(k\sum_{i=1}^{k}a_i^k-\sum_{i=1}^{k}a_i^{k-1}\right)\geq 0.$$

We have

$$\begin{aligned}\prod_{i=1}^{k}a_i+(k-1)a_{k+1}^k-a_{k+1}^{k-1}&=\prod_{i=1}^{k}(a_i-a_{k+1}+a_{k+1})+(k-1)a_{k+1}^k-a_{k+1}^{k-1}\\&\geq a_{k+1}^k+a_{k+1}^{k-1}\cdot\sum_{i=1}^{k}(a_i-a_{k+1})+(k-1)a_{k+1}^k-a_{k+1}^{k-1}\\&=0.\end{aligned}$$

The second inequality is equivalent to

$$k\sum_{i=1}^{k}a_i^{k+1}-\sum_{i=1}^{k}a_i^k\geq a_{k+1}\left(k\sum_{i=1}^{k}a_i^k-\sum_{i=1}^{k}a_i^{k-1}\right).$$

By *Chebishev's inequality* we have

$$k\sum_{i=1}^{k}a_i^k\geq\sum_{i=1}^{k}a_i\sum_{i=1}^{k}a_i^{k-1}=\sum_{i=1}^{k}a_i^{k-1},\quad\text{i.e.}\quad k\sum_{i=1}^{k}a_i^k-\sum_{i=1}^{k}a_i^{k-1}\geq 0,$$

and since $a_1 + a_2 + \cdots + a_{k+1} = 1$, by the assumptation that $a_1 \geq a_2 \geq \cdots \geq a_{k+1}$, we deduce that

$$a_{k+1} \leq \frac{1}{k}.$$

So it is enough to prove that

$$k\sum_{i=1}^{k} a_i^{k+1} - \sum_{i=1}^{k} a_i^k \geq \frac{1}{k}\left(k\sum_{i=1}^{k} a_i^k - \sum_{i=1}^{k} a_i^{k-1}\right),$$

which is equivalent to

$$k\sum_{i=1}^{k} a_i^{k+1} + \frac{1}{k}\sum_{i=1}^{k} a_i^{k-1} \geq 2\sum_{i=1}^{k} a_i^k.$$

Since $AM \geq GM$ inequality we have that

$$ka_i^{k+1} + \frac{1}{k}a_i^{k-1} \geq 2a_i^k \quad \text{for all } i.$$

Adding this inequalities for $i = 1, 2, \ldots, k$ we obtain the required inequality. □

Exercise 4.1 Let x, y be positive real numbers. Prove the inequality

$$x^y + y^x \geq 1.$$

Solution We'll show that for every real number $a, b \in (0, 1)$ we have

$$a^b \geq \frac{a}{a+b-ab}.$$

By *Bernoulli's inequality* we have

$$a^{1-b} = (1 + a - 1)^{1-b} \leq 1 + (a-1)(1-b) = a + b - ab,$$

i.e.

$$a^b \geq \frac{a}{a+b-ab}.$$

If $x \geq 1$ or $y \geq 1$ then the given inequality clearly holds.

So let $0 < x, y < 1$.

By the previous inequality we have

$$x^y + y^x \geq \frac{x}{x+y-xy} + \frac{y}{x+y-xy} = \frac{x+y}{x+y-xy} > \frac{x+y}{x+y} = 1.$$

Exercise 4.2 Let $a, b, c > 0$. Prove *Nesbitt's inequality*

$$\frac{a}{b+c} + \frac{b}{c+a} + \frac{c}{a+b} \geq \frac{3}{2}.$$

Solution 1 Applying the *Cauchy–Schwarz inequality* for

$$a_1 = \sqrt{b+c}, \qquad a_2 = \sqrt{c+a}, \qquad a_3 = \sqrt{a+b};$$
$$b_1 = \frac{1}{\sqrt{b+c}}, \qquad b_2 = \frac{1}{\sqrt{c+a}}, \qquad b_3 = \frac{1}{\sqrt{a+b}}$$

gives us

$$((b+c)+(c+a)+(a+b))\left(\frac{1}{b+c}+\frac{1}{c+a}+\frac{1}{a+b}\right) \geq (1+1+1)^2 = 9,$$

i.e.

$$2(a+b+c)\left(\frac{1}{b+c}+\frac{1}{c+a}+\frac{1}{a+b}\right) \geq 9$$
$$\Leftrightarrow \quad \frac{a+b+c}{b+c}+\frac{a+b+c}{c+a}+\frac{a+b+c}{a+b} \geq \frac{9}{2}$$
$$\Leftrightarrow \quad \frac{a}{b+c}+\frac{b}{c+a}+\frac{c}{a+b} \geq \frac{9}{2}-3 = \frac{3}{2}.$$

Equality occurs iff $(b+c)^2 = (c+a)^2 = (a+b)^2$, i.e. iff $a = b = c$.

Solution 2 We'll use *Chebishev's inequality*.

Assume that $a \geq b \geq c$; then $\frac{1}{b+c} \geq \frac{1}{c+a} \geq \frac{1}{a+b}$.

Now by *Chebishev's inequality* we get

$$(a+b+c)\left(\frac{1}{b+c}+\frac{1}{c+a}+\frac{1}{a+b}\right) \leq 3\left(\frac{a}{b+c}+\frac{b}{c+a}+\frac{c}{a+b}\right). \tag{4.10}$$

Note that

$$(a+b+c)\left(\frac{1}{b+c}+\frac{1}{c+a}+\frac{1}{a+b}\right) = \frac{1}{2}((b+c)+(c+a)+(a+b)) \times \left(\frac{1}{b+c}+\frac{1}{c+a}+\frac{1}{a+b}\right).$$

Since $AM \geq HM$ (the same thing in this case with *Cauchy–Schwarz*) we have

$$((b+c)+(c+a)+(a+b))\left(\frac{1}{b+c}+\frac{1}{c+a}+\frac{1}{a+b}\right) \geq 9.$$

Therefore

$$(a+b+c)\left(\frac{1}{b+c}+\frac{1}{c+a}+\frac{1}{a+b}\right) \geq \frac{9}{2}. \tag{4.11}$$

By (4.10) and (4.11) we obtain

$$3\left(\frac{a}{b+c}+\frac{b}{c+a}+\frac{c}{a+b}\right) \geq \frac{9}{2}, \quad \text{i.e.} \quad \frac{a}{b+c}+\frac{b}{c+a}+\frac{c}{a+b} \geq \frac{3}{2}.$$

Equality occurs iff $a = b = c$.

Exercise 4.3 Let a, b, c, d be positive real numbers. Prove the inequality

$$\frac{1}{a}+\frac{1}{b}+\frac{4}{c}+\frac{16}{d} \geq \frac{64}{a+b+c+d}.$$

Solution By Corollary 4.3 we obtain

$$\frac{1}{a}+\frac{1}{b}+\frac{4}{c}+\frac{16}{d} \geq \frac{(1+1+2+4)^2}{a+b+c+d} = \frac{64}{a+b+c+d},$$

as required.

Exercise 4.4 Let $a, b, c \in \mathbb{R}^+$. Prove the inequality

$$\frac{a^2}{3^3}+\frac{b^2}{4^3}+\frac{c^2}{5^3} \geq \frac{(a+b+c)^2}{6^3}.$$

Solution Note that $3^3+4^3+5^3=6^3$.

Taking

$$a_1=\frac{a}{\sqrt{3^3}}, \qquad a_2=\frac{b}{\sqrt{4^3}}, \qquad a_3=\frac{c}{\sqrt{5^3}};$$

$$b_1=\sqrt{3^3}, \qquad b_2=\sqrt{4^3}, \qquad b_3=\sqrt{5^3},$$

by the *Cauchy–Schwarz inequality* we obtain

$$\left(\frac{a^2}{3^3}+\frac{b^2}{4^3}+\frac{c^2}{5^3}\right)(3^3+4^3+5^3) \geq (a+b+c)^2,$$

as required.

Exercise 4.5 Let a, b, c be positive real numbers. Determine the minimal value of

$$\frac{3a}{b+c}+\frac{4b}{c+a}+\frac{5c}{a+b}.$$

Solution By the *Cauchy–Schwarz inequality* we have

$$\begin{aligned}
&\frac{3a}{b+c}+\frac{4b}{c+a}+\frac{5c}{a+b}+(3+4+5)\\
&\quad=(a+b+c)\left(\frac{3}{b+c}+\frac{4}{c+a}+\frac{5}{a+b}\right)\\
&\quad=\frac{1}{2}((b+c)+(c+a)+(a+b))\left(\frac{3}{b+c}+\frac{4}{c+a}+\frac{5}{a+b}\right)\\
&\quad\geq \frac{1}{2}\left(\sqrt{3}+\sqrt{4}+\sqrt{5}\right)^2.
\end{aligned}$$

Hence

$$\frac{3a}{b+c}+\frac{4b}{c+a}+\frac{5c}{a+b} \geq \frac{1}{2}(\sqrt{3}+\sqrt{4}+\sqrt{5})^2-12.$$

So the minimal value of the expression is $\frac{1}{2}(\sqrt{3}+\sqrt{4}+\sqrt{5})^2 - 12$, and it is reached if and only if $\frac{b+c}{\sqrt{3}} = \frac{c+a}{2} = \frac{a+b}{\sqrt{5}}$.

Exercise 4.6 Let a, b, c be positive real numbers. Prove the inequality

$$\frac{a^2+b^2}{a+b} + \frac{b^2+c^2}{b+c} + \frac{c^2+a^2}{c+a} \geq a+b+c.$$

Solution By the *Cauchy–Schwarz inequality* (Corollary 4.3) we have

$$\begin{aligned} &\frac{a^2+b^2}{a+b} + \frac{b^2+c^2}{b+c} + \frac{c^2+a^2}{c+a} \\ &\quad = \frac{a^2}{a+b} + \frac{b^2}{b+c} + \frac{c^2}{c+a} + \frac{b^2}{a+b} + \frac{c^2}{b+c} + \frac{a^2}{c+a} \\ &\quad \geq \frac{(2(a+b+c))^2}{4(a+b+c)} = a+b+c. \end{aligned}$$

Exercise 4.7 Let $a, b, c \in \mathbb{R}^+$. Prove the inequality

$$\frac{a}{b+2c} + \frac{b}{c+2a} + \frac{c}{a+2b} \geq 1.$$

Solution Applying the *Cauchy–Schwarz inequality* we get

$$\begin{aligned} &\left(\frac{a}{b+2c} + \frac{b}{c+2a} + \frac{c}{a+2b}\right)(a(b+2c) + b(c+2a) + c(a+2b)) \\ &\quad \geq (a+b+c)^2, \end{aligned}$$

hence

$$\frac{a}{b+2c} + \frac{b}{c+2a} + \frac{c}{a+2b} \geq \frac{(a+b+c)^2}{3(ab+bc+ca)}.$$

So it suffices to show that

$$\frac{(a+b+c)^2}{3(ab+bc+ca)} \geq 1, \quad \text{i.e.} \quad (a+b+c)^2 \geq 3(ab+bc+ca),$$

which is equivalent to $a^2+b^2+c^2 \geq ab+bc+ca$, and clearly holds.

Equality occurs iff $a = b = c$.

Exercise 4.8 Let a, b, c be positive real numbers. Prove the inequality

$$\frac{a}{b+1} + \frac{b}{c+1} + \frac{c}{a+1} \geq \frac{3(a+b+c)}{3+a+b+c}.$$

Solution By the *Cauchy–Schwarz inequality* (Corollary 4.3) we have

$$\begin{aligned}\frac{a}{b+1}+\frac{b}{c+1}+\frac{c}{a+1}&=\frac{a^2}{a(b+1)}+\frac{b^2}{b(c+1)}+\frac{c^2}{c(a+1)}\\&\geq\frac{(a+b+c)^2}{a(b+1)+b(c+1)+c(a+1)}\\&=\frac{(a+b+c)^2}{ab+bc+ca+a+b+c}.\end{aligned}\tag{4.12}$$

Also we have

$$ab+bc+ca\leq\frac{(a+b+c)^2}{3}.\tag{4.13}$$

By (4.12) and (4.13) we get

$$\frac{a}{b+1}+\frac{b}{c+1}+\frac{c}{a+1}\geq\frac{(a+b+c)^2}{\frac{(a+b+c)^2}{3}+a+b+c}=\frac{3(a+b+c)}{3+a+b+c}.$$

Equality occurs iff $a=b=c$.

Exercise 4.9 Let $a,b,c>0$ be real numbers such that $ab+bc+ca=1$. Prove the inequality

$$\frac{a^2}{b+c}+\frac{b^2}{c+a}+\frac{c^2}{a+b}\geq\frac{\sqrt{3}}{2}.$$

Solution By the *Cauchy–Schwarz inequality* we have

$$\left(\frac{a^2}{b+c}+\frac{b^2}{c+a}+\frac{c^2}{a+b}\right)((b+c)+(c+a)+(a+b))\geq(a+b+c)^2,$$

i.e.

$$\frac{a^2}{b+c}+\frac{b^2}{c+a}+\frac{c^2}{a+b}\geq\frac{a+b+c}{2}.\tag{4.14}$$

Furthermore

$$(a+b+c)^2=a^2+b^2+c^2+2(ab+bc+ca)\geq 3(ab+bc+ca)=3,$$

i.e.

$$a+b+c\geq\sqrt{3}.\tag{4.15}$$

Using (4.14) and (4.15) we obtain the required inequality.

Equality occurs iff $a=b=c=1/\sqrt{3}$.

Exercise 4.10 Let a,b,c be positive real numbers such that $abc=1$. Prove the inequality

$$\frac{a}{a+b^4+c^4}+\frac{b}{b+c^4+a^4}+\frac{c}{c+a^4+b^4}\leq 1.$$

Solution By the *Cauchy–Schwarz inequality* we have

$$\frac{a}{a+b^4+c^4} = \frac{a(a^3+2)}{(a+b^4+c^4)(a^3+1+1)} \le \frac{a(a^3+2)}{(a^2+b^2+c^2)^2}.$$

Similarly we get

$$\frac{b}{b+c^4+a^4} \le \frac{b(b^3+2)}{(a^2+b^2+c^2)^2} \quad \text{and} \quad \frac{c}{c+a^4+b^4} \le \frac{c(c^3+2)}{(a^2+b^2+c^2)^2}.$$

Hence

$$\frac{a}{a+b^4+c^4} + \frac{b}{b+c^4+a^4} + \frac{c}{c+a^4+b^4} \le \frac{a^4+b^4+c^4+2(a+b+c)}{(a^2+b^2+c^2)^2},$$

and we need to prove that

$$(a^2+b^2+c^2)^2 \ge a^4+b^4+c^4+2(a+b+c),$$

which is equivalent to

$$a^2b^2+b^2c^2+c^2a^2 \ge a+b+c.$$

By the well-known inequality $a^2b^2+b^2c^2+c^2a^2 \ge abc(a+b+c)$ and $abc=1$, we have

$$a^2b^2+b^2c^2+c^2a^2 \ge abc(a+b+c) = a+b+c,$$

as required.

Exercise 4.11 Let a, b, c be positive real numbers such that $a+b+c=1$. Prove the inequality

$$(a+b)^2(1+2c)(2a+3c)(2b+3c) \ge 54abc.$$

Solution The given inequality can be rewritten as follows

$$(a+b)^2(1+2c)\left(2+3\frac{c}{a}\right)\left(2+3\frac{c}{b}\right) \ge 54c.$$

By the *Cauchy–Schwarz inequality* and $AM \ge GM$ we have

$$\left(2+3\frac{c}{a}\right)\left(2+3\frac{c}{b}\right) \ge \left(2+\frac{3c}{\sqrt{ab}}\right)^2 \ge \left(2+\frac{6c}{a+b}\right)^2 = \frac{(2(a+b)+6c)^2}{(a+b)^2}$$
$$= \frac{(2(1-c)+6c)^2}{(a+b)^2} = \frac{4(1+2c)^2}{(a+b)^2}.$$

Then we have

$$(a+b)^2(1+2c)\left(2+3\frac{c}{a}\right)\left(2+3\frac{c}{b}\right) \ge (a+b)^2(1+2c)\frac{4(1+2c)^2}{(a+b)^2}$$
$$= 4(1+2c)^3,$$

and it remains to prove that

$$4(1+2c)^3 \geq 54c, \quad \text{i.e.} \quad (1+2c)^3 \geq \frac{27c}{2}.$$

By the $AM \geq GM$ inequality we have

$$(1+2c)^3 = \left(\frac{1}{2}+\frac{1}{2}+2c\right)^3 \geq 27\cdot\frac{1}{2}\cdot\frac{1}{2}\cdot 2c = \frac{27c}{2},$$

as required.

Equality occurs iff $a = b = \frac{3}{8}$, $c = \frac{1}{4}$.

Exercise 4.12 Let a, b, c, d, e, f be positive real numbers. Prove the inequality

$$\frac{a}{b+c}+\frac{b}{c+d}+\frac{c}{d+e}+\frac{d}{e+f}+\frac{e}{f+a}+\frac{f}{a+b} \geq 3.$$

Solution By the *Cauchy–Schwarz inequality* we have

$$\begin{aligned}
&\frac{a}{b+c}+\frac{b}{c+d}+\frac{c}{d+e}+\frac{d}{e+f}+\frac{e}{f+a}+\frac{f}{a+b}\\
&= \frac{a^2}{ab+ad}+\frac{b^2}{bc+bd}+\frac{c^2}{cd+ce}+\frac{d^2}{de+df}+\frac{e^2}{ef+ea}+\frac{f^2}{fa+fb}\\
&\geq \frac{(a+b+c+d+e+f)^2}{ab+ac+bc+bd+cd+ce+de+df+ef+ea+fa+fb}. \qquad (4.16)
\end{aligned}$$

Let

$$S = ab+ac+bc+bd+cd+ce+de+df+ef+ea+fa+fb.$$

Then

$$\begin{aligned}
2S &= (a+b+c+d+e+f)^2\\
&\quad - (a^2+b^2+c^2+d^2+e^2+f^2+2ad+2bd+2cf). \qquad (4.17)
\end{aligned}$$

Also we have

$$\begin{aligned}
&a^2+b^2+c^2+d^2+e^2+f^2+2ad+2be+2cf\\
&\quad = (a+d)^2+(b+e)^2+(c+f)^2\\
&\quad \overset{QM \geq AM}{\geq} \frac{1}{3}(a+b+c+d+e+f)^2. \qquad (4.18)
\end{aligned}$$

Using (4.17) and (4.18) we get

$$\begin{aligned}
2S &= (a+b+c+d+e+f)^2\\
&\quad - (a^2+b^2+c^2+d^2+e^2+f^2+2ad+2bd+2cf)\\
&\leq (a+b+c+d+e+f)^2 - \frac{1}{3}(a+b+c+d+e+f)^2\\
&= \frac{2}{3}(a+b+c+d+e+f)^2,
\end{aligned}$$

i.e.

$$\frac{(a+b+c+d+e+f)^2}{S} \geq 3. \tag{4.19}$$

Finally from (4.16) and (4.19) we obtain the required inequality.

Equality occurs iff $a=b=c=d=e=f$.

Exercise 4.13 Let $a,b,c \in \mathbb{R}^+$ such that $\frac{1}{a+b+1}+\frac{1}{b+c+1}+\frac{1}{c+a+1} \geq 1$. Prove the inequality

$$a+b+c \geq ab+bc+ca.$$

Solution We'll use the *Cauchy–Schwarz inequality*.

We have

$$(a+b+1)(a+b+c^2) \geq (a+b+c)^2, \quad \text{i.e.} \quad \frac{1}{a+b+1} \leq \frac{a+b+c^2}{(a+b+c)^2}.$$

Analogously

$$\frac{1}{b+c+1} \leq \frac{b+c+a^2}{(a+b+c)^2} \quad \text{and} \quad \frac{1}{c+a+1} \leq \frac{c+a+b^2}{(a+b+c)^2}.$$

By the given condition we have

$$1 \leq \frac{1}{a+b+1}+\frac{1}{b+c+1}+\frac{1}{c+a+1} \leq \frac{a+b+c^2+b+c+a^2+c+a+b^2}{(a+b+c)^2},$$

i.e.

$$2(a+b+c) \geq (a+b+c)^2-(a^2+b^2+c^2)$$
$$\Leftrightarrow \quad a+b+c \geq ab+bc+ca.$$

Exercise 4.14 Let a,b,c be positive real numbers such that $ab+bc+ca=3$. Prove the inequality

$$\frac{a(b^2+c^2)}{a^2+bc}+\frac{b(c^2+a^2)}{b^2+ca}+\frac{c(a^2+a^2)}{c^2+ab} \geq 3.$$

Solution Let $x=a(b^2+c^2)$, $y=b(c^2+a^2)$ and $z=c(a^2+b^2)$.

Then we have

$$\frac{x}{y+z}(b+c)=\frac{a(b^2+c^2)(b+c)}{b(c^2+a^2)+c(a^2+b^2)}=\frac{a(b^2+c^2)}{a^2+bc}.$$

Analogously we get

$$\frac{y}{z+x}(c+a)=\frac{b(c^2+a^2)}{b^2+ca} \quad \text{and} \quad \frac{z}{x+y}(a+b)=\frac{c(a^2+a^2)}{c^2+ab}.$$

By Corollary 4.5 and the previous identities we have

$$\frac{a(b^2+c^2)}{a^2+bc}+\frac{b(c^2+a^2)}{b^2+ca}+\frac{c(a^2+a^2)}{c^2+ab}$$
$$=\frac{x}{y+z}(b+c)+\frac{y}{z+x}(c+a)+\frac{z}{x+y}(a+b)\geq\sqrt{3(ab+bc+ca)}=3.$$

Exercise 4.15 Let $x, y, z \geq 0$ be real numbers. Prove the inequality

$$\sqrt{x^2+1}+\sqrt{y^2+1}+\sqrt{z^2+1}\geq\sqrt{6(x+y+z)}.$$

Solution According to Corollary 4.4 we have

$$\sqrt{x^2+1}+\sqrt{y^2+1}+\sqrt{z^2+1}\geq\sqrt{(x+y+z)^2+9}. \tag{4.20}$$

Applying $AM \geq GM$ we deduce

$$(x+y+z)^2+9\geq 2\sqrt{9(x+y+z)^2}=6(x+y+z). \tag{4.21}$$

From (4.20) and (4.21) we get the required inequality.

Equality occurs if and only if $x=y=z=1$.

Exercise 4.16 Let $a, b, c \in \mathbb{R}^+$. Prove the inequalities

(1) $2(a^8+b^8)\geq(a^3+b^3)(a^5+b^5)$;
(2) $3(a^8+b^8+c^8)\geq(a^3+b^3+c^3)(a^5+b^5+c^5)$.

Solution (1) Let $a \geq b$. Then $a^3 \geq b^3$ and $a^5 \geq b^5$.

Due to *Chebishev's inequality* we have

$$(a^3+b^3)(a^5+b^5)\leq 2(a^8+b^8).$$

(2) Similarly to (1).

Exercise 4.17 Let a, b and c be the lengths of the sides of a triangle, and α, β, γ be its angles (in radians), respectively. Let s be the semi-perimeter of the triangle. Prove the inequality

$$\frac{b+c}{\alpha}+\frac{c+a}{\beta}+\frac{a+b}{\gamma}\geq\frac{12s}{\pi}.$$

Solution Without loss of generality we may assume that $a \leq b \leq c$. Then clearly $\alpha \leq \beta \leq \gamma, a+b \leq a+c \leq b+c$ and $\frac{1}{\gamma} \leq \frac{1}{\beta} \leq \frac{1}{\alpha}$.

Now by *Chebishev's inequality* we have

$$((a+b)+(b+c)+(c+a))\left(\frac{1}{\alpha}+\frac{1}{\beta}+\frac{1}{\gamma}\right)$$
$$\leq 3\left((b+c)\frac{1}{\alpha}+(c+a)\frac{1}{\beta}+(a+b)\frac{1}{\gamma}\right),$$

i.e.

$$\frac{b+c}{\alpha}+\frac{c+a}{\beta}+\frac{a+b}{\gamma}\geq\frac{4s}{3}\left(\frac{1}{\alpha}+\frac{1}{\beta}+\frac{1}{\gamma}\right). \tag{4.22}$$

Using (4.22) and $AM \geq HM$ we obtain

$$\frac{b+c}{\alpha}+\frac{c+a}{\beta}+\frac{a+b}{\gamma}\geq\frac{4s}{3}\left(\frac{1}{\alpha}+\frac{1}{\beta}+\frac{1}{\gamma}\right)\geq\frac{4s}{3}\cdot\frac{9}{\alpha+\beta+\gamma}=\frac{12s}{\pi}.$$

Equality occurs iff $a=b=c$.

Exercise 4.18 Let $a,b,c,d\in\mathbb{R}^+$. Prove the inequality

$$\frac{a^3+b^3+c^3}{a+b+c}+\frac{a^3+b^3+d^3}{a+b+d}+\frac{a^3+c^3+d^3}{a+c+d}+\frac{b^3+c^3+d^3}{b+c+d}$$
$$\geq a^2+b^2+c^2+d^2.$$

Solution Without loss of generality we may assume that $a\geq b\geq c\geq d$. Then clearly $a^2\geq b^2\geq c^2\geq d^2$.

We'll use *Chebishev's inequality*, i.e. we have

$$(a+b+c)(a^2+b^2+c^2)\leq 3(a^3+b^3+c^3)$$
$$\Leftrightarrow\quad\frac{a^3+b^3+c^3}{a+b+c}\geq\frac{a^2+b^2+c^2}{3}.$$

Similarly we get

$$\frac{a^3+b^3+d^3}{a+b+d}\geq\frac{a^2+b^2+d^2}{3},\qquad\frac{a^3+c^3+d^3}{a+c+d}\geq\frac{a^2+c^2+d^2}{3},$$
$$\frac{b^3+c^3+d^3}{b+c+d}\geq\frac{b^2+c^2+d^2}{3}.$$

After adding these inequalities we get the required inequality.

Exercise 4.19 Let $a_1,a_2,\ldots,a_n\in\mathbb{R}^+$ such that $a_1+a_2+\cdots+a_n=1$. Prove the inequality

$$\frac{a_1}{2-a_1}+\frac{a_2}{2-a_2}+\cdots+\frac{a_n}{2-a_n}\geq\frac{n}{2n-1}.$$

Solution Without loss of generality we may assume that $a_1\geq a_2\geq\cdots\geq a_n$.

Then

$$\frac{1}{2-a_1}\geq\frac{1}{2-a_2}\geq\cdots\geq\frac{1}{2-a_n}.$$

Now by *Chebishev's inequality* we have

$$(a_1+a_2+\cdots+a_n)\left(\frac{1}{2-a_1}+\frac{1}{2-a_2}+\cdots+\frac{1}{2-a_n}\right)$$
$$\leq n\left(\frac{a_1}{2-a_1}+\frac{a_2}{2-a_2}+\cdots+\frac{a_n}{2-a_n}\right),$$

hence

$$\begin{aligned}\frac{a_1}{2-a_1}+\frac{a_2}{2-a_2}+\cdots+\frac{a_n}{2-a_n}&\geq\frac{1}{n}\left(\frac{1}{2-a_1}+\frac{1}{2-a_2}+\cdots+\frac{1}{2-a_n}\right)\\&\geq\frac{1}{n}\cdot\frac{n^2}{2n-(a_1+a_2+\cdots+a_n)}=\frac{n}{2n-1}.\end{aligned}$$

Equality occurs if and only if $a_1=a_2=\cdots=a_n=1/n$.

Exercise 4.20 Let a,b,c,d be positive real numbers such that $a+b+c+d=4$. Prove the inequality

$$\frac{1}{11+a^2}+\frac{1}{11+b^2}+\frac{1}{11+c^2}+\frac{1}{11+d^2}\leq\frac{1}{3}.$$

Solution Rewrite the given inequality as follows

$$\frac{1}{11+a^2}-\frac{1}{12}+\frac{1}{11+b^2}-\frac{1}{12}+\frac{1}{11+c^2}-\frac{1}{12}+\frac{1}{11+d^2}-\frac{1}{12}\leq 0,$$

i.e.

$$\frac{a^2-1}{11+a^2}+\frac{b^2-1}{11+b^2}+\frac{c^2-1}{11+c^2}+\frac{d^2-1}{11+d^2}\geq 0,$$

i.e.

$$(a-1)\frac{a+1}{11+a^2}+(b-1)\frac{b+1}{11+b^2}+(c-1)\frac{c+1}{11+c^2}+(d-1)\frac{d+1}{11+d^2}\geq 0. \tag{4.23}$$

Without loss of generality we may assume that $a\geq b\geq c\geq d$.

Then we have

$$a-1\geq b-1\geq c-1\geq d-1 \quad\text{and}\quad \frac{a+1}{11+a^2}\geq\frac{b+1}{11+b^2}\geq\frac{c+1}{11+c^2}\geq\frac{d+1}{11+d^2}.$$

Now inequality (4.23) is a direct consequences of *Chebishev's inequality*.

Equality occurs if and only if $a=b=c=d=1$.

Chapter 5
Inequalities Between Means (General Case)

In Chap. 2 we discussed *mean inequalities* of two and three variables. In this section we will develop their generalization, i.e. we'll present an analogous theorem for an arbitrary number of variables.

These inequalities are of particular importance because they are part of the basic apparatus for proving more complicated inequalities.

Theorem 5.1 (Mean inequalities) *Let $a_1, a_2, \ldots, a_n$ be positive real numbers. The numbers*

$$QM = \sqrt{\frac{a_1^2 + a_2^2 + \cdots + a_n^2}{n}}, \qquad AM = \frac{a_1 + a_2 + \cdots + a_n}{n},$$

$$GM = \sqrt[n]{a_1 a_2 \cdots a_n} \quad \text{and} \quad HM = \frac{n}{\frac{1}{a_1} + \frac{1}{a_2} + \cdots + \frac{1}{a_n}},$$

are called the quadratic, arithmetic, geometric and harmonic mean for the numbers $a_1, a_2, \ldots, a_n$, respectively, and we have

$$QM \geq AM \geq GM \geq HM.$$

Equalities occur if and only if $a_1 = a_2 = \cdots = a_n$.

Proof Firstly, we'll show that $AM \geq GM$, i.e.

$$\frac{a_1 + a_2 + \cdots + a_n}{n} \geq \sqrt[n]{a_1 a_2 \cdots a_n}. \tag{5.1}$$

Let

$$x_i = \frac{a_i}{\sqrt[n]{a_1 a_2 \cdots a_n}}, \qquad \text{for } i = 1, 2, \ldots, n. \tag{5.2}$$

Z. Cvetkovski, *Inequalities*,
DOI 10.1007/978-3-642-23792-8_5,

Then $x_i > 0$ for each $i = 1, 2, \ldots, n$ and we have

$$x_1 x_2 \cdots x_n = 1.$$

Inequality (5.1) is equivalent to

$$\frac{a_1}{\sqrt[n]{a_1 a_2 \cdots a_n}} + \frac{a_2}{\sqrt[n]{a_1 a_2 \cdots a_n}} + \cdots + \frac{a_n}{\sqrt[n]{a_1 a_2 \cdots a_n}} \geq n,$$

i.e. to

$$x_1 + x_2 + \cdots + x_n \geq n, \quad \text{when } x_1 x_2 \cdots x_n = 1, \tag{5.3}$$

with equality if and only if $x_1 = x_2 = \cdots = x_n = 1$.

We'll prove inequality (5.3) by induction.

For $n = 1$, inequality (5.3) is true; it becomes equality.

If $n = 2$ then $x_1 x_2 = 1$ and since $x_1 + x_2 \geq 2\sqrt{x_1 x_2}$ we get $x_1 + x_2 \geq 2$.

Hence (5.3) is true, and equality occurs iff $x_1 = x_2 = 1$.

Let assume that for $n = k$, and arbitrary positive real numbers $x_1, x_2, \ldots, x_k$ such that $x_1 x_2 \cdots x_k = 1$, we have $x_1 + x_2 + \cdots + x_k \geq k$, with equality if and only if $x_1 = x_2 = \cdots = x_k = 1$.

Let $n = k + 1$ and $x_1, x_2, \ldots, x_{k+1}$ be arbitrary positive real numbers such that

$$x_1 x_2 \cdots x_{k+1} = 1.$$

If $x_1 = x_2 = \cdots = x_{k+1} = 1$ then inequality (5.3) clearly holds.

Therefore, let us assume that there are numbers smaller then 1. Then clearly, there are also numbers which are greater then 1.

Without loss of generality we may assume that $x_1 < 1$ and $x_2 > 1$.

Then, for the sequences $x_1 x_2, x_3, \ldots, x_{k+1}$ which contain k terms we have $(x_1 x_2) x_3 \cdots x_{k+1} = 1$, and according to the induction hypothesis we have that $x_1 x_2 + x_3 + \cdots + x_{k+1} \geq k$, and equality occurs iff $x_1 x_2 = x_3 = \cdots = x_{k+1} = 1$.

Now we have

$$\begin{aligned} x_1 + x_2 + \cdots + x_{k+1} &\geq x_1 x_2 + x_3 + \cdots + x_{k+1} + 1 + (x_2 - 1)(1 - x_1) \\ &\geq k + 1 + (x_2 - 1)(1 - x_1) \geq k + 1, \end{aligned}$$

with equality if and only if $x_1 x_2 = x_3 = \cdots = x_{k+1} = 1$ and $(x_2 - 1)(1 - x_1) = 0$, i.e. iff $x_1 = x_2 = \cdots = x_{k+1} = 1$.

So, due to the *principle of mathematical induction*, we conclude that (5.3) is proved.

Thus by (5.2) we have $\frac{a_1}{\sqrt[n]{a_1 a_2 \ldots a_n}} = \frac{a_2}{\sqrt[n]{a_1 a_2 \ldots a_n}} = \cdots = \frac{a_n}{\sqrt[n]{a_1 a_2 \ldots a_n}}$, i.e.

$$a_1 = a_2 = \cdots = a_n.$$

Hence we have proved (5.1), and we are done.

We'll show that $GM \geq HM$, i.e.

$$\sqrt[n]{a_1 a_2 \cdots a_n} \geq \frac{n}{\frac{1}{a_1} + \frac{1}{a_2} + \cdots + \frac{1}{a_n}}.$$

By $AM \geq GM$ it follows that

$$\frac{1}{a_1} + \frac{1}{a_2} + \cdots + \frac{1}{a_n} \geq n \sqrt[n]{\frac{1}{a_1} \frac{1}{a_2} \cdots \frac{1}{a_n}} = \frac{n}{\sqrt[n]{a_1 a_2 \cdots a_n}},$$

i.e. we have

$$\sqrt[n]{a_1 a_2 \cdots a_n} \geq \frac{n}{\frac{1}{a_1} + \frac{1}{a_2} + \cdots + \frac{1}{a_n}},$$

and clearly equality holds if and only if $\frac{1}{a_1} = \frac{1}{a_2} = \cdots = \frac{1}{a_n}$, i.e. $a_1 = a_2 = \cdots = a_n$.

It is left to be shown that $QM \geq AM$, i.e.

$$\sqrt{\frac{a_1^2 + a_2^2 + \cdots + a_n^2}{n}} \geq \frac{a_1 + a_2 + \cdots + a_n}{n}.$$

We'll use the *Cauchy–Schwarz inequality* for the sequences $(a_1, a_2, \ldots, a_n)$ and $(1, 1, \ldots, 1)$.

So we have

$$\begin{aligned}
&(a_1^2 + a_2^2 + \cdots + a_n^2)(1^2 + 1^2 + \cdots + 1^2) \geq (a_1 + a_2 + \cdots + a_n)^2 \\
\Leftrightarrow \quad & n(a_1^2 + a_2^2 + \cdots + a_n^2) \geq (a_1 + a_2 + \cdots + a_n)^2 \\
\Leftrightarrow \quad & \frac{a_1^2 + a_2^2 + \cdots + a_n^2}{n} \geq \left(\frac{a_1 + a_2 + \cdots + a_n}{n}\right)^2 \\
\Leftrightarrow \quad & \sqrt{\frac{a_1^2 + a_2^2 + \cdots + a_n^2}{n}} \geq \frac{a_1 + a_2 + \cdots + a_n}{n}.
\end{aligned}$$

Equality holds if and only if $\frac{a_1}{1} = \frac{a_2}{1} = \cdots = \frac{a_n}{1}$, i.e. $a_1 = a_2 = \cdots = a_n$. □

Exercise 5.1 Let $a, b, c, d \in \mathbb{R}^+$ such that $abcd = 1$. Prove the inequality

$$a^2 + b^2 + c^2 + d^2 + ab + ac + ad + bc + bd + cd \geq 10.$$

Solution Since $AM \geq GM$ we have

$$a^2 + b^2 + c^2 + d^2 + ab + ac + ad + bc + bd + cd \geq 10 \sqrt[10]{a^5 b^5 c^5 d^5} = 10.$$

Equality holds if and only if $a = b = c = d = 1$.

Exercise 5.2 Let $a, b, c \in \mathbb{R}^+$. Prove the inequality

$$(a+b+c)^3 \geq a^3+b^3+c^3+24abc.$$

Solution We have

$$\begin{aligned}(a+b+c)^3 &= a^3+b^3+c^3+6abc+3(a^2b+a^2c+b^2a+b^2c+c^2a+c^2b) \\ &\geq a^3+b^3+c^3+6abc+3\cdot 6\sqrt[6]{a^6b^6c^6} = a^3+b^3+c^3+24abc.\end{aligned}$$

Equality holds if and only if $a=b=c$.

Exercise 5.3 Let $k \in \mathbb{N}$, and $a_1, a_2, \ldots, a_n$ be positive real numbers such that $a_1 + a_2 + \cdots + a_n = 1$. Prove the inequality

$$a_1^{-k}+a_2^{-k}+\cdots+a_n^{-k} \geq n^{k+1}.$$

Solution Since $AM \geq GM$ we have

$$\sqrt[n]{a_1a_2\cdots a_n} \leq \frac{a_1+a_2+\cdots+a_n}{n} = \frac{1}{n}$$

or

$$n \leq \sqrt[n]{\frac{1}{a_1}\frac{1}{a_2}\cdots\frac{1}{a_n}}.$$

Hence

$$n^k \leq \sqrt[n]{a_1^{-k}a_2^{-k}\cdots a_n^{-k}} \leq \frac{a_1^{-k}+a_2^{-k}+\cdots+a_n^{-k}}{n},$$

i.e.

$$a_1^{-k}+a_2^{-k}+\cdots+a_n^{-k} \geq n^{k+1},$$

as required.

Exercise 5.4 Let $a, b, c, d \in \mathbb{R}^+$. Prove the inequality

$$a^6+b^6+c^6+d^6 \geq abcd(ab+bc+cd+da).$$

Solution We have

$$\begin{aligned}a^6+b^6+c^6+d^6 &= \frac{1}{6}((2a^6+2b^6+c^6+d^6)+(2b^6+2c^6+d^6+a^6) \\ &\quad +(2c^6+2d^6+a^6+b^6)+(2d^6+2a^6+b^6+c^6)).\end{aligned}$$

Since $AM \geq GM$ we have

$$\frac{2a^6+2b^6+c^6+d^6}{6} = \frac{a^6+a^6+b^6+b^6+c^6+d^6}{6} \geq \sqrt[6]{a^{12}b^{12}c^6d^6} = a^2b^2cd.$$

Similarly we get

$$\frac{2b^6+2c^6+d^6+a^6}{6} \geq b^2c^2ad,$$

$$\frac{2c^6+2d^6+a^6+b^6}{6} \geq c^2d^2ab$$

and

$$\frac{2d^6+2a^6+b^6+c^6}{6} \geq d^2a^2bc.$$

Adding the last four inequalities we obtain the required inequality.

Equality holds if and only if $a=b=c=d$.

Exercise 5.5 Let $x, y, z \geq 2$ be real numbers. Prove the inequality

$$(y^3+x)(z^3+y)(x^3+z) \geq 125xyz.$$

Solution We have

$$y^3+x \geq 4y+x = y+y+y+y+x \geq 5\sqrt[5]{y^4x}.$$

Analogously

$$z^3+y \geq 5\sqrt[5]{z^4y} \quad \text{and} \quad x^3+z \geq 5\sqrt[5]{x^4z}.$$

Multiplying the last three inequalities gives us the required inequality.

5.1 Points of Incidence in Applications of the *AM–GM* Inequality

In this subsection we will consider characteristic examples in which we can use incorrectly the inequality $AM \geq GM$. Namely, a possible major route for the proper use of this inequality (the means inequalities) will be the fact that equality in these inequalities is achieved when all variables are equal. These points at which equality (all their coordinates are equal) of a given inequality is satisfied are called points of incidence. It is also important to note that symmetrical expressions achieve a minimum or maximum at a point of incidence.

Exercise 5.6 Let $x > 0$ be a real number. Find the minimum value of the expression

$$x+\frac{1}{x}.$$

Solution Since $AM \geq GM$ we have

$$x+\frac{1}{x} \geq 2\sqrt{x\cdot\frac{1}{x}} = 2,$$

with equality iff $x = \frac{1}{x}$, i.e. $x=1$.

Thus $\min\{x+\frac{1}{x}\}=2$.

Exercise 5.7 Let $x \geq 3$ be a real number. Find the minimum value of the expression

$$x + \frac{1}{x}.$$

Solution In this case we cannot directly use the inequality $AM \geq GM$ since the point $x = 1$ doesn't belongs to the domain $[3, +\infty)$.

We can easily show that the function $f(x) = x + \frac{1}{x}$ is an increasing function on $[3, +\infty)$, so it follows that $\min\{x + \frac{1}{x}\} = 3 + \frac{1}{3} = \frac{10}{3}$.

Now we will show how we can use $AM \geq GM$.

Since we have equality in $AM \geq GM$ if and only if all variables are equal, we deduce that we cannot use this inequality for the numbers x and $\frac{1}{x}$ at the point of incidence $x = 3$ since $3 \neq \frac{1}{3}$.

Assume that $AM \geq GM$ is used for the couple $(\frac{x}{\alpha}, \frac{1}{x})$ such that at the point of incidence $x = 3$, equality occurs, i.e. $\frac{x}{\alpha} = \frac{1}{x}$.

So it follows that $\alpha = x^2 = 3^2 = 9$.

According to this we transform $x + \frac{1}{x}$ as follows

$$A = x + \frac{1}{x} = \frac{x}{9} + \frac{1}{x} + \frac{8}{9}x \geq 2\sqrt{\frac{x}{9} \cdot \frac{1}{x}} + \frac{8}{9}x = \frac{2}{3} + \frac{8}{9} \cdot 3 = \frac{10}{3}.$$

Exercise 5.8 Let $a, b > 0$ be real numbers such that $a + b \leq 1$. Find the minimum value of the expression

$$A = ab + \frac{1}{ab}.$$

Solution If we use $AM \geq GM$ we get

$$A = ab + \frac{1}{ab} \geq 2\sqrt{ab \cdot \frac{1}{ab}} = 2,$$

and equality occurs if and only if $ab = \frac{1}{ab}$, i.e. $ab = 1$.

But then we have $a + b \geq 2\sqrt{ab} = 2$, contradicting $a + b \leq 1$.

If we take $x = \frac{1}{ab}$, then we have $x = \frac{1}{ab} \geq \frac{4}{(a+b)^2} \geq \frac{4}{1^2} = 4$.

Thus we may consider an equivalent problem of the given problem:

Find the minimum of the function $A = x + \frac{1}{x}$, with $x \geq 4$.

Point of incidence is $x = 4$.

So we have $\frac{x}{\alpha} = \frac{1}{x}$, from which it follows that $\alpha = x^2 = 16$.

Then we transform as follows

$$A = x + \frac{1}{x} = \frac{x}{16} + \frac{1}{x} + \frac{15}{16}x \geq 2\sqrt{\frac{x}{16} \cdot \frac{1}{x}} + \frac{15}{16}x \geq 2 \cdot \frac{1}{4} + \frac{15}{16} \cdot 4 = \frac{17}{4}.$$

Equality holds if and only if $x = 4$, i.e. $a = b = 1/2$.

Exercise 5.9 Let $a, b, c > 0$ be real numbers such that $a + b + c \leq \frac{3}{2}$. Find the minimum value of the expression

$$A = a + b + c + \frac{1}{a} + \frac{1}{b} + \frac{1}{c}.$$

Solution If we use $AM \geq GM$ we get

$$A = a + b + c + \frac{1}{a} + \frac{1}{b} + \frac{1}{c} \geq 6\sqrt[6]{abc \cdot \frac{1}{abc}} = 6,$$

with equality if and only if $a = b = c = 1$.

But then $a + b + c = 3 > \frac{3}{2}$, a contradiction.

Since A is a symmetrical expression on a, b and c we estimate that min A occurs at $a = b = c$, i.e. at $a = b = c = 1/2$.

Therefore for a point of incidence we have $\frac{1}{\alpha a} = \frac{1}{\alpha b} = \frac{1}{\alpha c} = a = b = c = 1/2$, and it follows that $\alpha = \frac{1}{a^2} = 4$.

Now we have

$$A = a + b + c + \frac{1}{a} + \frac{1}{b} + \frac{1}{c} = \left(a + b + c + \frac{1}{4a} + \frac{1}{4b} + \frac{1}{4c}\right) + \frac{3}{4}\left(\frac{1}{a} + \frac{1}{b} + \frac{1}{c}\right)$$

$$\geq 6\sqrt[6]{abc \cdot \frac{1}{(4a)(4b)(4c)}} + \frac{3}{4}\left(\frac{1}{a} + \frac{1}{b} + \frac{1}{c}\right) = 3 + \frac{3}{4}\left(\frac{1}{a} + \frac{1}{b} + \frac{1}{c}\right)$$

$$\geq 3 + \frac{3}{4} \cdot \frac{9}{a + b + c} \geq 3 + \frac{27}{4} \cdot \frac{1}{3/2} = \frac{15}{2}.$$

So min $A = \frac{15}{2}$, for $a = b = c = 1/2$.

Exercise 5.10 Let a, b, c be positive real numbers such that $a + b + c = 1$. Find the minimum value of the expression

$$abc + \frac{1}{a} + \frac{1}{b} + \frac{1}{c}.$$

Solution By the inequality $AM \geq GM$ we get

$$abc + \frac{1}{a} + \frac{1}{b} + \frac{1}{c} \geq 4\sqrt[4]{abc \cdot \frac{1}{a} \cdot \frac{1}{b} \cdot \frac{1}{c}} = 4,$$

with equality if and only if $abc = \frac{1}{a} = \frac{1}{b} = \frac{1}{c}$, from which we easily deduce that $a = b = c = 1$ and then $a + b + c = 3$, a contradiction since $a + b + c = 1$.

Since $abc + \frac{1}{a} + \frac{1}{b} + \frac{1}{c}$ is symmetrical with respect to a, b and c we estimate that the minimal value occurs when $a = b = c$, i.e. $a = b = c = 1/3$, since $a + b + c = 1$.

Let $abc = \frac{1}{\alpha a} = \frac{1}{\alpha b} = \frac{1}{\alpha c}$, from which we obtain $\alpha = \frac{1}{a^2 bc} = 81$.

Therefore let us rewrite the given expression as follows

$$abc+\frac{1}{a}+\frac{1}{b}+\frac{1}{c}=abc+\frac{1}{81a}+\frac{1}{81b}+\frac{1}{81c}+\frac{80}{81}\left(\frac{1}{a}+\frac{1}{b}+\frac{1}{c}\right). \tag{5.4}$$

By $AM \geq GM$ and $AM \geq HM$ we have

$$abc+\frac{1}{81a}+\frac{1}{81b}+\frac{1}{81c}\geq 4\sqrt[4]{abc\cdot\frac{1}{81a}\cdot\frac{1}{81b}\cdot\frac{1}{81c}}=\frac{4}{27} \tag{5.5}$$

and

$$\frac{1}{a}+\frac{1}{b}+\frac{1}{c}\geq\frac{9}{a+b+c}=9. \tag{5.6}$$

By (5.4), (5.5) and (5.6) we have

$$abc+\frac{1}{a}+\frac{1}{b}+\frac{1}{c}\geq\frac{4}{27}+\frac{80}{9}=\frac{244}{27},$$

with equality if and only if $a=b=c=\frac{1}{3}$.

Exercise 5.11 Let $a,b,c,d>0$ be real numbers. Find the minimum value of the expression

$$\frac{a}{b+c+d}+\frac{b}{c+d+a}+\frac{c}{d+a+b}+\frac{d}{a+b+c}+\frac{b+c+d}{a}+\frac{c+d+a}{b}$$
$$+\frac{a+b+d}{c}+\frac{a+b+c}{d}.$$

Solution Let us denote

$$A=\frac{a}{b+c+d}+\frac{b}{c+d+a}+\frac{c}{d+a+b}+\frac{d}{a+b+c}+\frac{b+c+d}{a}+\frac{c+d+a}{b}$$
$$+\frac{a+b+d}{c}+\frac{a+b+c}{d}.$$

If we use $AM \geq GM$ we get $A\geq 8$, with equality iff

$$\frac{a}{b+c+d}=\frac{b}{c+d+a}=\frac{c}{d+a+b}=\frac{d}{a+b+c}=\frac{b+c+d}{a}=\frac{c+d+a}{b}$$
$$=\frac{a+b+d}{c}=\frac{a+b+c}{d},$$

i.e.

$$a=b+c+d,\qquad b=c+d+a,\qquad c=d+a+b\quad\text{and}\quad d=a+b+c.$$

After adding the last identities we deduce $a+b+c+d=3(a+b+c+d)$, i.e. $3=1$, a contradiction.

Since A is a symmetrical expression with variables a, b, c, d, it follows that the minimum (maximum) will occur at the point of incidence $a = b = c = d > 0$.

Suppose $a = b = c = d > 0$.

We have

$$\frac{a}{b+c+d} = \frac{b}{c+d+a} = \frac{c}{d+a+b} = \frac{d}{a+b+c} = \frac{1}{3}$$

and

$$\frac{b+c+d}{\alpha a} = \frac{c+d+a}{\alpha b} = \frac{a+b+d}{\alpha c} = \frac{a+b+c}{\alpha d} = \frac{3}{\alpha},$$

i.e. $\frac{1}{3} = \frac{3}{\alpha}$, and it follows that $\alpha = 9$.

Therefore

$$\begin{aligned} A &= \frac{a}{b+c+d} + \frac{b}{c+d+a} + \frac{c}{d+a+b} + \frac{d}{a+b+c} + \frac{b+c+d}{9a} \\ &\quad + \frac{c+d+a}{9b} + \frac{a+b+d}{9c} + \frac{a+b+c}{9d} \\ &\quad + \frac{8}{9}\left(\frac{b+c+d}{a} + \frac{c+d+a}{b} + \frac{a+b+d}{c} + \frac{a+b+c}{d}\right) \\ &\geq \frac{8}{3} + \frac{8}{9}(2+2+2+2+2+2) = \frac{40}{3}. \end{aligned}$$

Exercise 5.12 Let $a, b, c \geq 0$ be real numbers such that $a + b + c = 1$. Find the maximum value of the expression $A = \sqrt[3]{a+b} + \sqrt[3]{b+c} + \sqrt[3]{c+a}$.

Solution Since $AM \geq GM$ we have

$$\sqrt[3]{a+b} = \sqrt[3]{(a+b)\cdot 1 \cdot 1} \leq \frac{a+b+1+1}{3} = \frac{a+b+2}{3}.$$

Similarly

$$\sqrt[3]{b+c} \leq \frac{b+c+2}{3} \quad \text{and} \quad \sqrt[3]{c+a} \leq \frac{c+a+2}{3}.$$

Thus it follows that

$$A \leq \frac{a+b+2}{3} + \frac{b+c+2}{3} + \frac{c+a+2}{3} = \frac{2(a+b+c)}{3} + 2 = \frac{8}{3},$$

with equality iff $a + b = b + c = c + a = 1$, i.e. $a = b = c = 1/2$.

But then $a + b + c = 3/2 \neq 1$, a contradiction.

Since A is symmetrical expression in a, b, c, we estimate that the minimum (maximum) will occur at the point of incidence $a = b = c$, i.e. $a = b = c = 1/3$.

Clearly $a + b = b + c = c + a = 2/3$.

Since $AM \geq GM$ we have

$$\sqrt[3]{a+b} = \sqrt[3]{(a+b) \cdot \frac{2}{3} \cdot \frac{2}{3}} \cdot \sqrt[3]{\frac{9}{4}} \leq \sqrt[3]{\frac{9}{4}} \cdot \frac{a+b+\frac{2}{3}+\frac{2}{3}}{3} = \sqrt[3]{\frac{9}{4}} \cdot \frac{3(a+b)+4}{9}.$$

Similarly we get

$$\sqrt[3]{b+c} \leq \sqrt[3]{\frac{9}{4}} \cdot \frac{3(b+c)+4}{9} \quad \text{and} \quad \sqrt[3]{c+a} \leq \sqrt[3]{\frac{9}{4}} \cdot \frac{3(c+a)+4}{9}.$$

Adding the last three inequalities gives us

$$\begin{aligned} A &\leq \sqrt[3]{\frac{9}{4}} \cdot \left(\frac{3(a+b)+4}{9} + \frac{3(b+c)+4}{9} + \frac{3(c+a)+4}{9} \right) \\ &= \sqrt[3]{\frac{9}{4}} \cdot \frac{6(a+b+c)+12}{9} = \sqrt[3]{18}. \end{aligned}$$

So $\max A = \sqrt[3]{18}$, and it occurs iff $a+b = b+c = c+a = 2/3$, i.e. $a = b = c = 1/3$.

Exercise 5.13 Let a, b, c be positive real numbers such that $a+b+c=6$. Prove the inequality

$$\sqrt[3]{ab+bc} + \sqrt[3]{bc+ca} + \sqrt[3]{ca+ab} \leq 6.$$

Solution 1 Since we have a symmetrical expression we estimate that the maximum value of $\sqrt[3]{ab+bc} + \sqrt[3]{bc+ca} + \sqrt[3]{ca+ab}$ will occur at the point of incidence $a = b = c = 2$ and then clearly we have $ab + bc = 8$.

By the inequality $AM \geq GM$ we get

$$\sqrt[3]{ab+bc} = \frac{\sqrt[3]{(ab+bc) \cdot 8 \cdot 8}}{4} \leq \frac{1}{4} \left(\frac{(ab+bc)+8+8}{3} \right).$$

Similarly we obtain

$$\sqrt[3]{bc+ca} \leq \frac{1}{4} \left(\frac{(bc+ca)+8+8}{3} \right) \quad \text{and} \quad \sqrt[3]{ca+ab} \leq \frac{1}{4} \left(\frac{(ca+ab)+8+8}{3} \right).$$

Adding the last three inequalities gives us

$$\sqrt[3]{ab+bc} + \sqrt[3]{bc+ca} + \sqrt[3]{ca+ab} \leq \frac{1}{4} \left(\frac{2(ab+bc+ca)+48}{3} \right). \tag{5.7}$$

Since $ab+bc+ca \leq \frac{(a+b+c)^2}{3} = 12$ by (5.7) we get

$$\sqrt[3]{ab+bc} + \sqrt[3]{bc+ca} + \sqrt[3]{ca+ab} \leq \frac{1}{4} \left(\frac{24+48}{3} \right) = 6.$$

Equality occurs if and only if $ab+bc = bc+ca = ca+ab = 8$, i.e. $a = b = c = 2$.

Solution 2 The given inequality is equivalent to

$$\sqrt[3]{b(a+c)}+\sqrt[3]{c(b+a)}+\sqrt[3]{a(c+b)}\le 6$$

i.e.

$$\sqrt[3]{b(6-b)}+\sqrt[3]{c(6-c)}+\sqrt[3]{a(6-a)}\le 6. \tag{5.8}$$

Since at the point of incidence $a=b=c=2$ we have $2a=6-a=4$ by $AM\ge GM$ we deduce

$$\sqrt[3]{a(6-a)}=\frac{\sqrt[3]{2a\cdot(6-a)\cdot 4}}{2}\le\frac{2a+6-a+4}{6}=\frac{a+10}{6}.$$

Analogously we obtain

$$\sqrt[3]{b(6-b)}\le\frac{b+10}{6}\quad \text{and} \sqrt[3]{c(6-c)}\le\frac{c+10}{6}.$$

After adding the last three inequalities we get

$$\sqrt[3]{b(6-b)}+\sqrt[3]{c(6-c)}+\sqrt[3]{a(6-a)}\le\frac{a+b+c+30}{6}=\frac{36}{6}=6.$$

Equality occurs if and only if $a=b=c=2$.

Exercise 5.14 Let a,b,c be positive real numbers such that $a^2+b^2+c^2=3$. Prove the inequality

$$\sqrt[3]{a^2+bc}+\sqrt[3]{b^2+ca}+\sqrt[3]{c^2+ab}\le 3\sqrt[3]{2}.$$

Solution Since we have a symmetrical expression we estimate that the maximum value will occur at the point of incidence $a=b=c=1$. Then we have $a^2+bc=2$.

By the inequality $AM\ge GM$ we get

$$\sqrt[3]{a^2+bc}=\frac{1}{\sqrt[3]{4}}\cdot\sqrt[3]{(a^2+bc)\cdot 2\cdot 2}\le\frac{1}{\sqrt[3]{4}}\left(\frac{a^2+bc+4}{3}\right).$$

Similarly we obtain

$$\sqrt[3]{b^2+ca}\le\frac{1}{\sqrt[3]{4}}\left(\frac{b^2+ca+4}{3}\right)\quad\text{and}\quad\sqrt[3]{c^2+ab}\le\frac{1}{\sqrt[3]{4}}\left(\frac{c^2+ab+4}{3}\right).$$

Adding the last three inequalities gives us

$$\begin{aligned}\sqrt[3]{a^2+bc}+\sqrt[3]{b^2+ca}+\sqrt[3]{c^2+ab}&\le\frac{1}{3\sqrt[3]{4}}(a^2+b^2+c^2+ab+bc+ca+12)\\&\le\frac{1}{3\sqrt[3]{4}}(2(a^2+b^2+c^2)+12)=\frac{18}{3\sqrt[3]{4}}=3\sqrt[3]{2}.\end{aligned}$$

Equality occurs if and only if $a=b=c=1$.

Exercise 5.15 Let a, b, c be positive real numbers such that $a^2+b^2+c^2=3$. Prove the inequality

$$\sqrt[4]{5a^2+4(b+c)+3}+\sqrt[4]{5b^2+4(c+a)+3}+\sqrt[4]{5c^2+4(a+b)+3}\leq 6.$$

Solution At the point of incidence $a=b=c=1$ we have $5a^2+4(b+c)+3=16$.
By the inequality $AM \geq GM$ we get

$$\begin{aligned}\sqrt[4]{5a^2+4(b+c)+3} &= \frac{\sqrt[4]{(5a^2+4(b+c)+3)\cdot 16^3}}{8}\\ &\leq \frac{1}{32}(5a^2+4(b+c)+3+3\cdot 16)\\ &= \frac{5a^2+4(b+c)+51}{32}.\end{aligned}$$

Similarly we obtain

$$\sqrt[4]{5b^2+4(c+a)+3}\leq \frac{5b^2+4(c+a)+51}{32}$$

and

$$\sqrt[4]{5c^2+4(a+b)+3}\leq \frac{5c^2+4(a+b)+51}{32}.$$

Adding the last three inequalities gives us

$$\begin{aligned}&\sqrt[4]{5a^2+4(b+c)+3}+\sqrt[4]{5b^2+4(c+a)+3}+\sqrt[4]{5c^2+4(a+b)+3}\\ &\quad\leq \frac{5(a^2+b^2+c^2)+8(a+b+c)+153}{32}.\end{aligned}$$

Since $a^2+b^2+c^2=3$ we have $a+b+c\leq\sqrt{3(a^2+b^2+c^2)}=3$, and by the last inequality we obtain

$$\begin{aligned}&\sqrt[4]{5a^2+4(b+c)+3}+\sqrt[4]{5b^2+4(c+a)+3}+\sqrt[4]{5c^2+4(a+b)+3}\\ &\quad\leq \frac{5\cdot 3+8\cdot 3+153}{32}=\frac{192}{32}=6.\end{aligned}$$

Equality occurs if and only if $a=b=c=1$.

Chapter 6
The Rearrangement Inequality

In this section we will introduce one really useful inequality called the *rearrangement inequality*. This inequality has a very broad and easy use in proving other inequalities.

Theorem 6.1 (Rearrangement inequality) *Let $a_1 \leq a_2 \leq \cdots \leq a_n$ and $b_1 \leq b_2 \leq \cdots \leq b_n$ be real numbers. For any permutation $(x_1, x_2, \ldots, x_n)$ of $(a_1, a_2, \ldots, a_n)$ we have the following inequalities*:

$$\begin{aligned} a_1b_1 + a_2b_2 + \cdots + a_nb_n &\geq x_1b_1 + x_2b_2 + \cdots + x_nb_n \\ &\geq a_nb_1 + a_{n-1}b_2 + \cdots + a_1b_n. \end{aligned}$$

In case when $a_1 < a_2 < \cdots < a_n$ and $b_1 < b_2 < \cdots < b_n$ there is a simple necessary and sufficient condition for equality in either of the inequalities. The left inequality becomes equality only if $(x_1, x_2, \ldots, x_n)$ matches $(a_1, a_2, \ldots, a_n)$, and the right inequality becomes equality only if $(x_1, x_2, \ldots, x_n)$ matches $(a_n, a_{n-1}, \ldots, a_1)$.

Corollary 6.1 *Let $a_1, a_2, \ldots, a_n$ be real numbers and let $(x_1, x_2, \ldots, x_n)$ be a permutation of $(a_1, a_2, \ldots, a_n)$. Then*

$$a_1^2 + a_2^2 + \cdots + a_n^2 \geq a_1x_1 + a_2x_2 + \cdots + a_nx_n.$$

Exercise 6.1 Let a, b and c be positive real numbers. Prove *Nesbitt's inequality*

$$\frac{a}{b+c} + \frac{b}{c+a} + \frac{c}{a+b} \geq \frac{3}{2}.$$

Z. Cvetkovski, *Inequalities*,
DOI 10.1007/978-3-642-23792-8_6, © Springer-Verlag Berlin Heidelberg 2012

Solution Without loss of generality we may assume that $a \geq b \geq c$. Then clearly

$$\frac{1}{b+c} \geq \frac{1}{c+a} \geq \frac{1}{a+b}.$$

By the *rearrangement inequality* we deduce

$$\frac{a}{b+c} + \frac{b}{c+a} + \frac{c}{a+b} \geq \frac{b}{b+c} + \frac{c}{c+a} + \frac{a}{a+b}$$

and

$$\frac{a}{b+c} + \frac{b}{c+a} + \frac{c}{a+b} \geq \frac{c}{b+c} + \frac{a}{c+a} + \frac{b}{a+b}.$$

Adding the last two inequalities gives us

$$2\left(\frac{a}{b+c} + \frac{b}{c+a} + \frac{c}{a+b}\right) \geq 3 \quad \text{or} \quad \frac{a}{b+c} + \frac{b}{c+a} + \frac{c}{a+b} \geq \frac{3}{2}.$$

Exercise 6.2 Let $a_1 \leq a_2 \leq \cdots \leq a_n$ and $b_1 \leq b_2 \leq \cdots \leq b_n$ be two sequences of real numbers and let $(c_1, c_2, \ldots, c_n)$ be a permutation of $(b_1, b_2, \ldots, b_n)$. Prove that

$$(a_1-b_1)^2+(a_2-b_2)^2+\cdots+(a_n-b_n)^2 \leq (a_1-c_1)^2+(a_2-c_2)^2+\cdots+(a_n-c_n)^2.$$

Solution Note that $b_1^2 + b_2^2 + \cdots + b_n^2 = c_1^2 + c_2^2 + \cdots + c_n^2$.

So it suffices to prove that

$$a_1c_1 + a_2c_2 + \cdots + a_nc_n \leq a_1b_1 + a_2b_2 + \cdots + a_nb_n,$$

which is true due to the *rearrangement inequality*.

Exercise 6.3 Let $a_1, a_2, \ldots, a_n$ be different positive integers. Prove the inequality

$$\frac{a_1}{1^2} + \frac{a_2}{2^2} + \cdots + \frac{a_n}{n^2} \geq 1 + \frac{1}{2} + \cdots + \frac{1}{n}.$$

Solution Let $(x_1, x_2, \ldots, x_n)$ be a permutation of $(a_1, a_2, \ldots, a_n)$ such that $x_1 \leq x_2 \leq \cdots \leq x_n$.

Then clearly $x_i \geq i$ for each $i = 1, 2, \ldots, n$ and $\frac{1}{1^2} \geq \frac{1}{2^2} \geq \cdots \geq \frac{1}{n^2}$.

By the *rearrangement inequality* and the previous conclusion we obtain

$$\frac{a_1}{1^2} + \frac{a_2}{2^2} + \cdots + \frac{a_n}{n^2} \geq \frac{x_1}{1^2} + \frac{x_2}{2^2} + \cdots + \frac{x_n}{n^2} \geq 1 + \frac{2}{2^2} + \cdots + \frac{n}{n^2} = 1 + \frac{1}{2} + \cdots + \frac{1}{n}.$$

Exercise 6.4 Let a, b, c be the lengths of the sides of a triangle. Prove the inequality

$$a^2(b+c-a) + b^2(c+a-b) + c^2(a+b-c) \leq 3abc.$$

Solution Without loss of generality we may assume that $a \geq b \geq c$. Then very easily we can verify that

$$c(a+b-c) \geq b(c+a-b) \geq a(b+c-a).$$

Applying the *rearrangement inequality* we obtain the following inequalities

$$\begin{aligned} &a^2(b+c-a)+b^2(c+a-b)+c^2(a+b-c) \\ &\quad \leq ba(b+c-a)+cb(c+a-b)+ac(a+b-c) \end{aligned}$$

and

$$\begin{aligned} &a^2(b+c-a)+b^2(c+a-b)+c^2(a+b-c) \\ &\quad \leq ca(b+c-a)+ab(c+a-b)+bc(a+b-c). \end{aligned}$$

Adding the last two inequalities gives us the required result.

Exercise 6.5 Let a, b, c be real numbers. Prove the inequality

$$a^5+b^5+c^5 \geq a^4b+b^4c+c^4a.$$

Solution 1 Without loss of generality we may assume that $a \geq b \geq c$, and then clearly $a^4 \geq b^4 \geq c^4$ (since the given inequality is cyclic we also need to consider the case when $c \geq b \geq a$, which is analogous).

Now by the *rearrangement inequality* we get the required inequality. Equality occurs iff $a=b=c$.

Solution 2 Since $AM \geq GM$ we obtain the following inequalities:

$$\begin{aligned} a^5+a^5+a^5+a^5+b^5 &\geq 5a^4b, \\ b^5+b^5+b^5+b^5+c^5 &\geq 5b^4c, \\ c^5+c^5+c^5+c^5+a^5 &\geq 5c^4a, \end{aligned}$$

and adding the previous three inequalities yields required inequality. Equality occurs iff $a=b=c$.

Exercise 6.6 Let a, b, c be positive real numbers such that $abc=1$. Prove the inequality

$$\frac{1}{a^3(b+c)}+\frac{1}{b^3(c+a)}+\frac{1}{c^3(a+b)} \geq \frac{3}{2}.$$

Solution Without loss of generality we may assume that $a \geq b \geq c$.

Let $x=\frac{1}{a}, y=\frac{1}{b}, z=\frac{1}{c}$. Then clearly $xyz=1$.

We have

$$\frac{1}{a^3(b+c)}+\frac{1}{b^3(c+a)}+\frac{1}{c^3(a+b)}=\frac{x^3}{1/y+1/z}+\frac{y^3}{1/z+1/x}+\frac{z^3}{1/x+1/y}$$
$$=\frac{x^2}{y+z}+\frac{y^2}{z+x}+\frac{z^2}{x+y}.$$

Since $c\le b\le a$ we have $x\le y\le z$.

So clearly $x+y\le z+x\le y+z$ and $\frac{x}{y+z}\le\frac{y}{z+x}\le\frac{z}{x+y}$.

Now by the *rearrangement inequality* we get the following inequalities

$$\frac{x^2}{y+z}+\frac{y^2}{z+x}+\frac{z^2}{x+y}\ge\frac{xy}{y+z}+\frac{yz}{z+x}+\frac{zx}{x+y},$$
$$\frac{x^2}{y+z}+\frac{y^2}{z+x}+\frac{z^2}{x+y}\ge\frac{xz}{y+z}+\frac{yx}{z+x}+\frac{zy}{x+y}.$$

So we obtain

$$2\left(\frac{1}{a^3(b+c)}+\frac{1}{b^3(c+a)}+\frac{1}{c^3(a+b)}\right)$$
$$=2\left(\frac{x^2}{y+z}+\frac{y^2}{z+x}+\frac{z^2}{x+y}\right)$$
$$\ge\frac{xy}{y+z}+\frac{yz}{z+x}+\frac{zx}{x+y}+\frac{xz}{y+z}+\frac{yx}{z+x}+\frac{zy}{x+y}$$
$$=x+y+z\ge 3\sqrt[3]{xyz}=3,$$

as required.

Exercise 6.7 Let a,b,c be positive real numbers. Prove the inequality

$$\frac{a^2+c^2}{b}+\frac{b^2+a^2}{c}+\frac{c^2+b^2}{a}\ge 2(a+b+c).$$

Solution Since the given inequality is symmetric, without loss of generality we may assume that $a\ge b\ge c$. Then clearly

$$a^2\ge b^2\ge c^2 \quad\text{and}\quad \frac{1}{c}\ge\frac{1}{b}\ge\frac{1}{a}.$$

By the *rearrangement inequality* we have

$$\frac{a^2}{b}+\frac{b^2}{c}+\frac{c^2}{a}=a^2\cdot\frac{1}{b}+b^2\cdot\frac{1}{c}+c^2\cdot\frac{1}{a}\ge a^2\cdot\frac{1}{a}+b^2\cdot\frac{1}{b}+c^2\cdot\frac{1}{c}=a+b+c \qquad (6.1)$$

and

$$\frac{a^2}{c}+\frac{b^2}{a}+\frac{c^2}{b}=a^2\cdot\frac{1}{c}+b^2\cdot\frac{1}{a}+c^2\cdot\frac{1}{b}\geq a^2\cdot\frac{1}{a}+b^2\cdot\frac{1}{b}+c^2\cdot\frac{1}{c}=a+b+c. \quad (6.2)$$

Adding (6.1) and (6.2) yields the required inequality.

Equality occurs if and only if $a=b=c$.

Exercise 6.8 Let $x, y, z > 0$ be real numbers. Prove the inequality

$$\frac{x^2-z^2}{y+z}+\frac{y^2-x^2}{z+x}+\frac{z^2-y^2}{x+y}\geq 0.$$

Solution We need to prove that $\frac{x^2}{y+z}+\frac{y^2}{z+x}+\frac{z^2}{x+y}\geq\frac{z^2}{y+z}+\frac{x^2}{z+x}+\frac{y^2}{x+y}$.

Without loss of generality we may assume that $x\geq y\geq z$(since the given inequality is cyclic we also will consider the case $z\geq y\geq x$).

Then clearly $x^2\geq y^2\geq z^2$ and $\frac{1}{y+z}\geq\frac{1}{z+x}\geq\frac{1}{x+y}$.

By the *rearrangement inequality* we have

$$\frac{x^2}{y+z}+\frac{y^2}{z+x}+\frac{z^2}{x+y}\geq\frac{z^2}{y+z}+\frac{x^2}{z+x}+\frac{y^2}{x+y},$$

as required.

If we assume that $z\geq y\geq x$, then $z^2\geq y^2\geq x^2$ and $\frac{1}{x+y}\geq\frac{1}{x+z}\geq\frac{1}{z+y}$.

By the *rearrangement inequality* we obtain

$$\begin{aligned}\frac{x^2}{y+z}+\frac{y^2}{z+x}+\frac{z^2}{x+y}&=z^2\cdot\frac{1}{x+y}+x^2\cdot\frac{1}{y+z}+y^2\cdot\frac{1}{z+x}\\&\geq z^2\cdot\frac{1}{y+z}+x^2\cdot\frac{1}{z+x}+y^2\cdot\frac{1}{x+y}\\&=\frac{z^2}{y+z}+\frac{x^2}{z+x}+\frac{y^2}{x+y}.\end{aligned}$$

Equality occurs if and only if $x=y=z$.

Exercise 6.9 Let x, y, z be positive real numbers. Prove the inequality

$$\frac{x^3}{yz}+\frac{y^3}{zx}+\frac{z^3}{xy}\geq x+y+z.$$

Solution Since the given inequality is symmetric we may assume that $x\geq y\geq z$.

Then

$$x^3\geq y^3\geq z^3 \quad\text{and}\quad \frac{1}{yz}\geq\frac{1}{zx}\geq\frac{1}{xy}.$$

By the *rearrangement inequality* we have

$$\begin{aligned}\frac{x^3}{yz}+\frac{y^3}{zx}+\frac{z^3}{xy}&=x^3\cdot\frac{1}{yz}+y^3\cdot\frac{1}{zx}+z^3\cdot\frac{1}{xy}\\&\geq x^3\cdot\frac{1}{xy}+y^3\cdot\frac{1}{yz}+z^3\cdot\frac{1}{zx}=\frac{x^2}{y}+\frac{y^2}{z}+\frac{z^2}{x}.\end{aligned}\tag{6.3}$$

We will prove that

$$\frac{x^2}{y}+\frac{y^2}{z}+\frac{z^2}{x}\geq x+y+z.\tag{6.4}$$

Let $x\geq y\geq z$.

Then $x^2\geq y^2\geq z^2$ and $\frac{1}{z}\geq\frac{1}{y}\geq\frac{1}{x}$ (since inequality (6.4) is cyclic we also need to consider the case $z\geq y\geq x$).

By the *rearrangement inequality* we obtain

$$\frac{x^2}{y}+\frac{y^2}{z}+\frac{z^2}{x}\geq\frac{x^2}{x}+\frac{y^2}{y}+\frac{z^2}{z}=x+y+z.$$

The case when $z\geq y\geq x$ is analogous to the previous case.

Now by (6.3) and (6.4) we obtain

$$\frac{x^3}{yz}+\frac{y^3}{zx}+\frac{z^3}{xy}\geq x+y+z.$$

Equality occurs if and only if $x=y=z$.

Exercise 6.10 Let a,b,c,d be positive real numbers such that $a+b+c+d=4$. Prove the inequality

$$a^2bc+b^2cd+c^2da+d^2ab\leq 4.$$

Solution Let (x,y,z,t) be a permutation of (a,b,c,d) such that $x\geq y\geq z\geq t$. Then clearly $xyz\geq xyt\geq xzt\geq yzt$.

By the *rearrangement inequality* we obtain

$$x\cdot xyz+y\cdot xyt+z\cdot xzt+t\cdot yzt\geq a^2bc+b^2cd+c^2da+d^2ab.\tag{6.5}$$

Since $AM\geq GM$ we deduce

$$x\cdot xyz+y\cdot xyt+z\cdot xzt+t\cdot yzt=(xy+zt)(xz+yt)\leq\frac{(xy+xz+yt+zt)^2}{4}.\tag{6.6}$$

Since

$$xy+xz+yt+zt=(x+z)(y+t)\leq\frac{(x+y+z+t)^2}{4}=4$$

by (6.6) we deduce that

$$x \cdot xyz + y \cdot xyt + z \cdot xzt + t \cdot yzt \leq 4.$$

Finally by (6.5) we obtain

$$a^2bc + b^2cd + c^2da + d^2ab \leq 4,$$

and we are done.

Equality holds iff $a = b = c = d = 1$ or $a = 2, b = c = 1, d = 0$ (up to permutation).

Chapter 7
Convexity, Jensen's Inequality

The main purpose of this section is to acquaint the reader with one of the most important theorems, that is widely used in proving inequalities, *Jensen's inequality*. This is an inequality regarding so-called convex functions, so firstly we will give some definitions and theorems whose proofs are subject to mathematical analysis, and therefore we'll present them here without proof.

Also we will consider that the reader has an elementary knowledge of differential calculus.

Definition 7.1 For the function $f : [a, b] \to \mathbb{R}$ we'll say that it is convex on the interval $[a, b]$ if for any $x, y \in [a, b]$ and any $\alpha \in (0, 1)$ we have

$$f(\alpha x + (1-\alpha)y) \leq \alpha f(x) + (1-\alpha)f(y). \tag{7.1}$$

If in (7.1) we have strict inequality then we'll say that f is strictly convex.

For the function f we'll say that it is concave if $-f$ is a convex function.

If the function f is defined on $\mathbb{R}$, it can happen that on some interval this function is a convex function, but on another interval it is a concave function. For this reason, we will consider functions defined on intervals.

Example 7.1 The function $f(x) = x^2$ is convex on $\mathbb{R}$, moreover $f(x) = x^n$ is convex on $\mathbb{R}$ for even n. Also $f(x) = x^n$ is convex on $\mathbb{R}^+$ for n odd, and it is concave on $\mathbb{R}^-$.

The function $f(x) = \sin x$ on $(\pi, 2\pi)$ is convex, but on $(0, \pi)$ it is concave.

Now we will state a theorem that will give a criterion for determining whether and when a function is convex, respectively concave.

Z. Cvetkovski, *Inequalities*,
DOI 10.1007/978-3-642-23792-8_7, © Springer-Verlag Berlin Heidelberg 2012

Theorem 7.1 *Let $f:(a,b)\to\mathbb{R}$ and for any $x\in(a,b)$ suppose there exists a second derivative $f''(x)$. The function $f(x)$ is convex on (a,b) if and only if for each $x\in(a,b)$ we have $f''(x)\geq 0$. If $f''(x)>0$ for each $x\in(a,b)$, then f is strictly convex on (a,b).*

Clearly, according to Definition 7.1 and Theorem 7.1 we have that the function $f(x)$ is concave on (a,b) if and only if $f''(x)\leq 0$, for all $x\in(a,b)$.

Example 7.2 Consider the power function $f:\mathbb{R}^+\to\mathbb{R}^+$ defined as $f(x)=x^\alpha$. For the second derivative we have $f''(x)=\alpha(\alpha-1)x^{\alpha-2}$, and clearly $f''(x)>0$ for $\alpha>1$ or $\alpha<0$ and $f''(x)<0$ for $0<\alpha<1$. So f is (strictly) convex for $\alpha>1$ or $\alpha<0$ and f is (strictly) concave for $0<\alpha<1$.

Example 7.3 For the function $f:\mathbb{R}\to\mathbb{R}$, $f(x)=\ln(1+e^x)$ we have $f'(x)=\frac{e^x}{1+e^x}$, and $f''(x)=\frac{e^x}{(1+e^x)^2}>0$ for $x\in\mathbb{R}$, and therefore f is convex on $\mathbb{R}$.

Example 7.4 For the function $f:\mathbb{R}^+\to\mathbb{R}^+$, $f(x)=(1+x^\alpha)^{\frac{1}{\alpha}}$ for $\alpha\neq 0$ we have $f''(x)=(\alpha-1)x^{\alpha-2}(1+x^\alpha)^{\frac{1}{\alpha}}$, from where it follows that for $\alpha<1$ the function f is strictly concave and for $\alpha>1$ the function f is strictly convex.

Theorem 7.2 *Let $f_1,f_2,\ldots,f_n$ be convex functions on (a,b). Then the function $c_1f_1+c_2f_2+\cdots+c_nf_n$ is also convex on (a,b), for any $c_1,c_2,\ldots,c_n\in(0,\infty)$.*

Theorem 7.3 (Jensen's inequality) *Let $f:(a,b)\to\mathbb{R}$ be a convex function on the interval (a,b). Let $n\in\mathbb{N}$ and $\alpha_1,\alpha_2,\ldots,\alpha_n\in(0,1)$ be real numbers such that $\alpha_1+\alpha_2+\cdots+\alpha_n=1$. Then for any $x_1,x_2,\ldots,x_n\in(a,b)$ we have*

$$f\left(\sum_{i=1}^n\alpha_ix_i\right)\leq\sum_{i=1}^n\alpha_if(x_i),$$

i.e.

$$f(\alpha_1x_1+\alpha_2x_2+\cdots+\alpha_nx_n)\leq\alpha_1f(x_1)+\alpha_2f(x_2)+\cdots+\alpha_nf(x_n).\quad(7.2)$$

Proof We'll prove inequality (7.2) by mathematical induction.

For $n=1$ we have $\alpha_1=1$ and since $f(x_1)=f(x_1)$ we get $f(\alpha_1x_1)=\alpha_1f(x_1)$, so (7.2) is true.

Let $n = 2$. Then (7.2) holds due to Definition 7.1.

Suppose that for $n = k$, and any real numbers $\alpha_1, \alpha_2, \ldots, \alpha_k \in [0, 1]$ such that $\alpha_1 + \alpha_2 + \cdots + \alpha_k = 1$ and any $x_1, x_2, \ldots, x_k \in (a, b)$, we have

$$f(\alpha_1 x_1 + \cdots + \alpha_k x_k) \le \alpha_1 f(x_1) + \cdots + \alpha_k f(x_k). \tag{7.3}$$

Let $n = k + 1$, and let $\alpha_1, \alpha_2, \ldots, \alpha_{k+1} \in [0, 1]$ such that $\alpha_1 + \alpha_2 + \cdots + \alpha_{k+1} = 1$.

Let $x_1, x_2, \ldots, x_{k+1} \in (a, b)$.

Then we have

$$\begin{aligned}
&\alpha_1 x_1 + \alpha_2 x_2 + \cdots + \alpha_{k+1} x_{k+1} \\
&\quad = (\alpha_1 x_1 + \cdots + \alpha_k x_k) + \alpha_{k+1} x_{k+1} \\
&\quad = (1 - \alpha_{k+1}) \left(\frac{\alpha_1}{1 - \alpha_{k+1}} x_1 + \frac{\alpha_2}{1 - \alpha_{k+1}} x_2 + \cdots + \frac{\alpha_k}{1 - \alpha_{k+1}} x_k \right) + \alpha_{k+1} x_{k+1}.
\end{aligned} \tag{7.4}$$

Let

$$\frac{\alpha_1}{1 - \alpha_{k+1}} x_1 + \frac{\alpha_2}{1 - \alpha_{k+1}} x_2 + \cdots + \frac{\alpha_k}{1 - \alpha_{k+1}} x_k = y_{k+1}.$$

Then since $x_1, x_2, \ldots, x_k \in (a, b)$ we deduce

$$\begin{aligned}
y_{k+1} &= \frac{\alpha_1}{1 - \alpha_{k+1}} x_1 + \frac{\alpha_2}{1 - \alpha_{k+1}} x_2 + \cdots + \frac{\alpha_k}{1 - \alpha_{k+1}} x_k \\
&< \frac{\alpha_1}{1 - \alpha_{k+1}} b + \frac{\alpha_2}{1 - \alpha_{k+1}} b + \cdots + \frac{\alpha_k}{1 - \alpha_{k+1}} b \\
&< \frac{b}{1 - \alpha_{k+1}} (\alpha_1 + \alpha_2 + \cdots + \alpha_k) = \frac{b}{1 - \alpha_{k+1}} (1 - \alpha_{k+1}) = b.
\end{aligned}$$

Similarly we deduce that $y_{k+1} > a$.

Thus $y_{k+1} \in (a, b)$.

According to Definition 7.1 and by (7.4) we obtain

$$\begin{aligned}
f(\alpha_1 x_1 + \cdots + \alpha_k x_k + \alpha_{k+1} x_{k+1}) &= f((1 - \alpha_{k+1}) y_{k+1} + \alpha_{k+1} x_{k+1}) \\
&\le (1 - \alpha_{k+1}) f(y_{k+1}) + \alpha_{k+1} f(x_{k+1}).
\end{aligned} \tag{7.5}$$

By inequality (7.3) and since

$$\frac{\alpha_1}{1 - \alpha_{k+1}} + \frac{\alpha_2}{1 - \alpha_{k+1}} + \cdots + \frac{\alpha_k}{1 - \alpha_{k+1}} = 1$$

we obtain

$$\begin{aligned}
f(y_{k+1}) &= f\left(\frac{\alpha_1}{1 - \alpha_{k+1}} x_1 + \frac{\alpha_2}{1 - \alpha_{k+1}} x_2 + \cdots + \frac{\alpha_k}{1 - \alpha_{k+1}} x_k \right) \\
&\le \frac{\alpha_1}{1 - \alpha_{k+1}} f(x_1) + \frac{\alpha_2}{1 - \alpha_{k+1}} f(x_2) + \cdots + \frac{\alpha_k}{1 - \alpha_{k+1}} f(x_k).
\end{aligned} \tag{7.6}$$

Finally according to (7.5) and (7.6) we deduce

$$f(\alpha_1 x_1 + \cdots + \alpha_{k+1} x_{k+1}) \leq \alpha_1 f(x_1) + \cdots + \alpha_{k+1} f(x_{k+1}).$$

So by the principle of mathematical induction inequality, (7.2) holds for any positive integer n, any $\alpha_1, \alpha_2, \ldots, \alpha_n \in [0, 1]$ such that $\alpha_1 + \alpha_2 + \cdots + \alpha_n = 1$, and arbitrary $x_1, x_2, \ldots, x_n \in (a, b)$. □

Remark If f is strictly convex then equality in *Jensen's inequality* occurs only for $x_1 = x_2 = \cdots = x_n$.

If the function $f(x)$ is concave then in *Jensen's inequality* we have the reverse inequality, i.e.

$$f(\alpha_1 x_1 + \cdots + \alpha_n x_n) \geq \alpha_1 f(x_1) + \cdots + \alpha_n f(x_n).$$

It is important to note that *Jensen's inequality* can also be written in the equivalent form:

If $f : I \to \mathbb{R}$ is convex on $I, x_1, x_2, \ldots, x_n \in I$ and $m_1, m_2, \ldots, m_n \geq 0$ are real numbers such that $m_1 + m_2 + \cdots + m_n > 0$. Then

$$f\left(\frac{m_1 x_1 + m_2 x_2 + \cdots + m_n x_n}{m_1 + m_2 + \cdots + m_n}\right) \leq \frac{m_1 f(x_1) + m_2 f(x_2) + \cdots + m_n f(x_n)}{m_1 + m_2 + \cdots + m_n}.$$

Example 7.5 Consider the function $f(x) = -\ln x$, on the interval $(0, +\infty)$. For the second derivative we have $f''(x) = \frac{1}{x^2} > 0$, which means that $f(x)$ is a strictly convex on $x \in (0, +\infty)$.

By *Jensen's inequality* for $\alpha_1 = \alpha_2 = \cdots = \alpha_n = \frac{1}{n}$, and $x_i \in (0, +\infty), i = 1, 2, \ldots, n$, we obtain

$$\begin{aligned} -\ln\left(\frac{x_1 + x_2 + \cdots + x_n}{n}\right) &\leq -\left(\frac{\ln x_1 + \ln x_2 + \cdots + \ln x_n}{n}\right) \\ \Leftrightarrow \quad \frac{\ln x_1 + \ln x_2 + \cdots + \ln x_n}{n} &\leq \ln\left(\frac{x_1 + x_2 + \cdots + x_n}{n}\right) \\ \Leftrightarrow \quad \ln(x_1 x_2 \cdots x_n)^{1/n} &\leq \ln\left(\frac{x_1 + x_2 + \cdots + x_n}{n}\right), \end{aligned}$$

i.e.

$$\sqrt[n]{x_1 x_2 \cdots x_n} \leq \frac{x_1 + x_2 + \cdots + x_n}{n},$$

which is the well-known inequality $AM \geq GM$.

Example 7.6 Let us consider the function $f(x) = x^2$. Since $f''(x) = 2 > 0$ it follows that f is convex on $\mathbb{R}$. Then by *Jensen's inequality*

$$f\left(\frac{m_1 x_1 + m_2 x_2 + \cdots + m_n x_n}{m_1 + m_2 + \cdots + m_n}\right) \leq \frac{m_1 f(x_1) + m_2 f(x_2) + \cdots + m_n f(x_n)}{m_1 + m_2 + \cdots + m_n},$$

we obtain

$$\left(\frac{m_1x_1+m_2x_2+\cdots+m_nx_n}{m_1+m_2+\cdots+m_n}\right)^2\leq\frac{m_1x_1^2+m_2x_2^2+\cdots+m_nx_n^2}{m_1+m_2+\cdots+m_n},$$

i.e.

$$(m_1x_1+m_2x_2+\cdots+m_nx_n)^2\leq(m_1x_1^2+m_2x_2^2+\cdots+m_nx_n^2)(m_1+m_2+\cdots+m_n).$$

By taking $m_i=b_i^2, x_i=\frac{a_i}{b_i}$ for $i=1,2,\dots,n$ in the last inequality, we obtain

$$(a_1b_1+a_2b_2+\cdots+a_nb_n)^2\leq(a_1^2+a_2^2+\cdots+a_n^2)(b_1^2+b_2^2+\cdots+b_n^2),$$

which is the well-known *Cauchy–Schwarz inequality*.

On this occasion we will present *Popoviciu's inequality*, which will be used in the same manner as *Jensen's inequality*. But we must note that this inequality is stronger then *Jensen's inequality*, i.e. in some cases this inequality can be a powerful tool for proving other inequalities, where *Jensen's inequality* does not work.

Theorem 7.4 (Popoviciu's inequality) *Let $f:[a,b]\to\mathbb{R}$ be a convex function on the interval $[a,b]$. Then for any $x,y,z\in[a,b]$ we have*

$$f\left(\frac{x+y+z}{3}\right)+\frac{f(x)+f(y)+f(z)}{3}\geq\frac{2}{3}\left(f\left(\frac{x+y}{2}\right)+f\left(\frac{y+z}{2}\right)+f\left(\frac{z+x}{2}\right)\right).$$

Proof Without loss of generality we assume that $x\leq y\leq z$.

If $y\leq\frac{x+y+z}{3}$ then $\frac{x+y+z}{3}\leq\frac{x+z}{2}\leq z$ and $\frac{x+y+z}{3}\leq\frac{y+z}{2}\leq z$.

Therefore there exist $s,t\in[0,1]$ such that

$$\frac{x+z}{2}=\left(\frac{x+y+z}{3}\right)s+z(1-s)\quad\text{and}\tag{7.7}$$

$$\frac{y+z}{2}=\left(\frac{x+y+z}{3}\right)t+z(1-t).\tag{7.8}$$

Summing (7.7) and (7.8) gives

$$\frac{x+y-2z}{2}=\frac{x+y-2z}{3}(s+t),$$

from which we obtain $s+t=\frac{3}{2}$.

Because the function f is convex, we have

$$f\left(\frac{x+z}{2}\right) \le s \cdot f\left(\frac{x+y+z}{3}\right) + (1-s)f(z),$$
$$f\left(\frac{y+z}{2}\right) \le t \cdot f\left(\frac{x+y+z}{3}\right) + (1-t)f(z)$$

and

$$f\left(\frac{x+y}{2}\right) \le \frac{1}{2}f(x) + \frac{1}{2}f(y).$$

After adding together the last three inequalities we obtain the required inequality.

The case when $\frac{x+y+z}{3} < y$ is considered similarly, bearing in mind that $x \le \frac{x+z}{2} \le \frac{x+y+z}{3}$ and $x \le \frac{y+z}{2} \le \frac{x+y+z}{3}$. □

Note If f is a concave function on $[a,b]$ then in *Popoviciu's inequality* for all $x, y, z \in [a,b]$ we have the reverse inequality, i.e. we have

$$f\left(\frac{x+y+z}{3}\right) + \frac{f(x)+f(y)+f(z)}{3}$$
$$\le \frac{2}{3}\left(f\left(\frac{x+y}{2}\right) + f\left(\frac{y+z}{2}\right) + f\left(\frac{z+x}{2}\right)\right).$$

Theorem 7.5 (Generalized Popoviciu's inequality) *Let $f : [a,b] \to \mathbb{R}$ be a convex function on the interval $[a,b]$ and $a_1, a_2, \ldots, a_n \in [a,b]$. Then*

$$f(a_1) + f(a_2) + \cdots + f(a_n) + n(n-2)f(a)$$
$$\ge (n-1)(f(b_1) + f(b_2) + \cdots + f(b_n)),$$

where $a = \frac{a_1+a_2+\cdots+a_n}{n}$, and $b_i = \frac{1}{n-1}\sum_{i \ne j} a_j$ for all i.

Theorem 7.6 (Weighted *AM–GM* inequality) *Let $a_i \in (0,\infty)$, $i = 1, 2, \ldots, n$, and $\alpha_i \in [0,1]$, $i = 1, 2, \ldots, n$, be such that $\alpha_1 + \alpha_2 + \cdots + \alpha_n = 1$. Then*

$$a_1^{\alpha_1} a_2^{\alpha_2} \cdots a_n^{\alpha_n} \le a_1\alpha_1 + a_2\alpha_2 + \cdots + a_n\alpha_n. \tag{7.9}$$

Proof For the function $f(x) = -\ln x$ we have $f'(x) = -\frac{1}{x}$ and $f''(x) = \frac{1}{x^2}$, i.e. $f''(x) > 0$, for $x \in (0,\infty)$.

So due to Theorem 7.1 we conclude that the function f is convex on $(0,\infty)$.

Let $a_i \in (0, \infty), i = 1, 2, \ldots, n$, and $\alpha_i \in [0, 1], i = 1, 2, \ldots, n$, be arbitrary real numbers such that $\alpha_1 + \alpha_2 + \cdots + \alpha_n = 1$.

By *Jensen's inequality* we deduce

$$
\begin{aligned}
&-\ln\left(\sum_{i=1}^{n} a_i\alpha_i\right) = f\left(\sum_{i=1}^{n} a_i\alpha_i\right) \leq \sum_{i=1}^{n} \alpha_i f(a_i) = -\sum_{i=1}^{n} \alpha_i \ln a_i \\
\Leftrightarrow\quad & -\ln(a_1\alpha_1 + a_2\alpha_2 + \cdots + a_n\alpha_n) \leq -\alpha_1 \ln a_1 - \alpha_2 \ln a_2 - \cdots - \alpha_n \ln a_n \\
\Leftrightarrow\quad & \alpha_1 \ln a_1 + \alpha_2 \ln a_2 + \cdots + \alpha_n \ln a_n \leq \ln(a_1\alpha_1 + a_2\alpha_2 + \cdots + a_n\alpha_n) \\
\Leftrightarrow\quad & \ln a_1^{\alpha_1} a_2^{\alpha_2} \cdots a_n^{\alpha_n} \leq \ln(a_1\alpha_1 + a_2\alpha_2 + \cdots + a_n\alpha_n) \\
\Leftrightarrow\quad & a_1^{\alpha_1} a_2^{\alpha_2} \cdots a_n^{\alpha_n} \leq a_1\alpha_1 + a_2\alpha_2 + \cdots + a_n\alpha_n,
\end{aligned}
$$

as required. □

Note By inequality (7.9) for $\alpha_1 = \alpha_2 = \cdots = \alpha_n = \frac{1}{n}$, we obtain the inequality $AM \geq GM$.

Exercise 7.1 Let α, β, γ be the angles of a triangle. Prove the inequality

$$\sin\alpha \sin\beta \sin\gamma \leq \frac{3\sqrt{3}}{8}.$$

Solution Since $\alpha, \beta, \gamma \in (0, \pi)$ it follows that $\sin\alpha, \sin\beta, \sin\gamma > 0$.

Therefore since $AM \geq GM$ we obtain

$$\sqrt[3]{\sin\alpha \sin\beta \sin\gamma} \leq \frac{\sin\alpha + \sin\beta + \sin\gamma}{3}. \tag{7.10}$$

Since $f(x) = \sin x$ is concave on $(0, \pi)$, by *Jensen's inequality* we deduce

$$\frac{\sin\alpha + \sin\beta + \sin\gamma}{3} \leq \sin\frac{\alpha + \beta + \gamma}{3} = \frac{\sqrt{3}}{2}. \tag{7.11}$$

Due to (7.10) and (7.11) we get

$$\sqrt[3]{\sin\alpha \sin\beta \sin\gamma} \leq \frac{\sqrt{3}}{2} \quad \Leftrightarrow \quad \sin\alpha \sin\beta \sin\gamma \leq \frac{3\sqrt{3}}{8}.$$

Equality occurs iff $\alpha = \beta = \gamma$, i.e. the triangle is equilateral.

Exercise 7.2 Let $a, b, c \in \mathbb{R}^+$. Prove the inequalities:

(1) $4(a^3 + b^3) \geq (a + b)^3$;
(2) $9(a^3 + b^3 + c^3) \geq (a + b + c)^3$.

Solution (1) The function $f(x)=x^3$ is convex on $(0,+\infty)$, thus from *Jensen's inequality* it follows that

$$\left(\frac{a+b}{2}\right)^3 \le \frac{a^3+b^3}{2} \quad \Leftrightarrow \quad 4(a^3+b^3)\ge (a+b)^3.$$

(2) Similarly as in (1) we deduce that

$$\left(\frac{a+b+c}{3}\right)^3 \le \frac{a^3+b^3+c^3}{3} \quad \Leftrightarrow \quad 9(a^3+b^3+c^3)\ge (a+b+c)^3.$$

Exercise 7.3 Let $\alpha_i>0, i=1,2,\ldots,n$, be real numbers such that $\alpha_1+\alpha_2+\cdots+\alpha_n=1$. Prove the inequality

$$\alpha_1^{\alpha_1}\alpha_2^{\alpha_2}\cdots\alpha_n^{\alpha_n}\ge\frac{1}{n}.$$

Solution If we take $a_i=\frac{1}{\alpha_i}, i=1,2,\ldots,n$, by the *Weighted AM–GM inequality* we get

$$\frac{1}{\alpha_1^{\alpha_1}}\frac{1}{\alpha_2^{\alpha_2}}\cdots\frac{1}{\alpha_n^{\alpha_n}}\le\frac{1}{\alpha_1}\alpha_1+\frac{1}{\alpha_2}\alpha_2+\cdots+\frac{1}{\alpha_n}\alpha_n=n,$$

i.e.

$$\frac{1}{n}\le\alpha_1^{\alpha_1}\alpha_2^{\alpha_2}\cdots\alpha_n^{\alpha_n}.$$

Exercise 7.4 Find the minimum value of k such that for arbitrary $a,b>0$ we have

$$\sqrt[3]{a}+\sqrt[3]{b}\le k\sqrt[3]{a+b}.$$

Solution Consider the function $f(x)=\sqrt[3]{x}$.

We have $f'(x)=\frac{1}{3}x^{-\frac{2}{3}}$ and $f''(x)=-\frac{2}{9}x^{-\frac{5}{3}}<0$, for any $x\in(0,\infty)$. Thus $f(x)$ is concave on the interval $(0,\infty)$.

By *Jensen's inequality* we deduce

$$\begin{aligned}&\frac{1}{2}f(a)+\frac{1}{2}f(b)\le f\left(\frac{a+b}{2}\right)\\ \Leftrightarrow\quad &\frac{\sqrt[3]{a}+\sqrt[3]{b}}{2}\le\sqrt[3]{\frac{a+b}{2}}\\ \Leftrightarrow\quad &\sqrt[3]{a}+\sqrt[3]{b}\le\frac{2}{\sqrt[3]{2}}\sqrt[3]{a+b}=\sqrt[3]{4}\cdot\sqrt[3]{a+b}.\end{aligned}$$

Therefore $k_{\min}=\sqrt[3]{4}$, and for instance we reach this value for $a=b$.

Exercise 7.5 Let $x, y, z \geq 0$ be real numbers. Prove the inequality

$$\sqrt{x^2+1}+\sqrt{y^2+1}+\sqrt{z^2+1} \geq \sqrt{6(x+y+z)}.$$

Solution Consider the function $f(t)=\sqrt{t^2+1}, t \geq 0$.

Since $f''(t)=\frac{1}{(\sqrt{t^2+1})^3}>0$, f is convex on $[0, \infty)$.

Therefore by *Jensen's inequality* we have

$$\frac{\sqrt{x^2+1}+\sqrt{y^2+1}+\sqrt{z^2+1}}{3} \geq \sqrt{\left(\frac{x+y+z}{3}\right)^2+1},$$

i.e.

$$\sqrt{x^2+1}+\sqrt{y^2+1}+\sqrt{z^2+1} \geq \sqrt{(x+y+z)^2+9}. \tag{7.12}$$

From the obvious inequality $((x+y+z)-3)^2 \geq 0$ it follows that

$$(x+y+z)^2+9 \geq 6(x+y+z). \tag{7.13}$$

By (7.12) and (7.13) we obtain

$$\sqrt{x^2+1}+\sqrt{y^2+1}+\sqrt{z^2+1} \geq \sqrt{(x+y+z)^2+9} \geq \sqrt{6(x+y+z)}.$$

Equality occurs if and only if $x=y=z=1$.

Exercise 7.6 Let x, y, z be positive real numbers. Prove the inequality

$$\frac{x+y}{z}+\frac{y+z}{x}+\frac{z+x}{y} \geq 4\left(\frac{z}{x+y}+\frac{x}{y+z}+\frac{y}{z+x}\right).$$

Solution Consider the function $f(x)=x+\frac{1}{x}$.

Since $f'(x)=1-\frac{1}{x^2}$ and $f''(x)=\frac{2}{x^3}>0$ for any $x>0$ it follows that f is convex on $\mathbb{R}^+$.

Now by *Popoviciu's inequality* we can easily obtain the required inequality.

Chapter 8
Trigonometric Substitutions and Their Application for Proving Algebraic Inequalities

Very often, for proving a given algebraic inequality we can use trigonometric substitutions that work amazingly well, and can almost always lead the solver to a solution.

Using such substitutions, a given inequality may simplify to the point, where the final part of the proof will be only routine, and will need previous results (usually *Jensen's inequality* and elements of trigonometry). Therefore it is necessary to possess a knowledge of trigonometry.

We will give some basic facts that must be known and which are of benefit when *Jensen's inequality* is being used. Namely, the function $\sin x$ is concave on $(0,\pi)$, the function $\cos x$ is concave on $(-\pi/2,\pi/2)$, hence also on $(0,\pi/2)$, $\tan x$ is convex on $(0,\pi/2)$, while the function $\cot x$ is convex on $(0,\pi/2)$.

Furthermore, without proof (the proofs are "pure" trigonometry, and some of them can be found in standard collections of problems in mathematics at secondary level) we will give several trigonometric identities relating the angles of a triangle, which the reader should certainly know.

Proposition 8.1 *Let α,β,γ be the angles of a given triangle. Then we have the following identities*:

I_1: $\cos\alpha+\cos\beta+\cos\gamma=1+4\sin\frac{\alpha}{2}\cdot\sin\frac{\beta}{2}\cdot\sin\frac{\gamma}{2}$
I_2: $\sin\alpha+\sin\beta+\sin\gamma=4\cos\frac{\alpha}{2}\cdot\cos\frac{\beta}{2}\cdot\cos\frac{\gamma}{2}$
I_3: $\sin 2\alpha+\sin 2\beta+\sin 2\gamma=4\sin\alpha\cdot\sin\beta\cdot\sin\gamma$
I_4: $\sin^2\alpha+\sin^2\beta+\sin^2\gamma=2+2\cos\alpha\cdot\cos\beta\cdot\cos\gamma$
I_4': $\sin^2\frac{\alpha}{2}+\sin^2\frac{\beta}{2}+\sin^2\frac{\gamma}{2}+2\sin\frac{\alpha}{2}\cdot\sin\frac{\beta}{2}\cdot\sin\frac{\gamma}{2}=1$
I_5: $\tan\frac{\alpha}{2}\cdot\tan\frac{\beta}{2}+\tan\frac{\beta}{2}\cdot\tan\frac{\gamma}{2}+\sin\frac{\gamma}{2}\cdot\sin\frac{\alpha}{2}=1$
I_6: $\tan\alpha+\tan\beta+\tan\gamma=\tan\alpha\cdot\tan\beta\cdot\tan\gamma$
I_7: $\cot\frac{\alpha}{2}+\cot\frac{\beta}{2}+\cot\frac{\gamma}{2}=\cot\frac{\alpha}{2}\cdot\cot\frac{\beta}{2}\cdot\cot\frac{\gamma}{2}$.

Z. Cvetkovski, *Inequalities*,
DOI 10.1007/978-3-642-23792-8_8, © Springer-Verlag Berlin Heidelberg 2012

Proposition 8.2 *Let α, β, γ be arbitrary real numbers. Then we have*:

I_8: $\sin\alpha + \sin\beta + \sin\gamma - \sin(\alpha+\beta+\gamma) = 4\sin\frac{\alpha+\beta}{2}\cdot\sin\frac{\beta+\gamma}{2}\cdot\sin\frac{\gamma+\alpha}{2}$

I_9: $\cos\alpha + \cos\beta + \cos\gamma + \cos(\alpha+\beta+\gamma) = 4\cos\frac{\alpha+\beta}{2}\cdot\cos\frac{\beta+\gamma}{2}\cdot\cos\frac{\gamma+\alpha}{2}$.

Now we will give several inequalities concerning the angles of a given triangle, which will be used in proving inequalities by using trigonometric substitutions, and which are of great importance. The method of introducing certain substitutions and knowledge of these inequalities are the essence of this way of proving algebraic inequalities.

Proposition 8.3 *Let α, β, γ be the angles of a given triangle. Then we have the following inequalities*:

N_1: $\sin\alpha + \sin\beta + \sin\gamma \le \frac{3\sqrt{3}}{2}$ N_2: $\sin\alpha\cdot\sin\beta\cdot\sin\gamma \le \frac{3\sqrt{3}}{8}$

N_3: $\sin\frac{\alpha}{2} + \sin\frac{\beta}{2} + \sin\frac{\gamma}{2} \le \frac{3}{2}$ N_4: $\sin\frac{\alpha}{2}\cdot\sin\frac{\beta}{2}\cdot\sin\frac{\gamma}{2} \le \frac{1}{8}$

N_5: $\cos\alpha + \cos\beta + \cos\gamma \le \frac{3}{2}$ N_6: $\cos\alpha\cdot\cos\beta\cdot\cos\gamma \le \frac{1}{8}$

N_7: $\cos\frac{\alpha}{2} + \cos\frac{\beta}{2} + \cos\frac{\gamma}{2} \le \frac{3\sqrt{3}}{2}$ N_8: $\cos\frac{\alpha}{2}\cdot\cos\frac{\beta}{2}\cdot\cos\frac{\gamma}{2} \le \frac{3\sqrt{3}}{8}$

N_9: $\sin^2\alpha + \sin^2\beta + \sin^2\gamma \le \frac{9}{4}$ N_{10}: $\sin^2\frac{\alpha}{2} + \sin^2\frac{\beta}{2} + \sin^2\frac{\gamma}{2} \ge \frac{3}{4}$

N_{11}: $\cos^2\alpha + \cos^2\beta + \cos^2\gamma \ge \frac{3}{4}$ N_{12}: $\cos^2\frac{\alpha}{2} + \cos^2\frac{\beta}{2} + \cos^2\frac{\gamma}{2} \le \frac{9}{4}$

N_{13}: $\tan\frac{\alpha}{2} + \tan\frac{\beta}{2} + \tan\frac{\gamma}{2} \ge \sqrt{3}$ N_{14}: $\cot\frac{\alpha}{2} + \cot\frac{\beta}{2} + \cot\frac{\gamma}{2} \ge 3\sqrt{3}$

N_{15}: $\cot\alpha + \cot\beta + \cot\gamma \ge \sqrt{3}$.

Proof N_1: The function $\sin x$ is concave on the interval $(0, \pi)$, thus from *Jensen's inequality* we obtain

$$\frac{\sin\alpha + \sin\beta + \sin\gamma}{3} \le \sin\left(\frac{\alpha+\beta+\gamma}{3}\right) = \sin\frac{\pi}{3} = \frac{\sqrt{3}}{2}$$

$$\Leftrightarrow \quad \sin\alpha + \sin\beta + \sin\gamma \le \frac{3\sqrt{3}}{2}.$$

N_2: Since $\sin x > 0$ for any $x \in (0, \pi)$ we can apply the inequality $AM \ge GM$, and we obtain

$$\sin\alpha\cdot\sin\beta\cdot\sin\gamma \le \left(\frac{\sin\alpha + \sin\beta + \sin\gamma}{3}\right)^3 \overset{N_1}{\le} \left(\frac{\sqrt{3}}{2}\right)^3 = \frac{3\sqrt{3}}{8}.$$

N_3: Similarly as in the proof of N_1 we have

$$\frac{\sin\frac{\alpha}{2} + \sin\frac{\beta}{2} + \sin\frac{\gamma}{2}}{3} \le \sin\left(\frac{\alpha+\beta+\gamma}{6}\right) = \sin\frac{\pi}{6} = \frac{1}{2}$$

or

$$\sin\frac{\alpha}{2}+\sin\frac{\beta}{2}+\sin\frac{\gamma}{2}\leq\frac{3}{2},$$

since the function $\sin x$ is concave on $(0,\pi/2)$.

N_4: Similarly as in proof of N_2 and since $AM \geq GM$ we have

$$\sqrt[3]{\sin\frac{\alpha}{2}\cdot\sin\frac{\beta}{2}\cdot\sin\frac{\gamma}{2}}\leq\frac{\sin\frac{\alpha}{2}+\sin\frac{\beta}{2}+\sin\frac{\gamma}{2}}{3}\overset{N_4}{\leq}\frac{1}{2},$$

i.e.

$$\sin\frac{\alpha}{2}\cdot\sin\frac{\beta}{2}\cdot\sin\frac{\gamma}{2}\leq\frac{1}{8}.$$

N_5: Since $\alpha+\beta=\pi-\gamma$ it follows that

$$\cos\gamma=-\cos(\alpha+\beta)=-\cos\alpha\cos\beta+\sin\alpha\sin\beta.$$

Thus

$$\begin{aligned}3-2(\cos\alpha+\cos\beta+\cos\gamma)&=3-2(\cos\alpha+\cos\beta-\cos\alpha\cos\beta+\sin\alpha\sin\beta)\\&=\sin^2\alpha+\sin^2\beta-2\sin\alpha\sin\beta+1+\cos^2\alpha\\&\quad+\cos^2\beta-2\cos\alpha-2\cos\beta+2\cos\alpha\cos\beta\\&=(\sin\alpha-\sin\beta)^2+(1-\cos\alpha-\cos\beta)^2\geq 0,\end{aligned}$$

which is equivalent to

$$\cos\alpha+\cos\beta+\cos\gamma\leq\frac{3}{2}.$$

N_6: Since $\cos(\alpha+\beta)=-\cos\gamma$, we have

$$\begin{aligned}\cos\alpha\cos\beta\cos\gamma&=\frac{1}{2}(\cos(\alpha+\beta)+\cos(\alpha-\beta))\cos\gamma\\&=\frac{1}{2}(\cos(\alpha-\beta)-\cos\gamma)\cos\gamma=\frac{1}{2}\cos(\alpha-\beta)\cos\gamma-\frac{\cos^2\gamma}{2}\\&=-\frac{1}{2}\left(\cos\gamma-\frac{\cos(\alpha-\beta)}{2}\right)^2+\frac{\cos^2(\alpha-\beta)}{8}\\&\leq\frac{\cos^2(\alpha-\beta)}{8}\leq\frac{1}{8}.\end{aligned}$$

N_7: Since $\alpha,\beta,\gamma\in(0,\pi)$ it follows that $\alpha/2,\beta/2,\gamma/2\in(0,\pi/2)$.
The function $\cos x$ is concave on the interval $(0,\pi/2)$.
Thus by *Jensen's inequality* we get

$$\frac{\cos\frac{\alpha}{2}+\cos\frac{\beta}{2}+\cos\frac{\gamma}{2}}{3}\leq\cos\frac{\alpha+\beta+\gamma}{6}=\cos\frac{\pi}{6}=\frac{\sqrt{3}}{2},$$

i.e.

$$\cos\frac{\alpha}{2}+\cos\frac{\beta}{2}+\cos\frac{\gamma}{2}\leq\frac{3\sqrt{3}}{2}.$$

N_8: Since $\alpha,\beta,\gamma\in(0,\pi)$ it follows that $\alpha/2,\beta/2,\gamma/2\in(0,\pi/2)$, i.e.

$$\cos\alpha,\cos\beta,\cos\gamma>0,$$

so we can apply $AM\geq GM$ to conclude that

$$\sqrt[3]{\cos\frac{\alpha}{2}\cos\frac{\beta}{2}\cos\frac{\gamma}{2}}\leq\frac{\cos\frac{\alpha}{2}+\cos\frac{\beta}{2}+\cos\frac{\gamma}{2}}{3}\overset{N_7}{\leq}\frac{\sqrt{3}}{2},$$

from which it follows that

$$\cos\frac{\alpha}{2}\cdot\cos\frac{\beta}{2}\cdot\cos\frac{\gamma}{2}\leq\frac{3\sqrt{3}}{8}.$$

N_9: By identity I_4 and inequality N_6 we obtain

$$\sin^2\alpha+\sin^2\beta+\sin^2\gamma=2+2\cos\alpha\cdot\cos\beta\cdot\cos\gamma\leq 2+2\cdot\frac{1}{8}=\frac{9}{4}.$$

N_{10}: By I_4' we have that

$$\sin^2\frac{\alpha}{2}+\sin^2\frac{\beta}{2}+\sin^2\frac{\gamma}{2}+2\sin\frac{\alpha}{2}\cdot\sin\frac{\beta}{2}\cdot\sin\frac{\gamma}{2}=1,$$

i.e.

$$\sin^2\frac{\alpha}{2}+\sin^2\frac{\beta}{2}+\sin^2\frac{\gamma}{2}=1-2\sin\frac{\alpha}{2}\cdot\sin\frac{\beta}{2}\cdot\sin\frac{\gamma}{2}.$$

According to N_4: $\sin\frac{\alpha}{2}\cdot\sin\frac{\beta}{2}\cdot\sin\frac{\gamma}{2}\leq\frac{1}{8}$ and the previous identity we obtain

$$\sin^2\frac{\alpha}{2}+\sin^2\frac{\beta}{2}+\sin^2\frac{\gamma}{2}\geq 1-\frac{2}{8}=\frac{3}{4}.$$

N_{11}: We have

$$\cos^2\alpha+\cos^2\beta+\cos^2\gamma=3-(\sin^2\alpha+\sin^2\beta+\sin^2\gamma)\overset{N_9}{\geq}3-\frac{9}{4}=\frac{3}{4}.$$

N_{12}: We have

$$\cos^2\frac{\alpha}{2}+\cos^2\frac{\beta}{2}+\cos^2\frac{\gamma}{2}=3-\left(\sin^2\frac{\alpha}{2}+\sin^2\frac{\beta}{2}+\sin^2\frac{\gamma}{2}\right)\overset{N_{10}}{\leq}3-\frac{3}{4}=\frac{9}{4}.$$

N_{13}: Since $\tan x$ is convex on the interval $(0,\pi/2)$, by *Jensen's inequality* we deduce

$$\frac{\tan\frac{\alpha}{2}+\tan\frac{\beta}{2}+\tan\frac{\gamma}{2}}{3}\geq\tan\frac{\alpha+\beta+\gamma}{6}=\tan\frac{\pi}{6}=\frac{1}{\sqrt{3}},$$

i.e.

$$\tan\frac{\alpha}{2}+\tan\frac{\beta}{2}+\tan\frac{\gamma}{2}\geq\sqrt{3}.$$

N_{14}: Due to the convexity of $\cot x$ on $(0,\pi/2)$ by *Jensen's inequality* we obtain

$$\cot\frac{\alpha}{2}+\cot\frac{\beta}{2}+\cot\frac{\gamma}{2}\geq 3\cot\frac{\alpha+\beta+\gamma}{6}=3\sqrt{3}.$$

N_{15}: Firstly we have

$$\cot\alpha+\cot\beta=\frac{\cos\alpha}{\sin\alpha}+\frac{\cos\beta}{\sin\beta}=\frac{\cos\alpha\sin\beta+\sin\alpha\cos\beta}{\sin\alpha\sin\beta}=\frac{\sin(\alpha+\beta)}{\sin\alpha\sin\beta}. \quad (8.1)$$

Also

$$1\geq\cos(\alpha-\beta)=\cos\alpha\cos\beta+\sin\alpha\sin\beta, \quad (8.2)$$

$$\cos\gamma=-\cos(\alpha+\beta)=-\cos\alpha\cos\beta+\sin\alpha\sin\beta. \quad (8.3)$$

Adding (8.2) and (8.3) gives us

$$\begin{aligned}
&2\sin\alpha\sin\beta\leq 1+\cos\gamma\\
&\Leftrightarrow\quad 2\sin\alpha\sin\beta\sin(\alpha+\beta)\leq(1+\cos\gamma)\sin(\alpha+\beta)\\
&\Leftrightarrow\quad 2\sin\alpha\sin\beta\sin\gamma\leq(1+\cos\gamma)\sin(\alpha+\beta)\\
&\Leftrightarrow\quad \frac{2\sin\alpha\sin\beta\sin\gamma}{\sin\alpha\sin\beta(1+\cos\gamma)}\leq\frac{(1+\cos\gamma)\sin(\alpha+\beta)}{\sin\alpha\sin\beta(1+\cos\gamma)}\\
&\Leftrightarrow\quad \frac{2\sin\gamma}{1+\cos\gamma}\leq\frac{\sin(\alpha+\beta)}{\sin\alpha\sin\beta}. \quad (8.4)
\end{aligned}$$

Therefore

$$\begin{aligned}
\cot\alpha+\cot\beta+\cot\gamma&\overset{(8.1)}{=}\frac{\sin(\alpha+\beta)}{\sin\alpha\sin\beta}+\frac{\cos\gamma}{\sin\gamma}\overset{(8.4)}{\geq}\frac{2\sin\gamma}{1+\cos\gamma}+\frac{\cos\gamma}{\sin\gamma}\\
&=\frac{1}{2}\left(\frac{4\sin^2\gamma+2\cos^2\gamma+2\cos\gamma}{(1+\cos\gamma)\sin\gamma}\right)\\
&=\frac{1}{2}\left(\frac{3\sin^2\gamma+(1+\cos\gamma)^2}{(1+\cos\gamma)\sin\gamma}\right)\\
&\geq\frac{1}{2}\left(\frac{2\sqrt{3\sin^2\gamma(1+\cos\gamma)^2}}{(1+\cos\gamma)\sin\gamma}\right)=\frac{2\sqrt{3}}{2}=\sqrt{3},
\end{aligned}$$

as required. □

Proposition 8.4 *Let α, β, γ be the angles of an acute triangle. Then*

$$N_{16}\text{: } \tan\alpha + \tan\beta + \tan\gamma \geq 3\sqrt{3}.$$

Proof Since the triangle is acute it follows that $\alpha, \beta, \gamma \in (0, \pi/2)$. The function $f(x) = \tan x$ is convex on $(0, \pi/2)$, so by *Jensen's inequality* we obtain

$$\tan\alpha + \tan\beta + \tan\gamma \geq 3\tan\frac{\alpha+\beta+\gamma}{3} = 3\tan\frac{\pi}{3} = 3\sqrt{3}.$$

Furthermore, we'll give two theorems that will be the basis for the introduction of trigonometric substitutions. □

Theorem 8.1 *Let $\alpha, \beta, \gamma \in (0, \pi)$. Then α, β and γ are the angles of a triangle if and only if*

$$\tan\frac{\alpha}{2}\tan\frac{\beta}{2} + \tan\frac{\beta}{2}\tan\frac{\gamma}{2} + \tan\frac{\alpha}{2}\tan\frac{\gamma}{2} = 1.$$

Proof Let α, β, γ be the angles of an arbitrary triangle. Then $\alpha + \beta + \gamma = \pi$, i.e. $\frac{\gamma}{2} = \frac{\pi}{2} - \frac{\alpha+\beta}{2}$.

Therefore

$$\tan\frac{\gamma}{2} = \tan\left(\frac{\pi}{2} - \frac{\alpha+\beta}{2}\right) = \cot\left(\frac{\alpha}{2} + \frac{\beta}{2}\right) = \frac{\cot\frac{\alpha}{2}\cot\frac{\beta}{2} - 1}{\cot\frac{\alpha}{2} + \cot\frac{\beta}{2}} = \frac{1 - \tan\frac{\alpha}{2}\tan\frac{\beta}{2}}{\tan\frac{\alpha}{2} + \tan\frac{\beta}{2}}$$

$$\Leftrightarrow \quad \tan\frac{\alpha}{2}\tan\frac{\beta}{2} + \tan\frac{\beta}{2}\tan\frac{\gamma}{2} + \tan\frac{\alpha}{2}\tan\frac{\gamma}{2} = 1.$$

Conversely, let us suppose that for some $\alpha, \beta, \gamma \in (0, \pi)$ we have

$$\tan\frac{\alpha}{2}\tan\frac{\beta}{2} + \tan\frac{\beta}{2}\tan\frac{\gamma}{2} + \tan\frac{\alpha}{2}\tan\frac{\gamma}{2} = 1. \tag{8.5}$$

If $\alpha = \beta = \gamma$ then $3\tan^2\frac{\alpha}{2} = 1$, and since $\tan\frac{\alpha}{2} > 0$ we get $\tan\frac{\alpha}{2} = \frac{1}{\sqrt{3}}$, i.e. $\alpha = \beta = \gamma = 60°$, from which it follows that $\alpha + \beta + \gamma = \pi$, i.e. α, β and γ are the angles of a triangle.

Without loss of generality let us assume that $\alpha \neq \beta$.

Since $0 < \alpha + \beta < 2\pi$ it follows that there exists $\gamma_1 \in (-\pi, \pi)$ such that $\alpha + \beta + \gamma_1 = \pi$.

Then by the previous part of this proof we must have

$$\tan\frac{\alpha}{2}\tan\frac{\beta}{2} + \tan\frac{\beta}{2}\tan\frac{\gamma_1}{2} + \tan\frac{\alpha}{2}\tan\frac{\gamma_1}{2} = 1. \tag{8.6}$$

We'll show that $\gamma=\gamma_1$, from which it will follow that $\alpha+\beta+\gamma=\pi$, i.e. α, β and γ are the angles of a triangle.

If we subtract (8.5) and (8.6) we get

$$\tan\frac{\gamma}{2}=\tan\frac{\gamma_1}{2}, \quad \text{i.e.} \quad \left|\frac{\gamma-\gamma_1}{2}\right|=k\pi, \quad \text{for some } k\geq 0, k\in\mathbb{Z}.$$

But $|\frac{\gamma-\gamma_1}{2}|\leq\frac{\gamma}{2}+\frac{\gamma_1}{2}<\frac{\pi}{2}+\frac{\pi}{2}=\pi$, so it follows that $k=0$, i.e. $\gamma=\gamma_1$, and the proof is finished. □

Theorem 8.2 *Let $\alpha, \beta, \gamma\in(0,\pi)$. Then α, β and γ are the angles of a triangle if and only if*

$$\sin^2\frac{\alpha}{2}+\sin^2\frac{\beta}{2}+\sin^2\frac{\gamma}{2}+2\sin\frac{\alpha}{2}\cdot\sin\frac{\beta}{2}\cdot\sin\frac{\gamma}{2}=1.$$

Proof Let α, β, γ be the angles of a triangle. Then we have

$$\begin{aligned}
&\sin^2\frac{\gamma}{2}+2\sin\frac{\alpha}{2}\cdot\sin\frac{\beta}{2}\cdot\sin\frac{\gamma}{2}\\
&=\cos^2\frac{\alpha+\beta}{2}+2\sin\frac{\alpha}{2}\cdot\sin\frac{\beta}{2}\cdot\cos\frac{\alpha+\beta}{2}\\
&=\cos\frac{\alpha+\beta}{2}\left(\cos\frac{\alpha+\beta}{2}+2\sin\frac{\alpha}{2}\cdot\sin\frac{\beta}{2}\right)\\
&=\cos\frac{\alpha+\beta}{2}\left(\cos\frac{\alpha+\beta}{2}+\cos\frac{\alpha-\beta}{2}-\cos\frac{\alpha+\beta}{2}\right)\\
&=\cos\frac{\alpha+\beta}{2}\cdot\cos\frac{\alpha-\beta}{2}=\frac{1}{2}(\cos\alpha+\cos\beta)\\
&=\frac{1-2\sin^2\frac{\alpha}{2}+1-\sin^2\frac{\beta}{2}}{2}=1-\sin^2\frac{\alpha}{2}-\sin^2\frac{\beta}{2},
\end{aligned}$$

i.e.

$$\sin^2\frac{\alpha}{2}+\sin^2\frac{\beta}{2}+\sin^2\frac{\gamma}{2}+2\sin\frac{\alpha}{2}\cdot\sin\frac{\beta}{2}\cdot\sin\frac{\gamma}{2}=1.$$

Conversely, let $\alpha, \beta, \gamma\in(0,\pi)$ be such that

$$\sin^2\frac{\alpha}{2}+\sin^2\frac{\beta}{2}+\sin^2\frac{\gamma}{2}+2\sin\frac{\alpha}{2}\cdot\sin\frac{\beta}{2}\cdot\sin\frac{\gamma}{2}=1. \tag{8.7}$$

We'll show that $\alpha+\beta+\gamma=\pi$.

Since $0<\alpha+\beta<2\pi$ it follows that there exists $\gamma_1\in(-\pi,\pi)$ such that $\alpha+\beta+\gamma_1=\pi$.

Then clearly

$$\sin^2\frac{\alpha}{2}+\sin^2\frac{\beta}{2}+\sin^2\frac{\gamma_1}{2}+2\sin\frac{\alpha}{2}\cdot\sin\frac{\beta}{2}\cdot\sin\frac{\gamma_1}{2}=1. \tag{8.8}$$

By subtracting (8.7) and (8.8) we obtain

$$\left(\sin\frac{\gamma}{2}-\sin\frac{\gamma_1}{2}\right)\left(\sin\frac{\gamma}{2}+\sin\frac{\gamma_1}{2}+2\sin\frac{\alpha}{2}\cdot\sin\frac{\beta}{2}\right)=0. \tag{8.9}$$

But

$$\begin{aligned}\sin\frac{\gamma}{2}+\sin\frac{\gamma_1}{2}+2\sin\frac{\alpha}{2}\cdot\sin\frac{\beta}{2}&=\sin\frac{\gamma}{2}+\sin\frac{\gamma_1}{2}+\cos\frac{\alpha-\beta}{2}-\cos\frac{\alpha+\beta}{2}\\&=\sin\frac{\gamma}{2}+\sin\frac{\gamma_1}{2}+\cos\frac{\alpha-\beta}{2}-\sin\frac{\gamma_1}{2}\\&=\sin\frac{\gamma}{2}+\cos\frac{\alpha-\beta}{2}.\end{aligned}$$

Since $\frac{\gamma}{2}\in(0,\pi/2)$ and $\frac{\alpha-\beta}{2}\in(-\pi/2,\pi/2)$ it follows that

$$\sin\frac{\gamma}{2}+\cos\frac{\alpha-\beta}{2}>0,\quad \text{i.e.}\quad \sin\frac{\gamma}{2}+\sin\frac{\gamma_1}{2}+2\sin\frac{\alpha}{2}\cdot\sin\frac{\beta}{2}>0.$$

From the last inequality and (8.9) we have

$$\sin\frac{\gamma}{2}=\sin\frac{\gamma_1}{2},\quad \text{i.e.}\quad \gamma=\gamma_1.$$

Thus $\alpha+\beta+\gamma=\pi$, as required. □

Now, based on these two theorems we will give basic cases, how a given algebraic inequality can be transformed by trigonometric substitutions. These substitutions, with the inequalities of Propositions 8.3 and 8.4 will be a powerful apparatus for proving algebraic inequalities.

8.1 The Most Usual Forms of Trigonometric Substitutions

Case 1. Let α, β and γ be the angles of an arbitrary triangle.
Let $A=\frac{\pi-\alpha}{2}$, $B=\frac{\pi-\beta}{2}$ and $C=\frac{\pi-\gamma}{2}$.
Then $A+B+C=\pi$; moreover $0<A,B,C<\pi/2$, i.e. this substitution allows us to transfer angles of an arbitrary triangle to angles of an acute triangle. (This is especially important when we use *Jensen's inequality*, since "Jensen" could not be used for the function $\cos x$ on the interval $(0,\pi)$, but only on the interval $(0,\pi/2)$.)
Observe that we have:

$$\sin\frac{\alpha}{2}=\cos A,\qquad \sin\frac{\beta}{2}=\cos B,\qquad \sin\frac{\gamma}{2}=\cos C.$$

Note: There are similar identities for the functions $\cos x$, $\tan x$ and $\cot x$.

Case 2. Let x, y and z be positive real numbers. Then there exist triangle with length-sides $a = x + y, b = y + z, c = z + x$.
Clearly $(x, y, z) = (s - b, s - c, s - a)$, where $s = \frac{a+b+c}{2} = x + y + z$.

Case 3. Let a, b, c be positive real numbers such that $ab + bc + ca = 1$.
Since $\tan x \in (0, \infty)$ for $x \in (0, \pi/2)$, and due to Theorem 8.1 we can use the substitutions

$$a = \tan\frac{\alpha}{2}, \qquad b = \tan\frac{\beta}{2}, \qquad c = \tan\frac{\gamma}{2},$$

where α, β and γ are the angles of a triangle, i.e. $\alpha, \beta, \gamma \in (0, \pi)$ and $\alpha + \beta + \gamma = \pi$.

Case 4. Let a, b, c be positive real numbers such that $ab + bc + ca = 1$.
Then according to *Case 3* and *Case 1* we can use the following substitutions

$$a = \cot\alpha, \qquad b = \cot\beta, \qquad c = \cot\gamma,$$

where α, β and γ are the angles of an acute triangle, i.e. $\alpha, \beta, \gamma \in (0, \pi/2)$ and $\alpha + \beta + \gamma = \pi$.

Case 5. Let a, b and c be positive real numbers such that

$$a + b + c = abc, \quad \text{i.e.} \quad \frac{1}{bc} + \frac{1}{ca} + \frac{1}{ab} = 1.$$

Then according to *Case 3* we can take

$$\frac{1}{a} = \tan\frac{\alpha}{2}, \qquad \frac{1}{b} = \tan\frac{\beta}{2}, \qquad \frac{1}{c} = \tan\frac{\gamma}{2},$$

$$\text{i.e.} \quad a = \cot\frac{\alpha}{2}, \qquad b = \cot\frac{\beta}{2}, \qquad c = \cot\frac{\gamma}{2},$$

where α, β and γ are the angles of an arbitrary triangle.

Case 6. Let a, b, c be positive real numbers such that $a + b + c = abc$.
Then according to *Case 5* and *Case 1* we can use the following substitutions

$$a = \tan\alpha, \qquad b = \tan\beta, \qquad c = \tan\gamma,$$

where α, β and γ are the angles of an acute triangle, i.e. $\alpha, \beta, \gamma \in (0, \pi/2)$ and $\alpha + \beta + \gamma = \pi$.

Case 7. Let a, b, c be positive real numbers such that

$$a^2 + b^2 + c^2 + 2abc = 1.$$

Note that since the numbers a, b, c are positive we must have $a, b, c < 1$. Therefore due to Theorem 8.2 we can use the substitutions

$$a = \sin\frac{\alpha}{2}, \qquad b = \sin\frac{\beta}{2}, \qquad c = \sin\frac{\gamma}{2},$$

where α, β and γ are the angles of an arbitrary triangle, i.e. $\alpha, \beta, \gamma \in (0, \pi/2)$ and $\alpha+\beta+\gamma=\pi$.

Case 8. Let a, b, c be positive real numbers such that

$$a^2+b^2+c^2+2abc=1.$$

Then according to *Case 7* and *Case 1* we can make the following substitutions

$$a=\cos\alpha, \qquad b=\cos\beta, \qquad c=\cos\gamma,$$

where α, β and γ are the angles of an acute triangle.

Case 9. Let x, y, z be positive real numbers.
Then the expressions:

$$\sqrt{\frac{yz}{(x+y)(x+z)}}, \sqrt{\frac{zx}{(y+z)(y+x)}}, \sqrt{\frac{xy}{(z+x)(z+y)}}$$

with the substitutions from *Case 2* become

$$\sqrt{\frac{(s-b)(s-c)}{bc}}, \sqrt{\frac{(s-c)(s-a)}{ca}}, \sqrt{\frac{(s-a)(s-b)}{ab}},$$

where a, b, c are the length-sides of a triangle.
But let us notice that

$$\sin\frac{\alpha}{2}=\sqrt{\frac{(s-b)(s-c)}{bc}}, \qquad \sin\frac{\beta}{2}=\sqrt{\frac{(s-c)(s-a)}{ca}},$$
$$\sin\frac{\gamma}{2}=\sqrt{\frac{(s-a)(s-b)}{ab}}.$$

Therefore for the given expressions we can simply make the substitutions: $\sin\frac{\alpha}{2}, \sin\frac{\beta}{2}, \sin\frac{\gamma}{2}$ (respectively), where α, β and γ are the angles of a triangle.

Case 10. Similarly as in *Case 9*, for the expressions:

$$\sqrt{\frac{x(x+y+z)}{(x+y)(x+z)}}, \sqrt{\frac{y(x+y+z)}{(y+z)(y+x)}}, \sqrt{\frac{z(x+y+z)}{(z+x)(z+y)}},$$

we can make the substitutions $\cos\frac{\alpha}{2}, \cos\frac{\beta}{2}, \cos\frac{\gamma}{2}$ (respectively), where α, β and γ are the angles of a triangle.

Now we will give practical applications of this material, through exercises that will demonstrate how it works, and how useful is this apparatus, which is based on the aforementioned substitutions in certain cases.

8.2 Characteristic Examples Using Trigonometric Substitutions

Exercise 8.1 Let $x, y, z > 0$ be real numbers. Prove the inequality

$$\sqrt{x(y+z)}+\sqrt{y(z+x)}+\sqrt{z(x+y)} \geq 2\sqrt{\frac{(x+y)(y+z)(z+x)}{x+y+z}}.$$

Solution The given inequality is equivalent to

$$\sqrt{\frac{x(x+y+z)}{(x+y)(x+z)}}+\sqrt{\frac{y(x+y+z)}{(y+z)(y+x)}}+\sqrt{\frac{z(x+y+z)}{(z+x)(z+y)}} \geq 2.$$

According to *Case 10*, it suffices to show that

$$\cos\frac{\alpha}{2}+\cos\frac{\beta}{2}+\cos\frac{\gamma}{2} \geq 2,$$

where α, β and γ are the angles of a triangle, i.e. $\alpha, \beta, \gamma \in (0, \pi)$ and $\alpha+\beta+\gamma=\pi$. Due to *Case 1*, it remains to prove that

$$\sin\alpha+\sin\beta+\sin\gamma \geq 2,$$

where α, β and γ are the angles of an acute triangle, i.e. $\alpha, \beta, \gamma \in (0, \pi/2)$ and $\alpha+\beta+\gamma=\pi$.

Since $\alpha \in (0, \pi/2]$ it follows that $0 < \sin\alpha \leq 1$, i.e. $\sin\alpha \geq \sin^2\alpha$, and equality occurs if and only if $\alpha = \pi/2$.

Similarly for $\beta, \gamma \in (0, \pi/2]$ we conclude that

$$\sin\beta \geq \sin^2\beta \quad \text{and} \quad \sin\gamma \geq \sin^2\gamma.$$

Thus we have

$$\begin{aligned}
&\sin\alpha+\sin\beta+\sin\gamma\\
&\quad\geq \sin^2\alpha+\sin^2\beta+\sin^2\gamma\\
&\quad= \frac{1-\cos 2\alpha}{2}+\frac{1-\cos 2\beta}{2}+\sin^2\gamma = 1-\frac{1}{2}(\cos 2\alpha+\cos 2\beta)+\sin^2\gamma\\
&\quad= 1-\frac{1}{2}2\cos(\alpha+\beta)\cos(\alpha-\beta)+1-\cos^2\gamma\\
&\quad= 2-\cos(\pi-\gamma)\cos(\alpha-\beta)-\cos^2\gamma = 2+\cos\gamma\cos(\alpha-\beta)-\cos^2\gamma\\
&\quad= 2+\cos\gamma(\cos(\alpha-\beta)-\cos\gamma) = 2+\cos\gamma[\cos(\alpha-\beta)-\cos(\pi-(\alpha+\beta))]\\
&\quad= 2+\cos\gamma(\cos(\alpha-\beta)+\cos(\alpha+\beta)) = 2+2\cos\gamma\cos\alpha\cos\beta > 2.
\end{aligned}$$

Exercise 8.2 Let a, b and c be positive real numbers such that $a+b+c=1$. Prove the inequality

$$a^2+b^2+c^2+2\sqrt{3abc} \leq 1.$$

Solution After taking $a = xy, b = yz, c = zx$, inequality becomes

$$x^2y^2 + y^2z^2 + z^2x^2 + 2\sqrt{3}xyz \le 1, \tag{8.10}$$

where x, y, z are positive real numbers such that

$$xy + yz + zx = 1. \tag{8.11}$$

Inequality (8.10) is equivalent to

$$(xy + yz + zx)^2 + 2\sqrt{3}xyz \le 1 + 2xyz(x + y + z)$$

or

$$\sqrt{3} \le x + y + z. \tag{8.12}$$

By (8.11) and according to *Case 3*, we can take

$$x = \tan\frac{\alpha}{2}, \qquad y = \tan\frac{\beta}{2}, \qquad z = \tan\frac{\gamma}{2},$$

where $\alpha, \beta, \gamma \in (0, \pi)$ and $\alpha + \beta + \gamma = \pi$.

Then inequality (8.12) is equivalent to $\tan\frac{\alpha}{2} + \tan\frac{\beta}{2} + \tan\frac{\gamma}{2} \ge \sqrt{3}$, which is N_{13}.

Exercise 8.3 Let $a, b, c \in (0, 1)$ be positive real numbers such that $ab + bc + ca = 1$. Prove the inequality

$$\frac{a}{1-a^2} + \frac{b}{1-b^2} + \frac{c}{1-c^2} \ge \frac{3}{4}\left(\frac{1-a^2}{a} + \frac{1-b^2}{b} + \frac{1-c^2}{c}\right).$$

Solution Since $ab + bc + ca = 1$ and by *Case 3*, we use the following substitutions

$$a = \tan\frac{\alpha}{2}, \qquad b = \tan\frac{\beta}{2}, \qquad c = \tan\frac{\gamma}{2},$$

where α, β and γ are the angles of a triangle.

Since $a, b, c \in (0, 1)$, it follows that $\tan\frac{\alpha}{2}, \tan\frac{\beta}{2}, \tan\frac{\gamma}{2} \in (0, 1)$, i.e. it follows that α, β and γ are the angles of an acute triangle.

Also we have

$$\frac{a}{1-a^2} = \frac{\tan\frac{\alpha}{2}}{1-\tan^2\frac{\alpha}{2}} = \frac{\sin\frac{\alpha}{2}\cdot\cos\frac{\alpha}{2}}{\cos\alpha} = \frac{\sin\alpha}{2\cos\alpha} = \frac{\tan\alpha}{2}.$$

Similarly

$$\frac{b}{1-b^2} = \frac{\tan\beta}{2} \quad\text{and}\quad \frac{c}{1-c^2} = \frac{\tan\gamma}{2}.$$

Therefore the given inequality becomes

$$\frac{\tan\alpha + \tan\beta + \tan\gamma}{2} \ge \frac{3}{4}\left(\frac{2}{\tan\alpha} + \frac{2}{\tan\beta} + \frac{2}{\tan\gamma}\right)$$

or

$$\tan\alpha + \tan\beta + \tan\gamma \geq 3\left(\frac{1}{\tan\alpha} + \frac{1}{\tan\beta} + \frac{1}{\tan\gamma}\right). \tag{8.13}$$

By I_6 we have that $\tan\alpha + \tan\beta + \tan\gamma = \tan\alpha \cdot \tan\beta \cdot \tan\gamma$. Thus it suffices to show that

$$(\tan\alpha + \tan\beta + \tan\gamma)^2 \geq 3(\tan\alpha\tan\beta + \tan\beta\tan\gamma + \tan\gamma\tan\alpha)$$
$$\Leftrightarrow \quad \frac{1}{2}((\tan\alpha - \tan\beta)^2 + (\tan\beta - \tan\gamma)^2 + (\tan\gamma - \tan\alpha)^2) \geq 0.$$

We are done.

Exercise 8.4 Let x, y, z be positive real numbers. Prove the inequality

$$\frac{x}{x + \sqrt{(x+y)(x+z)}} + \frac{y}{y + \sqrt{(y+z)(y+x)}} + \frac{z}{z + \sqrt{(z+x)(z+y)}} \leq 1.$$

Solution Rewrite the given inequality as follows

$$\frac{1}{1 + \sqrt{\frac{(x+y)(x+z)}{x^2}}} + \frac{1}{1 + \sqrt{\frac{(y+z)(y+x)}{y^2}}} + \frac{1}{1 + \sqrt{\frac{(z+x)(z+y)}{z^2}}} \leq 1. \tag{8.14}$$

Since this is homogenous we may take $xy + yz + zx = 1$.

Therefore by *Case 3*, we can take

$$a = \tan\frac{\alpha}{2}, \qquad b = \tan\frac{\beta}{2}, \qquad c = \tan\frac{\gamma}{2},$$

where α, β and γ are the angles of a triangle.

We have

$$\frac{(x+y)(x+z)}{x^2} = \frac{(\tan\frac{\alpha}{2} + \tan\frac{\beta}{2})(\tan\frac{\alpha}{2} + \tan\frac{\gamma}{2})}{\tan^2\frac{\alpha}{2}} = \frac{1}{\sin^2\frac{\alpha}{2}}.$$

Similarly

$$\frac{(y+z)(y+x)}{y^2} = \frac{1}{\sin^2\frac{\beta}{2}} \quad \text{and} \quad \frac{(z+x)(z+y)}{z^2} = \frac{1}{\sin^2\frac{\gamma}{2}}.$$

Thus inequality (8.14) becomes

$$\frac{\sin\frac{\alpha}{2}}{1 + \sin\frac{\alpha}{2}} + \frac{\sin\frac{\beta}{2}}{1 + \sin\frac{\beta}{2}} + \frac{\sin\frac{\gamma}{2}}{1 + \sin\frac{\gamma}{2}} \leq 1,$$

i.e.

$$\frac{1}{1 + \sin\frac{\alpha}{2}} + \frac{1}{1 + \sin\frac{\beta}{2}} + \frac{1}{1 + \sin\frac{\gamma}{2}} \geq 2. \tag{8.15}$$

Since $AM \geq HM$ we obtain

$$\frac{1}{1+\sin\frac{\alpha}{2}}+\frac{1}{1+\sin\frac{\beta}{2}}+\frac{1}{1+\sin\frac{\gamma}{2}} \geq \frac{9}{3+\sin\frac{\alpha}{2}+\sin\frac{\beta}{2}+\sin\frac{\gamma}{2}}. \tag{8.16}$$

According to N_3, we have that $\sin\frac{\alpha}{2}+\sin\frac{\beta}{2}+\sin\frac{\gamma}{2} \leq \frac{3}{2}$.

Finally, by the previous inequality and (8.16) we obtain

$$\frac{1}{1+\sin\frac{\alpha}{2}}+\frac{1}{1+\sin\frac{\beta}{2}}+\frac{1}{1+\sin\frac{\gamma}{2}} \geq \frac{9}{3+\frac{3}{2}}=2$$

as required.

Exercise 8.5 Let a, b, c $(a, b, c \neq 1)$ be non-negative real numbers such that $ab+bc+ca=1$. Prove the inequality

$$\frac{a}{1-a^2}+\frac{b}{1-b^2}+\frac{c}{1-c^2} \geq \frac{3\sqrt{3}}{2}.$$

Solution Since $ab+bc+ca=1$ (*Case 3*) we take:

$$a=\tan\frac{\alpha}{2}, \qquad b=\tan\frac{\beta}{2} \quad \text{and} \quad c=\tan\frac{\gamma}{2},$$

where $\alpha, \beta, \gamma \in (0, \pi)$ and $\alpha+\beta+\gamma=\pi$.

Using the well-known identity $\frac{\tan\frac{\alpha}{2}}{\tan^2\frac{\alpha}{2}-1}=\tan\alpha$, we get that the given inequality is equivalent to $\tan\alpha+\tan\beta+\tan\gamma \geq 3\sqrt{3}$, which is N_{16}.

Equality occurs if and only if $a=b=c=1/\sqrt{3}$.

Exercise 8.6 Let a, b, c be positive real numbers. Prove the inequality

$$(a^2+2)(b^2+2)(c^2+2) \geq 9(ab+bc+ac).$$

Solution Let $a=\sqrt{2}\tan\alpha, b=\sqrt{2}\tan\beta, c=\sqrt{2}\tan\gamma$ where $\alpha, \beta, \gamma \in (0, \pi/2)$. Then using the well-known identity $1+\tan^2 x=\frac{1}{\cos^2 x}$ the given inequality becomes

$$\frac{8}{\cos^2\alpha\cdot\cos^2\beta\cdot\cos^2\gamma} \geq 9\left(\frac{2}{\tan\alpha\tan\beta}+\frac{2}{\tan\beta\tan\gamma}+\frac{2}{\tan\gamma\tan\alpha}\right),$$

i.e.

$$\cos\alpha\cos\beta\cos\gamma(\cos\alpha\sin\beta\sin\gamma+\sin\alpha\cos\beta\sin\gamma+\sin\alpha\sin\beta\cos\gamma) \leq \frac{4}{9}. \tag{8.17}$$

Also since

$$\cos(\alpha+\beta+\gamma)=\cos\alpha\cos\beta\cos\gamma-\cos\alpha\sin\beta\sin\gamma-\sin\alpha\cos\beta\sin\gamma$$
$$-\sin\alpha\sin\beta\cos\gamma$$

inequality (8.17) is equivalent to

$$\cos\alpha\cos\beta\cos\gamma(\cos\alpha\cos\beta\cos\gamma-\cos(\alpha+\beta+\gamma))\leq\frac{4}{9}. \qquad (8.18)$$

Let $\theta=\frac{\alpha+\beta+\gamma}{3}$.

Since $\cos\alpha, \cos\beta, \cos\gamma>0$, and since the function $\cos x$ is concave on $(0,\pi/2)$ by the inequality $AM\geq GM$ and *Jensen's inequality*, we obtain

$$\cos\alpha\cos\beta\cos\gamma\leq\left(\frac{\cos\alpha+\cos\beta+\cos\gamma}{3}\right)^3\leq\cos^3\theta.$$

Therefore according to (8.18) we need to prove that

$$\cos^3\theta(\cos^3\theta-\cos 3\theta)\leq\frac{4}{9}. \qquad (8.19)$$

Using the trigonometric identity

$$\cos 3\theta=4\cos^3\theta-3\cos\theta, \quad \text{i.e.} \quad \cos^3\theta-\cos 3\theta=3\cos\theta-3\cos^3\theta$$

inequality (8.19) becomes

$$\cos^4\theta(1-\cos^2\theta)\leq\frac{4}{27},$$

which follows by the inequality $AM\geq GM$:

$$\left(\frac{\cos^2\theta}{2}\cdot\frac{\cos^2\theta}{2}\cdot(1-\cos^2\theta)\right)^3\leq\frac{1}{3}\left(\frac{\cos^2\theta}{2}+\frac{\cos^2\theta}{2}+(1-\cos^2\theta)\right)=\frac{1}{3}.$$

Equality occurs iff $\tan\alpha=\tan\beta=\tan\gamma=\frac{1}{\sqrt{2}}$, i.e. iff $a=b=c=1$.

Exercise 8.7 Let a, b, c be positive real numbers such that $a+b+c=1$. Prove the inequality

$$\frac{a}{a+bc}+\frac{b}{b+ca}+\frac{\sqrt{abc}}{c+ab}\leq 1+\frac{3\sqrt{3}}{4}.$$

Solution Since $a+b+c=1$ we use the following substitutions $a=xy, b=yz, c=zx$, where $x, y, z>0$ and the given inequality becomes

$$\frac{xy}{xy+(yz)(zx)}+\frac{yz}{yz+(zx)(xy)}+\frac{xyz}{zx+(xy)(yz)}\leq 1+\frac{3\sqrt{3}}{4},$$

i.e.

$$\frac{1}{1+z^2}+\frac{1}{1+x^2}+\frac{y}{1+y^2}\leq 1+\frac{3\sqrt{3}}{4} \qquad (8.20)$$

where $xy+yz+zx=1$.

Since $xy+yz+zx=1$ according to *Case 3* we may set $x=\tan\frac{\alpha}{2}$, $y=\tan\frac{\beta}{2}$, $z=\tan\frac{\gamma}{2}$ where $\alpha,\beta,\gamma\in(0,\pi)$, and $\alpha+\beta+\gamma=\pi$.

Then inequality (8.20) becomes

$$\frac{1}{1+\tan^2\frac{\gamma}{2}}+\frac{1}{1+\tan^2\frac{\alpha}{2}}+\frac{\tan\frac{\beta}{2}}{1+\tan^2\frac{\beta}{2}}\le 1+\frac{3\sqrt{3}}{4},$$

i.e.

$$\cos^2\frac{\gamma}{2}+\cos^2\frac{\alpha}{2}+\frac{\sin\beta}{2}\le 1+\frac{3\sqrt{3}}{4}.$$

Using the trigonometric identity $\cos x=2\cos^2\frac{x}{2}-1$ the last inequality becomes

$$\frac{\cos\gamma+1}{2}+\frac{\cos\alpha+1}{2}+\frac{\sin\beta}{2}\le 1+\frac{3\sqrt{3}}{4},$$

i.e.

$$\cos\gamma+\cos\alpha+\sin\beta\le\frac{3\sqrt{3}}{2}. \tag{8.21}$$

We have

$$\begin{aligned}
\cos\alpha+\cos\gamma+\sin\beta &= \cos\alpha+\cos\gamma+\sin(\pi-(\alpha+\gamma))\\
&= \frac{2}{\sqrt{3}}\left(\frac{\sqrt{3}}{2}\cos\alpha+\frac{\sqrt{3}}{2}\cos\gamma\right)\\
&\quad+\frac{1}{\sqrt{3}}(\sqrt{3}\sin\alpha\cos\gamma+\sqrt{3}\cos\alpha\sin\gamma)\\
&\le \frac{1}{\sqrt{3}}\left(\frac{3}{4}+\cos^2\alpha+\frac{3}{4}+\cos^2\gamma\right)\\
&\quad+\frac{1}{2\sqrt{3}}(3\sin^2\alpha+\cos^2\gamma+\cos^2\alpha+3\sin^2\gamma)\\
&= \frac{\sqrt{3}}{2}+\frac{\sqrt{3}}{2}(\cos^2\alpha+\sin^2\alpha)+\frac{\sqrt{3}}{2}(\cos^2\gamma+\sin^2\gamma)\\
&= \frac{3\sqrt{3}}{2}.
\end{aligned}$$

Chapter 9
Hölder's Inequality, Minkowski's Inequality and Their Variants

In this chapter we'll introduce two very useful inequalities with broad practical usage: *Hölder's inequality* and *Minkowski's inequality*. We'll also present few variants of these inequalities. For that purpose we will firstly introduce the following theorem.

Theorem 9.1 (Young's inequality) *Let $a, b > 0$ and $p, q > 1$ be real numbers such that $1/p + 1/q = 1$. Then $ab \leq \frac{a^p}{p} + \frac{b^q}{q}$. Equality occurs if and only if $a^p = b^q$.*

Proof For $f(x) = e^x$ we have $f'(x) = f''(x) = e^x > 0$, for any $x \in \mathbb{R}$.

Thus $f(x)$ is convex on $(0, \infty)$.

If we put $x = p \ln a$ and $y = q \ln b$ then due to *Jensen's inequality* we obtain

$$\begin{aligned} & f\left(\frac{x}{p} + \frac{y}{q}\right) \leq \frac{1}{p} f(x) + \frac{1}{q} f(y) \\ \Leftrightarrow \quad & e^{\frac{x}{p} + \frac{y}{q}} \leq \frac{e^x}{p} + \frac{e^y}{q} \\ \Leftrightarrow \quad & e^{\ln a + \ln b} \leq \frac{e^{p \ln a}}{p} + \frac{e^{q \ln b}}{q} \\ \Leftrightarrow \quad & e^{\ln ab} \leq \frac{e^{\ln a^p}}{p} + \frac{e^{\ln b^q}}{q} \\ \Leftrightarrow \quad & ab \leq \frac{a^p}{p} + \frac{b^q}{q}. \end{aligned}$$

Equality occurs iff $x = y$, i.e. iff $a^p = b^q$. □

Z. Cvetkovski, *Inequalities*,
DOI 10.1007/978-3-642-23792-8_9, © Springer-Verlag Berlin Heidelberg 2012

Theorem 9.2 (Hölder's inequality) *Let $a_1, a_2, \ldots, a_n; b_1, b_2, \ldots, b_n$ be positive real numbers and $p, q > 1$ be such that $1/p + 1/q = 1$.*
Then

$$\sum_{i=1}^{n} a_i b_i \leq \left(\sum_{i=1}^{n} a_i^p\right)^{\frac{1}{p}} \left(\sum_{i=1}^{n} b_i^q\right)^{\frac{1}{q}}.$$

Equality occurs if and only if $\frac{a_1^p}{b_1^q} = \frac{a_2^p}{b_2^q} = \cdots = \frac{a_n^p}{b_n^q}$.

Proof 1 By *Young's inequality* for

$$a = \frac{a_i}{(\sum_{i=1}^{n} a_i^p)^{\frac{1}{p}}}, \quad b = \frac{b_i}{(\sum_{i=1}^{n} b_i^q)^{\frac{1}{q}}}, \quad i = 1, 2, \ldots, n,$$

we obtain

$$\frac{a_i b_i}{(\sum_{i=1}^{n} a_i^p)^{\frac{1}{p}} (\sum_{i=1}^{n} b_i^q)^{\frac{1}{q}}} \leq \frac{1}{p} \frac{a_i^p}{\sum_{i=1}^{n} a_i^p} + \frac{1}{q} \frac{b_i^q}{\sum_{i=1}^{n} b_i^q}. \tag{9.1}$$

Adding the inequalities (9.1), for $i = 1, 2, \ldots, n$, we obtain

$$\frac{\sum_{i=1}^{n} a_i b_i}{(\sum_{i=1}^{n} a_i^p)^{\frac{1}{p}} (\sum_{i=1}^{n} b_i^q)^{\frac{1}{q}}} \leq \frac{1}{p} \frac{\sum_{i=1}^{n} a_i^p}{\sum_{i=1}^{n} a_i^p} + \frac{1}{q} \frac{\sum_{i=1}^{n} b_i^q}{\sum_{i=1}^{n} b_i^q} = \frac{1}{p} + \frac{1}{q} = 1,$$

i.e.

$$\sum_{i=1}^{n} a_i b_i \leq \left(\sum_{i=1}^{n} a_i^p\right)^{\frac{1}{p}} \left(\sum_{i=1}^{n} b_i^q\right)^{\frac{1}{q}}.$$

Obviously equality occurs if and only if $\frac{a_1^p}{b_1^q} = \frac{a_2^p}{b_2^q} = \cdots = \frac{a_n^p}{b_n^q}$. □

Proof 2 The function $f : \mathbb{R}^+ \to \mathbb{R}$, $f(x) = x^p$ for $p > 1$ and $p < 0$ is strictly convex, and for $0 < p < 1$, f is strictly concave (Example 7.2).

Let $p > 1$, then by *Jensen's inequality* we obtain

$$\left(\frac{m_1 x_1 + m_2 x_2 + \cdots + m_n x_n}{m_1 + m_2 + \cdots + m_n}\right)^p \leq \frac{m_1 x_1^p + m_2 x_2^p + \cdots + m_n x_n^p}{m_1 + m_2 + \cdots + m_n},$$

i.e.

$$\left(\sum_{i=1}^{n} m_i x_i\right)^p \leq \left(\sum_{i=1}^{n} m_i\right)^{p-1} \cdot \left(\sum_{i=1}^{n} m_i x_i^p\right),$$

i.e.

$$\sum_{i=1}^{n} m_i x_i \leq \left(\sum_{i=1}^{n} m_i\right)^{\frac{p-1}{p}} \cdot \left(\sum_{i=1}^{n} m_i x_i^p\right)^{\frac{1}{p}}.$$

Since $\frac{1}{p} + \frac{1}{q} = 1$ we obtain $\frac{p-1}{p} = \frac{1}{q}$ and the last inequality becomes

$$\sum_{i=1}^{n} m_i x_i \leq \left(\sum_{i=1}^{n} m_i\right)^{\frac{1}{q}} \cdot \left(\sum_{i=1}^{n} m_i x_i^p\right)^{\frac{1}{p}}.$$

By taking $m_i = b_i^q$ and $x_i = a_i b_i^{1-q}$, for $i = 1, 2, \ldots, n$ we obtain

$$\sum_{i=1}^{n} a_i b_i \leq \left(\sum_{i=1}^{n} b_i^q\right)^{\frac{1}{q}} \cdot \left(\sum_{i=1}^{n} a_i^p\right)^{\frac{1}{p}}.$$

□

Remark For $p = q = 2$ by *Hölder's inequality* we get the *Cauchy–Schwarz inequality*.

We'll introduce, without proof, two generalizations of *Hölder's inequality*.

Theorem 9.3 (Weighted Hölder's inequality) *Let $a_1, a_2, \ldots, a_n$; $b_1, b_2, \ldots, b_n$; $m_1, m_2, \ldots, m_n$ be three sequences of positive real numbers and $p, q > 1$ be such that $1/p + 1/q = 1$.*
Then

$$\sum_{i=1}^{n} a_i b_i m_i \leq \left(\sum_{i=1}^{n} a_i^p m_i\right)^{\frac{1}{p}} \left(\sum_{i=1}^{n} b_i^q m_i\right)^{\frac{1}{q}}.$$

Equality occurs iff $\frac{a_1^p}{b_1^q} = \frac{a_2^p}{b_2^q} = \cdots = \frac{a_n^p}{b_n^q}$.

Theorem 9.4 (Generalized Hölder's inequality) *Let a_{ij}, $i = 1, 2, \ldots, m$; $j = 1, 2, \ldots, n$, be positive real numbers, and $\alpha_1, \alpha_2, \ldots, \alpha_n$ be positive real numbers such that $\alpha_1 + \alpha_2 + \cdots + \alpha_n = 1$.*
Then

$$\sum_{i=1}^{m} \left(\prod_{j=1}^{m} a_{ij}^{\alpha_j}\right) \leq \prod_{j=1}^{n} \left(\sum_{i=1}^{m} a_{ij}\right)^{\alpha_j}.$$

A very useful and frequently used form of *Hölder's inequality* is given in the next corollary.

Corollary 9.1 *Let a_1, a_2, a_3; b_1, b_2, b_3; c_1, c_2, c_3 be positive real numbers. Then we have*

$$(a_1^3 + a_2^3 + a_3^3)(b_1^3 + b_2^3 + b_3^3)(c_1^3 + c_2^3 + c_3^3) \geq (a_1b_1c_1 + a_2b_2c_2 + a_3b_3c_3)^3.$$

Theorem 9.5 (First Minkowski's inequality) *Let $a_1, a_2, \ldots, a_n$; $b_1, b_2, \ldots, b_n$ be positive real numbers and $p > 1$. Then*

$$\left(\sum_{i=1}^{n}(a_i + b_i)^p\right)^{\frac{1}{p}} \leq \left(\sum_{i=1}^{n} a_i^p\right)^{\frac{1}{p}} + \left(\sum_{i=1}^{n} b_i^p\right)^{\frac{1}{p}}.$$

Equality occurs if and only if $\frac{a_1}{b_1} = \frac{a_2}{b_2} = \cdots = \frac{a_n}{b_n}$.

Proof For $p > 1$, we choose $q > 1$ such that $\frac{1}{p} + \frac{1}{q} = 1$, i.e. $q = \frac{p}{p-1}$.

By *Hölder's inequality* we have

$$\begin{aligned}
\sum_{i=1}^{n}(a_i + b_i)^p &= \sum_{i=1}^{n}(a_i + b_i)(a_i + b_i)^{p-1} \\
&= \sum_{i=1}^{n} a_i(a_i + b_i)^{p-1} + \sum_{i=1}^{n} b_i(a_i + b_i)^{p-1} \\
&\leq \left(\sum_{i=1}^{n} a_i^p\right)^{\frac{1}{p}} \left(\sum_{i=1}^{n}((a_i + b_i)^{p-1})^q\right)^{\frac{1}{q}} \\
&\quad + \left(\sum_{i=1}^{n} b_i^p\right)^{\frac{1}{p}} \left(\sum_{i=1}^{n}((a_i + b_i)^{p-1})^q\right)^{\frac{1}{q}} \\
&= \left(\sum_{i=1}^{n}((a_i + b_i)^{p-1})^q\right)^{\frac{1}{q}} \left(\left(\sum_{i=1}^{n} a_i^p\right)^{\frac{1}{p}} + \left(\sum_{i=1}^{n} b_i^p\right)^{\frac{1}{p}}\right) \\
&= \left(\sum_{i=1}^{n}((a_i + b_i)^{p-1})^{\frac{p}{p-1}}\right)^{\frac{p-1}{p}} \left(\left(\sum_{i=1}^{n} a_i^p\right)^{\frac{1}{p}} + \left(\sum_{i=1}^{n} b_i^p\right)^{\frac{1}{p}}\right) \\
&= \left(\sum_{i=1}^{n}(a_i + b_i)^p\right)^{\frac{p-1}{p}} \left(\left(\sum_{i=1}^{n} a_i^p\right)^{\frac{1}{p}} + \left(\sum_{i=1}^{n} b_i^p\right)^{\frac{1}{p}}\right),
\end{aligned}$$

i.e. we obtain

$$\sum_{i=1}^{n}(a_i+b_i)^p \le \left(\sum_{i=1}^{n}(a_i+b_i)^p\right)^{\frac{p-1}{p}}\left(\left(\sum_{i=1}^{n}a_i^p\right)^{\frac{1}{p}}+\left(\sum_{i=1}^{n}b_i^p\right)^{\frac{1}{p}}\right)$$

$$\Leftrightarrow \quad \left(\sum_{i=1}^{n}(a_i+b_i)^p\right)^{1-\frac{p-1}{p}} \le \left(\sum_{i=1}^{n}a_i^p\right)^{\frac{1}{p}}+\left(\sum_{i=1}^{n}b_i^p\right)^{\frac{1}{p}}$$

$$\Leftrightarrow \quad \left(\sum_{i=1}^{n}(a_i+b_i)^p\right)^{\frac{1}{p}} \le \left(\sum_{i=1}^{n}a_i^p\right)^{\frac{1}{p}}+\left(\sum_{i=1}^{n}b_i^p\right)^{\frac{1}{p}}.$$

Equality occurs if and only if $\frac{a_1}{b_1}=\frac{a_2}{b_2}=\cdots=\frac{a_n}{b_n}$. (Why?) □

Theorem 9.6 (Second Minkowski's inequality) *Let $a_1,a_2,\ldots,a_n;b_1,b_2,\ldots,b_n$ be positive real numbers and $p>1$. Then*

$$\left(\left(\sum_{i=1}^{n}a_i\right)^p+\left(\sum_{i=1}^{n}b_i\right)^p\right)^{\frac{1}{p}} \le \sum_{i=1}^{n}(a_i^p+b_i^p)^{\frac{1}{p}}.$$

Equality occurs if and only if $\frac{a_1}{b_1}=\frac{a_2}{b_2}=\cdots=\frac{a_n}{b_n}$.

Proof The function $f:\mathbb{R}^+\to\mathbb{R}^+$, $f(x)=(1+x^{\alpha})^{\frac{1}{\alpha}}$, $\alpha\neq 0$ for $\alpha>1$ is a strictly convex and for $\alpha<1$ is a strictly concave (Example 7.4).

By *Jensen's inequality* for $p>1$ we obtain

$$\left(1+\left(\frac{m_1x_1+m_2x_2+\cdots+m_nx_n}{m_1+m_2+\cdots+m_n}\right)^p\right)^{1/p} \le \frac{m_1(1+x_1^p)^{1/p}+m_2(1+x_2^p)^{1/p}+\cdots+m_n(1+x_n^p)^{1/p}}{m_1+m_2+\cdots+m_n},$$

i.e.

$$\left(\left(\sum_{i=1}^{n}m_i\right)^p+\left(\sum_{i=1}^{n}m_ix_i\right)^p\right)^{1/p} \le \sum_{i=1}^{n}(m_i^p+(m_ix_i)^p)^{1/p}.$$

If we take $m_i=a_i$ and $x_i=\frac{b_i}{a_i}$ for $i=1,2,\ldots,n$, by the last inequality we obtain

$$\left(\left(\sum_{i=1}^{n}a_i\right)^p+\left(\sum_{i=1}^{n}b_i\right)^p\right)^{\frac{1}{p}} \le \sum_{i=1}^{n}(a_i^p+b_i^p)^{\frac{1}{p}}.$$

□

Theorem 9.7 (Third Minkowski's inequality) *Let $a_1, a_2, \ldots, a_n$ and $b_1, b_2, \ldots, b_n$ be positive real numbers. Then*

$$\sqrt[n]{a_1a_2\cdots a_n}+\sqrt[n]{b_1b_2\cdots b_n}\leq\sqrt[n]{(a_1+b_1)(a_2+b_2)\cdots(a_n+b_n)}.$$

Equality occurs if and only if $\frac{a_1}{b_1}=\frac{a_2}{b_2}=\cdots=\frac{a_n}{b_n}$.

Proof The proof is a direct consequence of *Jensen's inequality* for the convex function $f(x)=\ln(1+e^x)$ (Example 7.3), with $x_i=\ln\frac{b_i}{a_i}$, $i=1,2,\ldots,n$. □

Theorem 9.8 (Weighted Minkowski's inequality) *Let $a_1, a_2, \ldots, a_n$; $b_1, b_2, \ldots, b_n$; $m_1, m_2, \ldots, m_n$ be three sequences of positive real numbers and let $p>1$. Then*

$$\left(\sum_{i=1}^n (a_i+b_i)^p m_i\right)^{\frac{1}{p}}\leq\left(\sum_{i=1}^n a_i^p m_i\right)^{\frac{1}{p}}+\left(\sum_{i=1}^n b_i^p m_i\right)^{\frac{1}{p}}.$$

Equality occurs if and only if $\frac{a_1}{b_1}=\frac{a_2}{b_2}=\cdots=\frac{a_n}{b_n}$.

Remark If $0<p<1$ then in Theorem 9.5, Theorem 9.6 and Theorem 9.8 the inequality is reversed.

Exercise 9.1 Let a, b, c be positive real numbers. Prove the inequality

$$3(a^3+b^3+c^3)^2\geq(a^2+b^2+c^2)^3.$$

Solution By Corollary 9.1 (or simply *Hölder's inequality*) we obtain

$$(a^3+b^3+c^3)(a^3+b^3+c^3)(1+1+1)\geq(a^2+b^2+c^2)^3,$$

i.e.

$$3(a^3+b^3+c^3)^2\geq(a^2+b^2+c^2)^3.$$

Exercise 9.2 Let $a, b, c, x, y, z\in\mathbb{R}^+$. Prove the inequality

$$\frac{a^3}{x}+\frac{b^3}{y}+\frac{c^3}{z}\geq\frac{(a+b+c)^3}{3(x+y+z)}.$$

Solution By the *generalized Hölder's inequality* (or simply *Hölder's inequality*) we have

$$\left(\frac{a^3}{x}+\frac{b^3}{y}+\frac{c^3}{z}\right)^{\frac{1}{3}}(1+1+1)^{\frac{1}{3}}(x+y+z)^{\frac{1}{3}}\geq a+b+c,$$

and the conclusion follows.

Exercise 9.3 Let a, b, c be positive real numbers such that $a+b+c=1$. Prove the inequality

$$(a^a+b^a+c^a)(a^b+b^b+c^b)(a^c+b^c+c^c) \geq (\sqrt[3]{a}+\sqrt[3]{b}+\sqrt[3]{c})^3.$$

Solution By *Hölder's inequality* we obtain

$$(a^a+b^a+c^a)^{\frac{1}{3}}(a^b+b^b+c^b)^{\frac{1}{3}}(a^c+b^c+c^c)^{\frac{1}{3}} \geq a^{\frac{a+b+c}{3}}+b^{\frac{a+b+c}{3}}+c^{\frac{a+b+c}{3}}.$$

Since $a+b+c=1$, the conclusion follows.

Exercise 9.4 Let a, b, c be positive real numbers. Prove the inequality

$$3(a^2b+b^2c+c^2a)(ab^2+bc^2+ca^2) \geq (ab+bc+ca)^3.$$

Solution By *Hölder's inequality* for the triples:

$$(a_1,a_2,a_3)=(1,1,1), \qquad (b_1,b_2,b_3)=\left(\sqrt[3]{a^2b}, \sqrt[3]{b^2c}, \sqrt[3]{c^2a}\right),$$

$$(c_1,c_2,c_3)=\left(\sqrt[3]{b^2a}, \sqrt[3]{c^2b}, \sqrt[3]{a^2c}\right),$$

we obtain the given inequality.

Exercise 9.5 Let $a, b, c \in \mathbb{R}^+$. Prove the inequality

$$\frac{abc}{(1+a)(a+b)(b+c)(c+16)} \leq \frac{1}{81}.$$

Solution By *Hölder's inequality* we have

$$\begin{aligned}(1+a)(a+b)(b+c)(c+16) &\geq (\sqrt[4]{1\cdot a\cdot b\cdot c}+\sqrt[4]{a\cdot b\cdot c\cdot 16})^4\\ &= (3\sqrt[4]{abc})^4 = 81abc.\end{aligned}$$

Equality occurs if and only if $\frac{1}{a}=\frac{a}{b}=\frac{b}{c}=\frac{c}{16}$, i.e. $a=2, b=4, c=8$.

Exercise 9.6 Let x, y, z be positive real numbers such that $xy+yz+zx+xyz=4$. Prove the inequality

$$\sqrt{x+2}+\sqrt{y+2}+\sqrt{z+2} \geq 3\sqrt{3}.$$

Solution Let us denote $x+2=a$, $y+2=b$ and $z+2=c$. Then the condition $xy+yz+zx+xyz=4$ becomes

$$abc=ab+bc+ca,$$

i.e.

$$\frac{1}{a}+\frac{1}{b}+\frac{1}{c}=1.$$

By *Hölder's inequality* we have

$$(\sqrt{a}+\sqrt{b}+\sqrt{c})^2\left(\frac{1}{a}+\frac{1}{b}+\frac{1}{c}\right)\geq 3^3,$$

and since $\frac{1}{a}+\frac{1}{b}+\frac{1}{c}=1$ we get

$$(\sqrt{a}+\sqrt{b}+\sqrt{c})^2\geq 3^3,$$

i.e.

$$\sqrt{a}+\sqrt{b}+\sqrt{c}\geq 3\sqrt{3},$$

as required.

Exercise 9.7 Let a, b, c be positive real numbers such that $abc = 1$. Prove the inequality

$$\frac{a}{\sqrt{7+b+c}}+\frac{b}{\sqrt{7+c+a}}+\frac{c}{\sqrt{7+a+b}}\geq 1.$$

Solution Let us denote

$$A=\frac{a}{\sqrt{7+b+c}}+\frac{b}{\sqrt{7+c+a}}+\frac{c}{\sqrt{7+a+b}}$$

and

$$B=a(7+b+c)+b(7+c+a)+c(7+a+b).$$

By *Hölder's inequality* we obtain

$$A^2B\geq (a+b+c)^3.$$

It remains to prove that

$$(a+b+c)^3\geq B=7(a+b+c)+2(ab+bc+ca).$$

Since $AM\geq GM$ we deduce that

$$a+b+c\geq 3\sqrt[3]{abc}=3,$$

so it follows that

$$\begin{aligned}(a+b+c)^3\geq 3(a+b+c)^2&=\frac{7}{3}(a+b+c)^2+\frac{2}{3}(a+b+c)^2\\&\geq\frac{7}{3}\cdot 3(a+b+c)+2(ab+bc+ca)\\&=7(a+b+c)+2(ab+bc+ca).\end{aligned}$$

Exercise 9.8 Let a, b, c be positive real numbers such that $a+b+c=1$. Prove the inequality

$$\frac{a}{\sqrt[3]{a+2b}}+\frac{b}{\sqrt[3]{b+2c}}+\frac{c}{\sqrt[3]{c+2a}}\geq 1.$$

Solution Let us denote

$$A=\frac{a}{\sqrt[3]{a+2b}}+\frac{b}{\sqrt[3]{b+2c}}+\frac{c}{\sqrt[3]{c+2a}}$$

and

$$B=a(a+2b)+b(b+2c)+c(c+2a)=(a+b+c)^2=1.$$

By *Hölder's inequality* we have

$$A^3B\geq (a+b+c)^4, \quad \text{i.e.} \quad A^3\geq (a+b+c)^2=1,$$

from which it follows that $A\geq 1$.

Equality occurs iff $a=b=c=1/3$.

Exercise 9.9 Let $p\geq 1$ be an arbitrary real number. Prove that for any positive integer n we have

$$1^p+2^p+\cdots+n^p\geq n\cdot\left(\frac{n+1}{2}\right)^p.$$

Solution If $p=1$ then the given inequality is true, i.e. it becomes equality.

So let $p>1$.

We take $x_1=1, x_2=2,\ldots,x_n=n$ and $y_1=n, y_2=n-1,\ldots,y_n=1$.

By *Minkowski's inequality* we have

$$((1+n)^p+(1+n)^p+\cdots+(1+n)^p)^{\frac{1}{p}}\leq 2(1^p+2^p+\cdots+n^p)^{\frac{1}{p}},$$

i.e.

$$n(1+n)^p\leq 2^p(1^p+2^p+\cdots+n^p)$$

or

$$1^p+2^p+\cdots+n^p\geq n\cdot\left(\frac{n+1}{2}\right)^p,$$

as required.

Equality occurs iff $n=1$. (Why?)

Exercise 9.10 Let x, y, z be positive real numbers. Prove the inequality

$$\frac{x}{x+\sqrt{(x+y)(x+z)}}+\frac{y}{y+\sqrt{(y+z)(y+x)}}+\frac{z}{z+\sqrt{(z+y)(z+x)}}\leq 1.$$

Solution By *Hölder's inequality* for $n=2$ and $p=q=2$, we obtain

$$\begin{aligned}\sqrt{(x+y)(x+z)} &= \left((\sqrt{x})^2+(\sqrt{y})^2\right)^{\frac{1}{2}}\left((\sqrt{z})^2+(\sqrt{x})^2\right)^{\frac{1}{2}}\\ &\ge \sqrt{x}\cdot\sqrt{z}+\sqrt{y}\cdot\sqrt{x}=\sqrt{xz}+\sqrt{xy},\end{aligned}$$

i.e.

$$\frac{1}{\sqrt{(x+y)(x+z)}}\le\frac{1}{\sqrt{xz}+\sqrt{xy}}=\frac{1}{\sqrt{x}(\sqrt{y}+\sqrt{z})}.$$

So it follows that

$$\frac{x}{x+\sqrt{(x+y)(x+z)}}\le\frac{x}{x+\sqrt{x}(\sqrt{y}+\sqrt{z})}=\frac{\sqrt{x}}{\sqrt{x}+\sqrt{y}+\sqrt{z}}.$$

Similarly

$$\frac{y}{y+\sqrt{(y+z)(y+x)}}\le\frac{\sqrt{y}}{\sqrt{x}+\sqrt{y}+\sqrt{z}}\quad\text{and}$$

$$\frac{z}{z+\sqrt{(z+y)(z+x)}}\le\frac{\sqrt{z}}{\sqrt{x}+\sqrt{y}+\sqrt{z}}.$$

Adding the last three inequalities yields

$$\begin{aligned}&\frac{x}{x+\sqrt{(x+y)(x+z)}}+\frac{y}{y+\sqrt{(y+z)(y+x)}}+\frac{z}{z+\sqrt{(z+y)(z+x)}}\\ &\le\frac{\sqrt{x}+\sqrt{y}+\sqrt{z}}{\sqrt{x}+\sqrt{y}+\sqrt{z}}=1,\end{aligned}$$

as required. Equality occurs iff $x=y=z$.

Exercise 9.11 Let $x,y,z>0$ be real numbers. Prove the inequality

$$\begin{aligned}&\sqrt{x^2+xy+y^2}+\sqrt{y^2+yz+z^2}+\sqrt{z^2+zx+x^2}\\ &\ge 3\sqrt{xy+yz+zx}.\end{aligned}$$

Solution By *Hölder's inequality* we have

$$\begin{aligned}xy+yz+zx &= (x^2)^{1/3}(xy)^{1/3}(y^2)^{1/3}+(y^2)^{1/3}(yz)^{1/3}(z^2)^{1/3}\\ &\quad+(z^2)^{1/3}(zx)^{1/3}(x^2)^{1/3}\\ &\le(x^2+xy+y^2)^{1/3}(y^2+yz+z^2)^{1/3}(z^2+zx+x^2)^{1/3},\end{aligned}$$

i.e.

$$(xy+yz+zx)^3\le(x^2+xy+y^2)(y^2+yz+z^2)(z^2+zx+x^2). \tag{9.2}$$

Since $AM \geq GM$ and by (9.2) we obtain

$$\left(\frac{1}{3}(\sqrt{x^2+xy+y^2}+\sqrt{y^2+yz+z^2}+\sqrt{z^2+zx+x^2})\right)^3$$
$$\geq \sqrt{x^2+xy+y^2}\cdot\sqrt{y^2+yz+z^2}\sqrt{z^2+zx+x^2} \geq \sqrt{(xy+yz+zx)^3},$$

i.e. we have

$$\sqrt{x^2+xy+y^2}+\sqrt{y^2+yz+z^2}+\sqrt{z^2+zx+x^2} \geq 3\sqrt{xy+yz+zx},$$

as required. Equality occurs iff $x=y=z$.

Chapter 10
Generalizations of the Cauchy–Schwarz Inequality, Chebishev's Inequality and the Mean Inequalities

In Chap. 4 we presented the *Cauchy–Schwarz inequality*, *Chebishev's inequality* and the *mean inequalities*. In this section we will give their generalizations. The proof of first theorem is left to the reader, since it is similar to the proof of *Cauchy–Schwarz inequality*.

Theorem 10.1 (Weighted Cauchy–Schwarz inequality) *Let $a_i, b_i \in \mathbb{R}, i = 1, 2, \ldots, n$, be real numbers and let $m_i \in \mathbb{R}^+, i = 1, 2, \ldots, n$. Then we have the inequality*

$$\left(\sum_{i=1}^{n} a_i b_i m_i\right)^2 \leq \left(\sum_{i=1}^{n} a_i^2 m_i\right)\left(\sum_{i=1}^{n} b_i^2 m_i\right).$$

Equality occurs iff $\frac{a_1}{b_1} = \frac{a_2}{b_2} = \cdots = \frac{a_n}{b_n}$.

Theorem 10.2 *Let $a_1, a_2, \ldots, a_n$ and $b_1, b_2, \ldots, b_n$ be two sequences of non-negative real numbers and $c_i > 0, i = 1, 2, \ldots, n$, such that $\frac{a_1}{c_1} \geq \frac{a_2}{c_2} \geq \cdots \geq \frac{a_n}{c_n}$ and $\frac{b_1}{c_1} \geq \frac{b_2}{c_2} \geq \cdots \geq \frac{b_n}{c_n}$ (the sequences $(\frac{a_i}{c_i})$ and $(\frac{b_i}{c_i})$ have the same orientation). Then*

$$\sum_{i=1}^{n} \frac{a_i b_i}{c_i} \geq \frac{\sum_{i=1}^{n} a_i \cdot \sum_{i=1}^{n} b_i}{\sum_{i=1}^{n} c_i} \tag{10.1}$$

i.e.

$$\frac{a_1 b_1}{c_1} + \frac{a_2 b_2}{c_2} + \cdots + \frac{a_n b_n}{c_n} \geq \frac{(a_1 + a_2 + \cdots + a_n)(b_1 + b_2 + \cdots + b_n)}{c_1 + c_2 + \cdots + c_n}.$$

Z. Cvetkovski, *Inequalities*,
DOI 10.1007/978-3-642-23792-8_10, © Springer-Verlag Berlin Heidelberg 2012

Proof In the proof we shall use the following lemma which can be easily proved using the principle of mathematical induction.

Lemma 10.1 *Let $a_1, a_2, \ldots, a_n$ be non-negative real numbers, and $c_i > 0$, $i = 1, 2, \ldots, n$, such that $\frac{a_1}{c_1} \geq \frac{a_2}{c_2} \geq \cdots \geq \frac{a_n}{c_n}$. Then*

$$\frac{a_1 + a_2 + \cdots + a_k}{c_1 + c_2 + \cdots + c_k} \geq \frac{a_n}{c_n}, \quad \textit{for any } k = 1, 2, \ldots, n.$$

We shall prove inequality (10.1) by mathematical induction.

For $n = 1$, we have equality in (10.1).

For $n = 2$ we need to prove that

$$\frac{a_1 b_1}{c_1} + \frac{a_2 b_2}{c_2} \geq \frac{(a_1 + a_2)(b_1 + b_2)}{c_1 + c_2},$$

which is equivalent to $(a_1 c_2 - a_2 c_1)(b_1 c_2 - b_2 c_1) \geq 0$.

The last inequality holds since we have $\frac{a_1}{c_1} \geq \frac{a_2}{c_2}$ and $\frac{b_1}{c_1} \geq \frac{b_2}{c_2}$.

Let us assume that for non-negative real numbers $a_1, a_2, \ldots, a_k; b_1, b_2, \ldots, b_k$ and $c_i > 0, i = 1, 2, \ldots, k$, such that $\frac{a_1}{c_1} \geq \frac{a_2}{c_2} \geq \cdots \geq \frac{a_k}{c_k}$ and $\frac{b_1}{c_1} \geq \frac{b_2}{c_2} \geq \cdots \geq \frac{b_k}{c_k}$ inequality (10.1) holds for $n = k$, i.e.

$$\frac{a_1 b_1}{c_1} + \frac{a_2 b_2}{c_2} + \cdots + \frac{a_k b_k}{c_k} \geq \frac{(a_1 + a_2 + \cdots + a_k)(b_1 + b_2 + \cdots + b_k)}{c_1 + c_2 + \cdots + c_k}. \quad (10.2)$$

For $n = k + 1$, for non-negative real numbers $a_1, a_2, \ldots, a_{k+1}; b_1, b_2, \ldots, b_{k+1}$ and $c_i > 0$, $i = 1, 2, \ldots, k + 1$ such that

$$\frac{a_1}{c_1} \geq \frac{a_2}{c_2} \geq \cdots \geq \frac{a_{k+1}}{c_{k+1}} \quad \text{and} \quad \frac{b_1}{c_1} \geq \frac{b_2}{c_2} \geq \cdots \geq \frac{b_{k+1}}{c_{k+1}},$$

we have

$$\begin{aligned}
&\frac{a_1 b_1}{c_1} + \frac{a_2 b_2}{c_2} + \cdots + \frac{a_k b_k}{c_k} + \frac{a_{k+1} b_{k+1}}{c_{k+1}} \\
&\overset{(10.2)}{\geq} \frac{(a_1 + a_2 + \cdots + a_k)(b_1 + b_2 + \cdots + b_k)}{c_1 + c_2 + \cdots + c_k} + \frac{a_{k+1} b_{k+1}}{c_{k+1}} \\
&\geq \frac{(a_1 + a_2 + \cdots + a_k + a_{k+1})(b_1 + b_2 + \cdots + b_{k+1})}{c_1 + c_2 + \cdots + c_{k+1}},
\end{aligned}$$

where the last inequality is true according to the case $n = 2$ and Lemma 10.1. □

Remark 10.1 If the sequences $(\frac{a_i}{c_i})$ and $(\frac{b_i}{c_i})$ have opposite orientation then in Theorem 10.1 we have the reverse inequality, i.e., we have $\sum_{i=1}^{n} \frac{a_i b_i}{c_i} \leq \frac{\sum_{i=1}^{n} a_i \cdot \sum_{i=1}^{n} b_i}{\sum_{i=1}^{n} c_i}$.

Remark 10.2 For $a_1 \geq a_2 \geq \cdots \geq a_n \geq 0, b_1 \geq b_2 \geq \cdots \geq b_n \geq 0$ and $0 < c_1 \leq c_2 \leq \cdots \leq c_n$ the required condition from Theorem 10.2 is satisfied, so we also have that

$$\sum_{i=1}^{n} \frac{a_i b_i}{c_i} \geq \frac{\sum_{i=1}^{n} a_i \cdot \sum_{i=1}^{n} b_i}{\sum_{i=1}^{n} c_i}.$$

If in Theorem 10.2 we put $a_i = c_i x_i, b_i = c_i y_i$ and $m_i = \frac{c_i}{\sum_{i=1}^{n} c_i}, i = 1, 2, \ldots, n$, then clearly $\sum_{i=1}^{n} m_i = 1$ and the following theorem is obtained:

Theorem 10.3 (Weighted Chebishev's inequality) *Let $a_1 \leq a_2 \leq \cdots \leq a_n$; $b_1 \leq b_2 \leq \cdots \leq b_n$ be real numbers and let $m_1, m_2, \ldots, m_n$ be non-negative real numbers such that $m_1 + m_2 + \cdots + m_n = 1$.*
Then

$$\left(\sum_{i=1}^{n} a_i m_i\right)\left(\sum_{i=1}^{n} b_i m_i\right) \leq \sum_{i=1}^{n} a_i b_i m_i.$$

Equality occurs iff $a_1 = a_2 = \cdots = a_n$ or $b_1 = b_2 = \cdots = b_n$.

Note If in the above Theorem 10.1 (Theorem 10.3) we choose $m_1 = m_2 = \cdots = m_n$ ($m_1 = m_2 = \cdots = m_n = \frac{1}{n}$), we get the *Cauchy–Schwarz inequality*, and *Chebishev's inequality*, respectively.

Exercise 10.1 Let $a_1, a_2, \ldots, a_n$ be the lengths of the sides of a given n-gon ($n \geq 3$) and let $s = a_1 + a_2 + \cdots + a_n$. Prove the inequality

$$\frac{a_1}{s - 2a_1} + \frac{a_2}{s - 2a_2} + \cdots + \frac{a_n}{s - 2a_n} \geq \frac{n}{n-2}.$$

Solution Without loss of generality we may assume that $a_1 \geq a_2 \geq \cdots \geq a_n$. Then clearly $0 < s - 2a_1 \leq s - 2a_2 \leq \cdots \leq s - 2a_n$.

According to Theorem 10.2 we obtain

$$\begin{aligned} \frac{a_1}{s - 2a_1} + \frac{a_2}{s - 2a_2} + \cdots + \frac{a_n}{s - 2a_n} &= \frac{a_1 \cdot 1}{s - 2a_1} + \frac{a_2 \cdot 1}{s - 2a_2} + \cdots + \frac{a_n \cdot 1}{s - 2a_n} \\ &\geq \frac{(a_1 + a_2 + \cdots + a_n)n}{ns - 2(a_1 + a_2 + \cdots + a_n)} \\ &= \frac{ns}{s(n-2)} = \frac{n}{n-2}. \end{aligned}$$

Exercise 10.2 Let M be the centroid of the triangle ABC, and let k be its circumscribed circle. Let $MA \cap k = \{A_1\}$, $MB \cap k = \{B_1\}$ and $MC \cap k = \{C_1\}$. Prove the inequality

$$\overline{MA} + \overline{MB} + \overline{MC} \leq \overline{MA_1} + \overline{MB_1} + \overline{MC_1}.$$

Solution Denote $\overline{BC} = a$, $\overline{AC} = b$ and $\overline{AB} = c$. Let A', B' and C' be the midpoints of the sides BC, AC and AB, respectively.

Without loss of generality we may assume that $a \leq b \leq c$, and then we may easily conclude that $\overline{MC} \leq \overline{MB} \leq \overline{MA}$.

Also by the power of a point we have $\frac{3}{2}\overline{MA} \cdot \overline{A'A_1} = \frac{1}{4}a^2$ from which it follows that

$$\overline{A'A_1} = \frac{a^2}{6\overline{MA}}, \quad \text{i.e.} \quad \overline{MA_1} = \frac{1}{2}\overline{MA} + \overline{A'A_1} = \frac{1}{2}\overline{MA} + \frac{a^2}{6\overline{MA}}.$$

Analogously we obtain

$$\overline{MB_1} = \frac{1}{2}\overline{MB} + \frac{b^2}{6\overline{MB}} \quad \text{and} \quad \overline{MC_1} = \frac{1}{2}\overline{MC} + \frac{c^2}{6\overline{MC}}.$$

So it suffices to prove the inequality

$$\frac{a^2}{3\overline{MA}} + \frac{b^2}{3\overline{MB}} + \frac{c^2}{3\overline{MC}} \geq \overline{MA} + \overline{MB} + \overline{MC}.$$

According to Theorem 10.2 we have

$$\begin{aligned}\frac{a^2}{3\overline{MA}} + \frac{b^2}{3\overline{MB}} + \frac{c^2}{3\overline{MC}} &= \frac{a^2 \cdot 1}{3\overline{MA}} + \frac{b^2 \cdot 1}{3\overline{MB}} + \frac{c^2 \cdot 1}{3\overline{MC}} \geq \frac{3(a^2+b^2+c^2)}{3(\overline{MA}+\overline{MB}+\overline{MC})}\\ &= \frac{a^2+b^2+c^2}{\overline{MA}+\overline{MB}+\overline{MC}} \geq \frac{3(\overline{MA}^2+\overline{MB}^2+\overline{MC}^2)}{\overline{MA}+\overline{MB}+\overline{MC}}\\ &\geq \overline{MA}+\overline{MB}+\overline{MC},\end{aligned}$$

as required.

Before introducing the *power mean inequality* we'll give following definition.

Definition 10.1 Let $a = (a_1, a_2, \dots, a_n)$ be a sequence of positive real numbers and $r \neq 0$ be real number. Then the *power mean* $M_r(a)$, of order r, is defined as follows: $M_r(a) = (\frac{a_1^r + a_2^r + \cdots + a_n^r}{n})^{\frac{1}{r}}$.

For $r = 1, r = 2, r = -1$ we get $M_1(a), M_2(a), M_{-1}(a)$, which represent the arithmetic, quadratic and harmonic means of the numbers $a_1, a_2, \dots, a_n$, respectively.

If r tends to 0 then it may be shown that $M_r(a)$ tends to the geometric mean of the numbers $a_1, a_2, \dots, a_n$, i.e. $M_0(a) = \sqrt[n]{a_1 a_2 \cdots a_n}$.

Also if $r \to -\infty$ then $M_r(a) \to \min\{a_1, a_2, \dots, a_n\}$, and if $r \to \infty$ then $M_r(a) \to \max\{a_1, a_2, \dots, a_n\}$.

Theorem 10.4 (Power mean inequality) *Let $a=(a_1,a_2,\dots,a_n)$ be a sequence of positive real numbers and $r\neq 0$ be real number. Then $M_r(a)\leq M_s(a)$, for any real numbers $r\leq s$.*

Exercise 10.3 Let a,b,c be positive real numbers. Prove the inequality

$$(a^2+b^2+c^2)^3\leq 3(a^3+b^3+c^3)^2.$$

Solution By the *power mean inequality* we have that

$$\begin{aligned} & M_2(a,b,c)\leq M_3(a,b,c) \\ \Leftrightarrow\quad & \sqrt{\frac{a^2+b^2+c^2}{3}}\leq\sqrt[3]{\frac{a^3+b^3+c^3}{3}} \\ \Leftrightarrow\quad & (a^2+b^2+c^2)^3\leq 3(a^3+b^3+c^3)^2. \end{aligned}$$

Definition 10.2 Let $m=(m_1,m_2,\dots,m_n)$ be a sequence of non-negative real numbers such that $m_1+m_2+\cdots+m_n=1$. Then the *weighted power mean $M_r^m(a)$, of order r* $(r\neq 0)$, for the sequence $a=(a_1,a_2,\dots,a_n)$ is defined as $M_r^m(a)=(a_1^r m_1+a_2^r m_2+\cdots+a_n^r m_n)^{\frac{1}{r}}$.

Example 10.1 If $m_1=m_2=\cdots=m_n=\frac{1}{n}$ then $M_r^m(x)=M_r(x)$.

Example 10.2 If $n=3, r=4$; $m_1=\frac{1}{2}, m_2=\frac{1}{3}, m_3=\frac{1}{6}$, then

$$M_4^m(x,y,z)=\left(\frac{1}{2}\cdot x^4+\frac{1}{3}\cdot y^4+\frac{1}{6}\cdot z^4\right)^{\frac{1}{4}}.$$

Theorem 10.5 (Weighted power mean inequality) *Let $a=(a_1,a_2,\dots,a_n)$ be a sequence of positive real numbers, and let $m=(m_1,m_2,\dots,m_n)$ also be a sequence of positive real numbers such that $m_1+m_2+\cdots+m_n=1$. Then for each $r\leq s$ we have*

$$M_r^m(a)\leq M_s^m(a),$$

i.e.

$$(m_1a_1^r+m_2a_2^r+\cdots+m_na_n^r)^{\frac{1}{r}}\leq(m_1a_1^s+m_2a_2^s+\cdots+m_na_n^s)^{\frac{1}{s}}.$$

Proof We shall use the fact that the power function $f(x)=x^{\alpha}$ is convex for $\alpha>1$ or $\alpha<0$, and it is concave for $0<\alpha<1$.

First we prove the inequality in the case $r<s$ where both s and r are different from 0.

Three sub-cases may to be considered: 1° $0<r<s$, 2° $r<0<s$ and 3° $r<s<0$.

1° $0<r<s$. Since $\frac{s}{r}>1$ we conclude that $f(x)=x^{\frac{s}{r}}$ is convex, so according to *Jensen's inequality*:

$$f(m_1x_1+m_2x_2+\cdots+m_nx_n)\le m_1f(x_1)+m_2f(x_2)+\cdots+m_nf(x_n),$$

where $m_1+m_2+\cdots+m_n=1$ we have

$$(m_1x_1+m_2x_2+\cdots+m_nx_n)^{s/r}\le m_1x_1^{s/r}+m_2x_2^{s/r}+\cdots+m_nx_n^{s/r}.$$

For $x_i=a_i^r, i=1,2,\dots,n$, from the last inequality we obtain

$$(m_1a_1^r+m_2a_2^r+\cdots+m_na_n^r)^{s/r}\le m_1a_1^s+m_2a_2^s+\cdots+m_na_n^s,$$

i.e.

$$(m_1a_1^r+m_2a_2^r+\cdots+m_na_n^r)^{1/r}\le (m_1a_1^s+m_2a_2^s+\cdots+m_na_n^s)^{1/s},$$

so inequality holds in this case.

2° $r<0<s$. Then since $\frac{s}{r}<0$ we have that $f(x)=x^{\frac{s}{r}}$ is a convex function. The rest of the proof in this case is the same as in case 1°.

3° $r<s<0$. Then since $0<\frac{s}{r}<1$ we have that $f(x)=x^{\frac{s}{r}}$ is a concave function and according to *Jensen's inequality* for concave functions we obtain

$$(m_1x_1+m_2x_2+\cdots+m_nx_n)^{s/r}\ge m_1x_1^{s/r}+m_2x_2^{s/r}+\cdots+m_nx_n^{s/r}.$$

For $x_i=a_i^r, i=1,2,\dots,n$, from the last inequality we obtain

$$(m_1a_1^r+m_2a_2^r+\cdots+m_na_n^r)^{s/r}\ge m_1a_1^s+m_2a_2^s+\cdots+m_na_n^s,$$

and since $r<s<0$ we obtain

$$(m_1a_1^r+m_2a_2^r+\cdots+m_na_n^r)^{1/r}\le (m_1a_1^s+m_2a_2^s+\cdots+m_na_n^s)^{1/s}.$$

The cases when some values of s and r equal 0 are covered by the fact that the function $t\to M_t^m(a)$ is a continuous function. □

Exercise 10.4 Let $a,b,c\in\mathbb{R}^+$. Prove the inequality

$$\frac{(a+2b+3c)^2}{a^2+2b^2+3c^2}\le 6.$$

Solution For $m_1=\frac{1}{6}, m_2=\frac{2}{6}, m_3=\frac{3}{6}, n=3$ by the inequality

$$M_1^m(a,b,c)\le M_2^m(a,b,c),$$

which is true due to the *weighted power mean inequality*, we obtain

$$\frac{a+2b+3c}{6}\le\sqrt{\frac{a^2+2b^2+3c^2}{6}},\quad \text{i.e.}\quad \frac{(a+2b+3c)^2}{a^2+2b^2+3c^2}\le 6.$$

Exercise 10.5 Let a,b,c be positive real numbers such that $a+b+c=6$. Prove the inequality

$$\sqrt[3]{ab+bc}+\sqrt[3]{bc+ca}+\sqrt[3]{ca+ab}\le 6.$$

Solution By the *power mean inequality* we have

$$\frac{\sqrt[3]{ab+bc}+\sqrt[3]{bc+ca}+\sqrt[3]{ca+ab}}{3}\le\sqrt[3]{\frac{(ab+bc)+(bc+ca)+(ca+ab)}{3}},$$

i.e.

$$\sqrt[3]{ab+bc}+\sqrt[3]{bc+ca}+\sqrt[3]{ca+ab}\le\sqrt[3]{18(ab+bc+ca)}. \tag{10.3}$$

Since $ab+bc+ca\le\frac{(a+b+c)^2}{3}=12$ by (10.3) we obtain

$$\sqrt[3]{ab+bc}+\sqrt[3]{bc+ca}+\sqrt[3]{ca+ab}\le\sqrt[3]{18\cdot 12}=6.$$

Equality occurs if and only if $a=b=c=2$.

Exercise 10.6 Let a,b,c be positive real numbers such that $a^2+b^2+c^2=3$. Prove the inequality

$$\sqrt[3]{a^2+bc}+\sqrt[3]{b^2+ca}+\sqrt[3]{c^2+ab}\le 3\sqrt[3]{2}.$$

Solution By the *power mean inequality* and the well-known inequality $ab+bc+ca\le a^2+b^2+c^2$ we have

$$\begin{aligned}\sqrt[3]{a^2+bc}+\sqrt[3]{b^2+ca}+\sqrt[3]{c^2+ab}&\le\sqrt[3]{9(a^2+b^2+c^2+ab+bc+ca)}\\&\le\sqrt[3]{18(a^2+b^2+c^2)}=\sqrt[3]{18\cdot 3}=3\sqrt[3]{2}.\end{aligned}$$

Equality occurs if and only if $a=b=c=1$.

Exercise 10.7 Let a,b,c be positive real numbers such that $a^2+b^2+c^2=3$. Prove the inequality

$$\sqrt[4]{5a^2+4(b+c)+3}+\sqrt[4]{5b^2+4(c+a)+3}+\sqrt[4]{5c^2+4(a+b)+3}\le 6.$$

Solution By the *power mean inequality* we have

$$\begin{aligned}&\sqrt[4]{5a^2+4(b+c)+3}+\sqrt[4]{5b^2+4(c+a)+3}+\sqrt[4]{5c^2+4(a+b)+3}\\&\quad\leq\sqrt[4]{27\left(5(a^2+b^2+c^2)+8(a+b+c)+9\right)}.\end{aligned}$$

Since $a^2+b^2+c^2=3$ we have $a+b+c\leq\sqrt{3(a^2+b^2+c^2)}=3$ and therefore

$$\begin{aligned}&\sqrt[4]{5a^2+4(b+c)+3}+\sqrt[4]{5b^2+4(c+a)+3}+\sqrt[4]{5c^2+4(a+b)+3}\\&\quad\leq\sqrt[4]{27(5\cdot 3+8\cdot 3+9)}=6.\end{aligned}$$

Equality occurs if and only if $a=b=c=1$.

Exercise 10.8 Let x, y, z be non-negative real numbers. Prove the inequality

$$8(x^3+y^3+z^3)^2\geq 9(x^2+yz)(y^2+xz)(z^2+xy).$$

Solution If one of the numbers x, y, z is zero, let us say $z=0$, then the above inequality is equivalent to

$$8(x^3+y^3)^2\geq 9x^3y^3\quad\text{or}\quad 8(x^6+y^6)+7x^3y^3\geq 0,$$

which clearly holds.

Equality occurs iff $x=y=0$.

So let us assume that $x, y, z>0$.

Then

$$x^2+yz\leq x^2+\frac{y^2+z^2}{2}=\frac{2x^2+y^2+z^2}{2}.$$

Similarly

$$y^2+xz\leq\frac{2y^2+x^2+z^2}{2}\quad\text{and}\quad z^2+xy\leq\frac{2z^2+x^2+y^2}{2}.$$

Hence

$$\begin{aligned}&9(x^2+yz)(y^2+xz)(z^2+xy)\\&\quad\leq\frac{9}{8}(2x^2+y^2+z^2)(2y^2+x^2+z^2)(2z^2+x^2+y^2)\\&\quad\leq\frac{9}{8}\left(\frac{(2x^2+y^2+z^2)+(2y^2+x^2+z^2)+(2z^2+x^2+y^2)}{3}\right)^3\\&\quad=\frac{9}{8}\left(\frac{4(x^2+y^2+z^2)}{3}\right)^3=\frac{9\cdot 4^3}{8}\left(\frac{x^2+y^2+z^2}{3}\right)^3.\end{aligned}\tag{10.4}$$

By the *power mean inequality* we have

$$\sqrt{\frac{x^2+y^2+z^2}{3}} \leq \sqrt[3]{\frac{x^3+y^3+z^3}{3}},$$

i.e.

$$\left(\frac{x^2+y^2+z^2}{3}\right)^3 \leq \left(\frac{x^3+y^3+z^3}{3}\right)^2. \tag{10.5}$$

Finally, by (10.4) and (10.5) it follows that

$$\begin{aligned} 9(x^2+yz)(y^2+xz)(z^2+xy) &\leq \frac{9\cdot 4^3}{8}\left(\frac{x^2+y^2+z^2}{3}\right)^3 \\ &\leq \frac{9\cdot 4^3}{8}\left(\frac{x^3+y^3+z^3}{3}\right)^2 = 8(x^3+y^3+z^3)^2. \end{aligned}$$

Exercise 10.9 Let a, b, c be positive real numbers. Prove the inequality

$$a^b b^c c^a \leq \left(\frac{a+b+c}{3}\right)^{a+b+c}.$$

Solution By the *weighted power mean inequality* we have

$$\begin{aligned} (a^b b^c c^a)^{\frac{1}{a+b+c}} &= a^{\frac{b}{a+b+c}} \cdot b^{\frac{c}{a+b+c}} \cdot c^{\frac{a}{a+b+c}} \leq \frac{ba+cb+ac}{a+b+c} \leq \frac{(a+b+c)^2}{3(a+b+c)} \\ &= \frac{a+b+c}{3}. \end{aligned}$$

Exercise 10.10 Let a, b, c be the lengths of the sides of a triangle. Prove the inequality

$$(a+b-c)^a(b+c-a)^b(c+a-b)^c \leq a^a b^b c^c.$$

Solution By the *weighted power mean inequality* we have

$$\begin{aligned} &\sqrt[a+b+c]{\left(\frac{a+b-c}{a}\right)^a\left(\frac{b+c-a}{b}\right)^b\left(\frac{c+a-b}{c}\right)^c} \\ &\quad\leq \frac{1}{a+b+c}\left(a\cdot\frac{a+b-c}{a}+b\cdot\frac{b+c-a}{b}+c\cdot\frac{c+a-b}{c}\right) = 1, \end{aligned}$$

i.e.

$$(a+b-c)^a(b+c-a)^b(c+a-b)^c \leq a^a b^b c^c.$$

Equality occurs iff $a=b=c$.

Exercise 10.11 Let $a, b \in \mathbb{R}^+$ and $n \in \mathbb{N}$. Prove the inequality

$$(a+b)^n(a^n+b^n) \leq 2^n(a^{2n}+b^{2n}).$$

Solution By the *power mean inequality*, for any $x, y \in \mathbb{R}^+, n \in \mathbb{N}$, we have

$$\left(\frac{x+y}{2}\right)^n \leq \frac{x^n+y^n}{2}.$$

Therefore

$$\begin{aligned}
(a+b)^n(a^n+b^n) &= 2^n\left(\frac{a+b}{2}\right)^n(a^n+b^n) \\
&\leq 2^n\left(\frac{a^n+b^n}{2}\right)(a^n+b^n) = 2^n\frac{(a^n+b^n)^2}{2} \\
&\leq 2^n\frac{2(a^{2n}+b^{2n})}{2} = 2^n(a^{2n}+b^{2n}).
\end{aligned}$$

Exercise 10.12 Let $a, b, c \in \mathbb{R}^+, n \in \mathbb{N}$. Prove the inequality

$$a^n+b^n+c^n \geq \left(\frac{a+2b}{3}\right)^n + \left(\frac{b+2c}{3}\right)^n + \left(\frac{c+2a}{3}\right)^n.$$

Solution By the *power mean inequality* for any $a, b, c \in \mathbb{R}^+$ and $n \in \mathbb{N}$, we have

$$\frac{a^n+b^n+c^n}{3} \geq \left(\frac{a+b+c}{3}\right)^n.$$

So it follows that

$$\frac{a^n+b^n+b^n}{3} \geq \left(\frac{a+b+b}{3}\right)^n = \left(\frac{a+2b}{3}\right)^n.$$

Similarly we obtain

$$\frac{b^n+c^n+c^n}{3} \geq \left(\frac{b+2c}{3}\right)^n \quad \text{and} \quad \frac{c^n+a^n+a^n}{3} \geq \left(\frac{c+2a}{3}\right)^n.$$

After adding these we get the required inequality.

Chapter 11
Newton's Inequality, Maclaurin's Inequality

Let $a_1, a_2, \ldots, a_n$ be arbitrary real numbers.

Consider the polynomial

$$P(x) = (x + a_1)(x + a_2) \cdots (x + a_n) = c_0 x^n + c_1 x^{n-1} + \cdots + c_{n-1} x + c_n.$$

Then the coefficients $c_0, c_1, \ldots, c_n$ can be expressed as functions of $a_1, a_2, \ldots, a_n$, i.e. we have

$$\begin{aligned} c_0 &= 1, \\ c_1 &= a_1 + a_2 + \cdots + a_n, \\ c_2 &= a_1 a_2 + a_1 a_3 + \cdots + a_{n-1} a_n, \\ c_3 &= a_1 a_2 a_3 + a_1 a_2 a_4 + \cdots + a_{n-2} a_{n-1} a_n, \\ &\ldots \\ c_n &= a_1 a_2 \cdots a_n. \end{aligned}$$

For each $k = 1, 2, \ldots, n$ we define $p_k = \frac{c_k}{\binom{n}{k}} = \frac{k!(n-k)!}{n!} c_k$.

Theorem 11.1 (Newton's inequality) *Let $a_1, a_2, \ldots, a_n > 0$ be arbitrary real numbers. Then for each $k = 1, 2, \ldots, n - 1$, we have*

$$p_{k-1} p_{k+1} \leq p_k^2.$$

Equality occurs if and only if $a_1 = a_2 = \cdots = a_n$.

Proof By induction. □

Z. Cvetkovski, *Inequalities*,
DOI 10.1007/978-3-642-23792-8_11, © Springer-Verlag Berlin Heidelberg 2012

Example 11.1 For $n=3$ we have

$$p_1p_3 \le p_2^2 \quad \Leftrightarrow \quad \frac{c_1}{\binom{3}{1}}\frac{c_3}{\binom{3}{3}} \le \frac{c_2^2}{\binom{3}{2}^2} \quad \Leftrightarrow \quad \frac{c_1c_3}{3} \le \frac{c_2^2}{9}$$

$$\Leftrightarrow \quad 3c_1c_3 \le c_2^2,$$

i.e.

$$3abc(a+b+c) \le (ab+ac+bc)^2.$$

Equality occurs iff $a=b=c$.

Theorem 11.2 (Maclaurin's inequality) *Let $a_1, a_2, \ldots, a_n > 0$. Then*

$$p_1 \ge p_2^{\frac{1}{2}} \ge \cdots \ge p_k^{\frac{1}{k}} \ge \cdots \ge p_n^{\frac{1}{n}}.$$

Equality occurs if and only if $a_1 = a_2 = \cdots = a_n$.

Proof By *Newton's inequality*. □

Exercise 11.1 Let $a,b,c,d > 0$ be real numbers. Let $u = ab+ac+ad+bc+bd+cd$ and $v = abc+abd+acd+bcd$. Prove the inequality

$$2u^3 \ge 27v^2.$$

Solution We have $p_2 = \frac{u}{\binom{4}{2}} = \frac{u}{6}$ and $p_3 = \frac{v}{\binom{4}{3}} = \frac{v}{4}$.

By *Maclaurin's inequality* we have

$$p_2^{\frac{1}{2}} \ge p_3^{\frac{1}{3}} \quad \Leftrightarrow \quad p_2^3 \ge p_3^2 \quad \Leftrightarrow \quad \left(\frac{u}{6}\right)^3 \ge \left(\frac{v}{4}\right)^2 \quad \Leftrightarrow \quad 2u^3 \ge 27v^2.$$

Equality occurs iff $a=b=c=d$.

Exercise 11.2 Let $a,b,c,d > 0$ be real numbers. Prove the inequality

$$\left(\frac{1}{ab}+\frac{1}{ac}+\frac{1}{ad}+\frac{1}{bc}+\frac{1}{bd}+\frac{1}{cd}\right) \le \frac{3}{8}\left(\frac{1}{a}+\frac{1}{b}+\frac{1}{c}+\frac{1}{d}\right)^2.$$

Solution If we multiply both sides by $(abcd)^2$ the above inequality becomes

$$abcd(cd+bd+bc+ad+ac+ab) \le \frac{3}{8}(bcd+acd+abd+abc)^2$$

$$\Leftrightarrow \quad abcd\left(\frac{cd+bd+bc+ad+ac+ab}{6}\right) \leq \left(\frac{bcd+acd+abd+abc}{4}\right)^2$$

$$\Leftrightarrow \quad p_4 p_2 \leq p_3^2.$$

The last inequality is true, due to *Newton's inequality*.

Equality occurs iff $a=b=c=d$.

Chapter 12
Schur's Inequality, Muirhead's Inequality and Karamata's Inequality

In this chapter we will present three very important theorems, which have broad usage in solving symmetric inequalities. In that way we'll start with following definition.

Definition 12.1 Let $x_1, x_2, \dots, x_n$ be a sequence of positive real numbers and let $\alpha_1, \alpha_2, \dots, \alpha_n$ be arbitrary real numbers.
Let us denote $F(x_1, x_2, \dots, x_n) = x_1^{\alpha_1} \cdot x_2^{\alpha_2} \cdots x_n^{\alpha_n}$, and by $T[\alpha_1, \alpha_2, \dots, \alpha_n]$ we'll denote the sum of all possible products $F(x_1, x_2, \dots, x_n)$, over all permutations of $\alpha_1, \alpha_2, \dots, \alpha_n$.

Example 12.1

$$T[1, 0, \dots, 0] = (n-1)! \cdot (x_1 + x_2 + \cdots + x_n),$$
$$T[a, a, \dots, a] = n! x_1^a x_2^a \cdots x_n^a, \qquad T[1, 2] = x^2 y + x y^2,$$
$$T[1, 2, 1] = 2x^2 yz + 2y^2 xz + 2z^2 yx, \qquad T[3, 0, 0] = 2(x^3 + y^3 + z^3),$$
$$T[2, 1, 0] = x^2 y + x^2 z + y^2 x + y^2 z + z^2 x + z^2 y.$$

Theorem 12.1 (Schur's inequality) *Let $\alpha \in \mathbb{R}$ and $\beta > 0$. Then we have*

$$T[\alpha + 2\beta, 0, 0] + T[\alpha, \beta, \beta] \geq 2T[\alpha + \beta, \beta, 0].$$

Proof Let (x, y, z) be the sequence of variables.

By Definition 12.1, and with elementary algebraic transformations we have

$$\frac{1}{2} T\left[\alpha + 2\beta, 0, 0\right] + \frac{1}{2} T\left[\alpha, \beta, \beta\right] - T\left[\alpha + \beta, \beta, 0\right]$$
$$= x^\alpha (x^\beta - y^\beta)(x^\beta - z^\beta) + y^\alpha (y^\beta - x^\beta)(y^\beta - z^\beta) + z^\alpha (z^\beta - x^\beta)(z^\beta - y^\beta).$$

Z. Cvetkovski, *Inequalities*,
DOI 10.1007/978-3-642-23792-8_12, © Springer-Verlag Berlin Heidelberg 2012

Thus the given inequality is equivalent to

$$x^\alpha(x^\beta - y^\beta)(x^\beta - z^\beta) + y^\alpha(y^\beta - x^\beta)(y^\beta - z^\beta) + z^\alpha(z^\beta - x^\beta)(z^\beta - y^\beta) \geq 0.$$

Without loss of generality we may assume that $x \geq y \geq z$.

Then clearly only the second term can be negative.

If $\alpha \geq 0$ then we have

$$\begin{aligned} x^\alpha(x^\beta - y^\beta)(x^\beta - z^\beta) &\geq x^\alpha(x^\beta - y^\beta)(y^\beta - z^\beta) \\ &\geq y^\alpha(x^\beta - y^\beta)(y^\beta - z^\beta) \\ &= -y^\alpha(y^\beta - x^\beta)(y^\beta - z^\beta), \end{aligned}$$

i.e.

$$x^\alpha(x^\beta - y^\beta)(x^\beta - z^\beta) + y^\alpha(y^\beta - x^\beta)(y^\beta - z^\beta) \geq 0,$$

and since $z^\alpha(z^\beta - x^\beta)(z^\beta - y^\beta) \geq 0$ we get the required result.

Similarly we consider the case when $\alpha < 0$. □

Let us notice that for $\beta = 1$ we get a special form of *Schur's inequality*, which is very useful. Therefore we have the next theorem.

Theorem 12.2 *Let $x, y, z \geq 0$ be real numbers and let $t \in \mathbb{R}$. Then we have*

$$x^t(x - y)(x - z) + y^t(y - x)(y - z) + z^t(z - x)(z - y) \geq 0,$$

with equality if and only if $x = y = z$ or $x = y, z = 0$ (up to permutation).

Proof Without loss of generality let us assume that $x \geq y \geq z$.

Suppose that $t > 0$.

Then we have

$$(z - x)(z - y) \geq 0, \quad \text{i.e.} \quad z^t(z - x)(z - y) \geq 0 \tag{12.1}$$

and

$$x^t(x - z) - y^t(y - z) = (x^{t+1} - y^{t+1}) + z(x^t - y^t) \geq 0$$

i.e.

$$x^t(x - y)(x - z) + y^t(y - x)(y - z) \geq 0. \tag{12.2}$$

By (12.1) and (12.2) clearly we have

$$x^t(x - y)(x - z) + y^t(y - x)(y - z) + z^t(z - x)(z - y) \geq 0.$$

Let $t \leq 0$. Then we have

$$(x - y)(x - z) \geq 0 \quad \text{i.e.} \quad x^t(x - y)(x - z) \geq 0 \tag{12.3}$$

and

$$z^t(x-z) - y^t(x-y) \geq z^t(x-y) - y^t(x-y) = (z^t - y^t)(x-y) \geq 0,$$

i.e.

$$y^t(y-x)(y-z) + z^t(z-x)(z-y) \geq 0. \tag{12.4}$$

By adding (12.3) and (12.4) we get

$$x^t(x-y)(x-z) + y^t(y-x)(y-z) + z^t(z-x)(z-y) \geq 0.$$

Equality occurs if and only if $x = y = z$ or $x = y, z = 0$ (up to permutation). □

Corollary 12.1 *Let x, y, z and a, b, c be positive real numbers such that $a \geq b \geq c$ or $a \leq b \leq c$. Then we have*

$$a(x-y)(x-z) + b(y-x)(y-z) + c(z-x)(z-y) \geq 0.$$

Proof Similar to the proof of Theorem 12.1. □

Example 12.2 If we take $\alpha = \beta = 1$ in *Schur's inequality* we get

$$T[3,0,0] + T[1,1,1] \geq 2T[2,1,0],$$

i.e.

$$2(x^3 + y^3 + z^3) + 6xyz \geq 2(x^2y + x^2z + y^2x + y^2z + z^2x + z^2y),$$

i.e.

$$x^3 + y^3 + z^3 + 3xyz \geq x^2y + x^2z + y^2x + y^2z + z^2x + z^2y.$$

Note that this inequality is a direct consequence of *Surányi's inequality* for $n = 3$.

Corollary 12.2 *Let $x, y, z > 0$. Then* $3xyz + x^3 + y^3 + z^3 \geq 2((xy)^{3/2} + (yz)^{3/2} + (zx)^{3/2})$.

Proof By *Schur's inequality* and *AM* $\geq$ *GM* we obtain

$$\begin{aligned} x^3 + y^3 + z^3 + 3xyz &\geq (x^2y + y^2x) + (z^2y + y^2z) + (x^2z + z^2x) \\ &\geq 2((xy)^{3/2} + (yz)^{3/2} + (zx)^{3/2}). \end{aligned}$$

□

Corollary 12.3 *Let $k \in (0,3]$. Then for any $a, b, c \in \mathbb{R}^+$ we have*

$$(3-k)+k(abc)^{2/k}+a^2+b^2+c^2 \geq 2(ab+bc+ca).$$

Proof After setting $x=a^{2/3}$, $y=b^{2/3}$, $z=c^{2/3}$, the given inequality becomes

$$(3-k)+k(xyz)^{3/k}+x^3+y^3+z^3 \geq 2((xy)^{3/2}+(yz)^{3/2}+(zx)^{3/2}),$$

and due to Corollary 12.1, it suffices to show that

$$(3-k)+k(xyz)^{3/k} \geq 3xyz.$$

By the *weighted power mean inequality* we have

$$\frac{3-k}{3}\cdot 1+\frac{k}{3}(xyz)^{3/k} \geq 1^{(3-k)/3}((xyz)^{3/k})^{k/3}=xyz,$$

i.e.

$$(3-k)+k(xyz)^{3/k} \geq 3xyz,$$

as required. □

Definition 12.2 We'll say that the sequence $(\beta_i)_{i=1}^n$ is majorized by $(\alpha_i)_{i=1}^n$, denoted $(\beta_i) \prec (\alpha_i)$, if we can rearrange the terms of the sequences (α_i) and (β_i) in such a way as to satisfy the following conditions:

(1) $\beta_1+\beta_2+\cdots+\beta_n=\alpha_1+\alpha_2+\cdots+\alpha_n$
(2) $\beta_1 \geq \beta_2 \geq \cdots \geq \beta_n$ and $\alpha_1 \geq \alpha_2 \geq \cdots \geq \alpha_n$
(3) $\beta_1+\beta_2+\cdots+\beta_s \leq \alpha_1+\alpha_2+\cdots+\alpha_s$ for any $1 \leq s < n$.

Without proofs we'll give the following two very important theorems.

Theorem 12.3 (Muirhead's theorem) *Let $x_1, x_2, \ldots, x_n$ be a sequence of non-negative real numbers and let (α_i) and (β_i) be sequences of positive real numbers such that $(\beta_i) \prec (\alpha_i)$. Then*

$$T[\beta_i] \leq T[\alpha_i].$$

Equality occurs iff $(\alpha_i)=(\beta_i)$ or $x_1=x_2=\cdots=x_n$.

Example 12.3 Let (x, y, z) be the sequence of variables.
Consider the sequences $(2, 2, 1), (3, 1, 1)$. Then clearly $(2, 2, 1) \prec (3, 1, 1)$.
So by *Muirhead's theorem* we obtain

$$T[2, 2, 1] \le T[3, 1, 1],$$

i.e.

$$2(x^2y^2z + x^2z^2y + y^2z^2x) \le 2(x^3yz + y^3zx + z^3yx),$$

i.e.

$$x^2y^2z + x^2z^2y + y^2z^2x \le x^3yz + y^3zx + z^3yx,$$

i.e.

$$xy + yz + zx \le x^2 + y^2 + z^2,$$

which clearly holds.

Theorem 12.4 (Karamata's inequality) *Let $f : I \to \mathbb{R}$ be a convex function on the interval $I \subseteq \mathbb{R}$ and let $(a_i)_{i=1}^n, (b_i)_{i=1}^n$, where $a_i, b_i \in I, i = 1, 2, \ldots, n$, are two sequences, such that $(a_i) \succ (b_i)$. Then*

$$f(a_1) + f(a_2) + \cdots + f(a_n) \ge f(b_1) + f(b_2) + \cdots + f(b_n).$$

Remark If $f : I \to \mathbb{R}$ is strictly convex on the interval $I \subseteq \mathbb{R}$, and $(a_i) \ne (b_i)$ are such that $(a_i) \succ (b_i)$ then in *Karamata's inequality* we have strict inequality, i.e.

$$f(a_1) + f(a_2) + \cdots + f(a_n) > f(b_1) + f(b_2) + \cdots + f(b_n).$$

Also if $f : I \to \mathbb{R}$ is concave (strictly concave) in *Karamata's inequality* we have the reverse inequalities.

Exercise 12.1 Let a, b, c be the lengths of the sides of a triangle. Prove the inequality

$$a^3(s-a) + b^3(s-b) + c^3(s-c) \le abcs.$$

Solution The given inequality is equivalent to

$$a^2(a-b)(a-c) + b^2(b-c)(b-a) + c^2(c-a)(c-b) \ge 0,$$

which clearly holds by *Schur's inequality*.

Exercise 12.2 Let a, b, c be positive real numbers. Prove the inequality

$$27 + \left(2 + \frac{a^2}{bc}\right)\left(2 + \frac{b^2}{ca}\right)\left(2 + \frac{c^2}{ab}\right) \ge 6(a+b+c)\left(\frac{1}{a} + \frac{1}{b} + \frac{1}{c}\right).$$

Solution The given inequality is equivalent to

$$2abc(a^3+b^3+c^3+3abc-a^2b-a^2c-b^2a-b^2c-c^2a-c^2b)$$
$$+(a^3b^3+b^3c^3+c^3a^3+3a^2b^2c^2-ab^3c^2-ab^2c^3-a^2b^1c^3-a^3b^1c^2)\geq 0,$$

which is true due to *Schur's inequality*, for variables a, b, c and ab, bc, ca.

Exercise 12.3 Let a, b, c be positive real numbers. Prove the inequality

$$\frac{a}{b+c}+\frac{b}{c+a}+\frac{c}{a+b}\geq\frac{a^2+bc}{(a+b)(a+c)}+\frac{b^2+ca}{(b+a)(b+c)}+\frac{c^2+ab}{(c+a)(c+b)}.$$

Solution The given inequality is equivalent to

$$\frac{a^3+b^3+c^3+3abc-ab(a+b)-bc(b+c)-ca(c+a)}{(a+b)(b+c)(c+a)}\geq 0,$$

i.e.

$$a(a-b)(a-c)+b(b-a)(b-c)+c(c-a)(c-b)\geq 0,$$

which is *Schur's inequality*.

Exercise 12.4 Let a, b, c be non-negative real numbers. Prove the inequality

$$\frac{a}{4b^2+bc+4c^2}+\frac{b}{4c^2+ca+4a^2}+\frac{c}{4a^2+ab+4b^2}\geq\frac{1}{a+b+c}.$$

Solution By the *Cauchy–Schwarz inequality* we have

$$\frac{a}{4b^2+bc+4c^2}+\frac{b}{4c^2+ca+4a^2}+\frac{c}{4a^2+ab+4b^2}$$
$$\geq\frac{(a+b+c)^2}{4a(b^2+c^2)+4b(c^2+a^2)+4c(a^2+b^2)+3abc}.$$

So we need to prove that

$$\frac{(a+b+c)^2}{4a(b^2+c^2)+4b(c^2+a^2)+4c(a^2+b^2)+3abc}\geq\frac{1}{a+b+c},$$

which is equivalent to

$$(a+b+c)^3\geq 4a(b^2+c^2)+4b(c^2+a^2)+4c(a^2+b^2)+3abc,$$

i.e.

$$a^3+b^3+c^3+3abc\geq a(b^2+c^2)+b(c^2+a^2)+c(a^2+b^2),$$

which is *Schur's inequality*.

Exercise 12.5 Let a, b, c be positive real numbers. Prove the inequality

$$a^2+b^2+c^2+2abc+1 \geq 2(ab+bc+ac).$$

Solution By *Schur's inequality* we deduce

$$2(ab+bc+ac)-(a^2+b^2+c^2) \leq \frac{9abc}{a+b+c}.$$

So it remains to prove that

$$\frac{9abc}{a+b+c} \leq 2abc+1.$$

Since $AM \geq GM$ we have

$$2abc+1 = abc+abc+1 \geq 3\sqrt[3]{(abc)^2}.$$

Therefore we only need to prove that $3\sqrt[3]{(abc)^2} \geq \frac{9abc}{a+b+c}$, which is equivalent to $a+b+c \geq 3\sqrt[3]{abc}$, and clearly holds.

Exercise 12.6 Let a, b, c be positive real numbers. Prove the inequality

$$\frac{a^2+bc}{(b+c)^2}+\frac{b^2+ca}{(c+a)^2}+\frac{c^2+ab}{(a+b)^2} \geq \frac{3}{2}.$$

Solution To begin we'll show that

$$\frac{a^2+bc}{(b+c)^2}+\frac{b^2+ca}{(c+a)^2}+\frac{c^2+ab}{(a+b)^2} \geq \frac{a}{b+c}+\frac{b}{c+a}+\frac{c}{a+b}. \tag{12.5}$$

We have

$$\frac{a^2+bc}{(b+c)^2}-\frac{a}{b+c}=\frac{(a-b)(a-c)}{(b+c)^2};$$

similarly we get

$$\frac{b^2+ca}{(c+a)^2}-\frac{b}{c+a}=\frac{(b-c)(b-a)}{(c+a)^2} \quad \text{and} \quad \frac{c^2+ab}{(a+b)^2}-\frac{c}{a+b}=\frac{(c-a)(c-b)}{(a+b)^2}.$$

Let

$$x=\frac{1}{(b+c)^2}, \qquad y=\frac{1}{(c+a)^2} \quad \text{and} \quad z=\frac{1}{(a+b)^2}.$$

Then we can rewrite inequality (12.5) as follows

$$x(a-b)(a-c)+y(b-c)(b-a)+z(c-a)(c-b) \geq 0. \tag{12.6}$$

Without loss of generality we may assume that $a \geq b \geq c$ from which it follows that $x \geq y \geq z$, and now inequality (12.6) i.e. inequality (12.5), will follow due to Corollary 12.1 from *Schur's inequality*. Equality occurs iff $a = b = c$.

Exercise 12.7 Let $a, b, c \in \mathbb{R}^+$. Prove the inequality

$$(a^2+2)(b^2+2)(c^2+2) \geq 9(ab+ac+bc).$$

Solution The given inequality is equivalent to

$$8+(abc)^2+2(a^2b^2+b^2c^2+c^2a^2)+4(a^2+b^2+c^2) \geq 9(ab+ac+bc). \quad (12.7)$$

From the obvious inequality

$$(ab-1)^2+(bc-1)^2+(ca-1)^2 \geq 0$$

we deduce that

$$6+2(a^2b^2+b^2c^2+c^2a^2) \geq 4(ab+ac+bc) \quad (12.8)$$

and clearly

$$3(a^2+b^2+c^2) \geq 3(ab+ac+bc). \quad (12.9)$$

For $k = 1$ by Corollary 12.2, we obtain

$$2+(abc)^2+a^2+b^2+c^2 \geq 2(ab+ac+bc). \quad (12.10)$$

By adding (12.8), (12.9) and (12.10) we obtain inequality (12.7), as required.

Exercise 12.8 Let $a, b, c \in \mathbb{R}^+$. Prove the inequality

$$a^4+b^4+c^4 \geq abc(a+b+c).$$

Solution We have

$$\begin{aligned} & a^4+b^4+c^4 \geq abc(a+b+c) \\ \Leftrightarrow \quad & a^4+b^4+c^4 \geq a^2bc+b^2ac+c^2ab \\ \Leftrightarrow \quad & \frac{T[4,0,0]}{2} \geq \frac{T[2,1,1]}{2}, \end{aligned}$$

i.e.

$$T[4,0,0] \geq T[2,1,1],$$

which is true according to *Muirhead's theorem*.

Exercise 12.9 Let a, b, c be positive real numbers. Prove the inequality

$$\frac{1}{a^3+b^3+abc}+\frac{1}{b^3+c^3+abc}+\frac{1}{c^3+a^3+abc}\leq\frac{1}{abc}.$$

Solution After multiplying both sides by

$$abc(a^3+b^3+abc)(b^3+c^3+abc)(c^3+a^3+abc),$$

above inequality becomes

$$\frac{3}{2}T[4,4,1]+2T[5,2,2]+\frac{1}{2}T[7,1,1]+\frac{1}{2}T[3,3,3]$$
$$\leq\frac{1}{2}T[3,3,3]+T[6,3,0]+\frac{3}{2}T[4,4,1]+\frac{1}{2}T[7,1,1]+T[5,2,2],$$

i.e.

$$T[5,2,2]\leq T[6,3,0],$$

which is true according to *Muirhead's theorem.*

Equality occurs iff $a=b=c$.

Exercise 12.10 Let a, b, c be positive real numbers such that $a+b+c=1$. Prove the inequality

$$0\leq ab+bc+ca-2abc\leq\frac{7}{27}.$$

Solution The left-hand inequality follows from the identity

$$\begin{aligned}ab+bc+ca-2abc&=(a+b+c)(ab+bc+ca)-2abc\\&=a^2b+a^2c+b^2a+b^2c+c^2a+c^2b+abc\\&=T[2,1,0]+\frac{1}{6}T[1,1,1],\end{aligned}$$

since $T[2,1,0]+\frac{1}{6}T[1,1,1]\geq 0$.

We have

$$\frac{7}{27}=\frac{7}{27}(x+y+z)^3=\frac{7}{27}\left(\frac{1}{2}T[3,0,0]+3T[2,1,0]+T[1,1,1]\right).$$

Therefore the given inequality is equivalent to

$$T[2,1,0]+\frac{1}{6}T[1,1,1]\leq\frac{7}{27}\left(\frac{1}{2}T[3,0,0]+3T[2,1,0]+T[1,1,1]\right),$$

i.e.

$$12T[2,1,0]\leq 7T[3,0,0]+5T[1,1,1]. \tag{12.11}$$

Muirhead's theorem we have

$$2T[2,1,0] \le 2T[3,0,0], \tag{12.12}$$

and by *Schur's inequality* for $\alpha = \beta = 1$ (third degree) we get

$$10T[2,1,0] \le 5T[3,0,0] + 5T[1,1,1]. \tag{12.13}$$

Adding (12.12) and (12.13) gives us inequality (12.11), as required.

Exercise 12.11 Let $a, b, c \in \mathbb{R}^+$ such that $abc = 1$. Prove the inequality

$$\frac{1}{a^3(b+c)} + \frac{1}{b^3(c+a)} + \frac{1}{c^3(a+b)} \ge \frac{3}{2}.$$

Solution If we divide both sides by $(abc)^{4/3} = 1$, and after clearing the denominators, the given inequality will be equivalent to

$$2T\left[\frac{16}{3}, \frac{13}{3}, \frac{7}{3}\right] + T\left[\frac{16}{3}, \frac{16}{3}, \frac{4}{3}\right] + T\left[\frac{13}{3}, \frac{13}{3}, \frac{10}{3}\right] \ge 3T[5,4,3] + T[4,4,4].$$

Now according to *Muirhead's inequality* we have

$$2T\left[\frac{16}{3}, \frac{13}{3}, \frac{7}{3}\right] \ge 2T[5,4,3], \qquad T\left[\frac{16}{3}, \frac{16}{3}, \frac{4}{3}\right] \ge T[5,4,3],$$

$$T\left[\frac{13}{3}, \frac{13}{3}, \frac{10}{3}\right] \ge T[4,4,4].$$

If we add the last three inequalities we obtain the required result.

Equality occurs iff $a = b = c = 1$.

Exercise 12.12 (Schur's inequality) Let a, b, c be positive real numbers. Prove the inequality

$$a^3 + b^3 + c^3 + 3abc \ge a^2b + a^2c + b^2a + b^2c + c^2a + c^2b.$$

Solution Since the given inequality is symmetric, without loss of generality we can assume that $a \ge b \ge c$.

After taking $x = \ln a$, $y = \ln b$ and $z = \ln c$ the given inequality becomes

$$e^{3x} + e^{3y} + e^{3z} + e^{x+y+z} + e^{x+y+z} + e^{x+y+z}$$
$$\ge e^{2x+y} + e^{2x+z} + e^{2y+x} + e^{2y+z} + e^{2z+x} + e^{2z+y}.$$

The function $f(x) = e^x$ is convex on $\mathbb{R}$, so by *Karamata's inequality* it suffices to prove that the sequence $a = (3x, 3y, 3z, x+y+z, x+y+z, x+y+z)$ majorizes the sequence $b = (2x+y, 2x+z, 2y+x, 2y+z, 2z+x, 2z+y)$.

Since $a \geq b \geq c$ it follows that $x \geq y \geq z$ and clearly $3x \geq x+y+z \geq 3z$.

If $x+y+z \geq 3y$ (the case when $3y \geq x+y+z$ is analogous) then we obtain the following inequalities

$$3x \geq x+y+z \geq 3y \geq 3z,$$

$$2x+y \geq 2x+z \geq 2y+x \geq 2z+x \geq 2y+z \geq 2z+y,$$

which means that $a \succ b$, and we are done.

Exercise 12.13 Let $a_1, a_2, \ldots, a_n$ be positive real numbers. Prove the inequality

$$\frac{a_1^3}{a_2}+\frac{a_2^3}{a_3}+\cdots+\frac{a_n^3}{a_1} \geq a_1^2+a_2^2+\cdots+a_n^2.$$

Solution Let $x_i = \ln a_i$. Then the given inequality becomes

$$e^{3x_1-x_2}+e^{3x_2-x_3}+\cdots+e^{3x_n-x_1} \geq e^{2x_1}+e^{2x_2}+\cdots+e^{2x_n}.$$

Let us consider the sequences $a: 3x_1-x_2, 3x_2-x_3, \ldots, 3x_n-x_1$ and $b: 2x_1, 2x_2, \ldots, 2x_n$.

Since $f(x)=e^x$ is a convex function on $\mathbb{R}$ by *Karamata's inequality* it suffices to prove that a (ordered in some way) majorizes the sequences b (ordered in some way).

For that purpose, let us assume that

$$3x_{m_1}-x_{m_1+1} \geq 3x_{m_2}-x_{m_2+1} \geq \cdots \geq 3x_{m_n}-x_{m_n+1} \quad \text{and}$$

$$2x_{k_1} \geq 2x_{k_2} \geq \cdots \geq 2x_{k_n},$$

for some indexes $m_i, k_i \in \{1,2,\ldots,n\}$.

Clearly

$$3x_{m_1}-x_{m_1+1} \geq 3x_{k_1}-x_{k_1+1} \geq 2x_{k_1}$$

and

$$(3x_{m_1}-x_{m_1+1})+(3x_{m_2}-x_{m_2+1}) \geq (3x_{k_1}-x_{k_1+1})+(3x_{k_2}-x_{k_2+1}) \geq 2x_{k_1}+2x_{k_2}.$$

Analogously the sum of the first s terms of (a) is not less than the sum of an arbitrary s terms of (a), hence it is not less than $(3x_{k_1}-x_{k_1+1})+(3x_{k_2}-x_{k_2+1})+\cdots+(3x_{k_s}-x_{k_s+1})$, which, on the other hand, is not less than $2x_{k_1}+2x_{k_2}+\cdots+2x_{k_s}$.

So $a \succ b$, and we are done.

Exercise 12.14 (Turkevicius inequality) Let a, b, c, d be positive real numbers. Prove the inequality

$$a^4+b^4+c^4+d^4+2abcd \geq a^2b^2+a^2c^2+a^2d^2+b^2c^2+b^2d^2+c^2d^2.$$

Solution Because of symmetry without loss of generality we can assume $a \geq b \geq c \geq d$.

Let $x = \ln a, y = \ln b, z = \ln c, t = \ln d$; then clearly $x \geq y \geq z \geq t$ and given inequality becomes

$$e^{4x} + e^{4y} + e^{4z} + e^{4t} + e^{x+y+z+t} + e^{x+y+z+t}$$
$$\geq e^{2(x+y)} + e^{2(x+z)} + e^{2(x+t)} + e^{2(y+z)} + e^{2(y+t)} + e^{2(z+t)}.$$

The function $f(x) = e^x$ is a convex on $\mathbb{R}$, so according to *Karamata's inequality* it suffices to prove that $(4x, 4y, 4z, 4t, x + y + z + t, x + y + z + t)$ (ordered in some way) majorizes the sequences $(2(x + y), 2(x + z), 2(x + t), 2(y + z), 2(y + t), 2(z + t))$ (ordered in some way).

Clearly $4x \geq 4y \geq 4z$ and $4x \geq x + y + z + t \geq 4t$.

We need to consider four cases:

If $4z \geq x + y + z + t$ then we can easily show that

$$2(x + y) \geq 2(x + z) \geq 2(y + z) \geq 2(x + t) \geq 2(y + t) \geq 2(z + t)$$

and we can check that the sequence $(4x, 4y, 4z, x + y + z + t, x + y + z + t, 4t)$ majorize the sequence $(2(x + y), 2(x + z), 2(y + z), 2(x + t), 2(y + t), 2(z + t))$.

The cases when $x + y + z + t \geq 4z, 4y \geq x + y + z + t$ or $x + y + z + t \geq 4y$ are analogous as the first case and therefore are left to the reader.

Chapter 13
Two Theorems from Differential Calculus, and Their Applications for Proving Inequalities

In this section we'll give two theorems (without proof), whose origins are part of differential calculus, and which are widely used in proving certain inequalities. We assume that the reader has basic knowledge of differential calculus.

Definition 13.1 For the function $f : (a, b) \to \mathbb{R}$ we'll say that it is a monotone increasing function on the interval (a, b) if for all $x, y \in (a, b)$ such that $x \geq y$ we have $f(x) \geq f(y)$.
If we have strict inequalities, i.e. if for all $x, y \in (a, b)$ such that $x > y$ we have $f(x) > f(y)$ then we'll say that f is strictly increasing on (a, b).

Similarly we define a monotone decreasing function and a strictly decreasing function. Therefore we have the following definition.

Definition 13.2 For the function $f : (a, b) \to \mathbb{R}$ we'll say that it is a monotone decreasing function on the interval (a, b) if for all $x, y \in (a, b)$ such that $x \geq y$ we have $f(x) \leq f(y)$.
If we have strict inequalities, i.e. if for all $x, y \in (a, b)$ such that $x > y$ we have $f(x) < f(y)$ then we'll say that f is strictly increasing on (a, b).

Theorem 13.1 (Characterization of monotonic functions) *Let $f : (a, b) \to \mathbb{R}$ be a differentiable function on (a, b).*
If, for all $x \in (a, b)$, $f'(x) \geq 0$, then f is a monotone increasing function on the interval (a, b).

Z. Cvetkovski, *Inequalities*,
DOI 10.1007/978-3-642-23792-8_13, © Springer-Verlag Berlin Heidelberg 2012

If, for all $x \in (a, b)$, we have $f'(x) \le 0$, then f is a monotone decreasing function on the interval (a, b).
If we have strict inequalities then f is a strictly increasing, respectively, strictly decreasing function on (a, b).

Theorem 13.2 *Let $f : [a, b] \to \mathbb{R}$ and $g : [a, b] \to \mathbb{R}$ be functions such that:*

(i) *f and g are continuous on $[a, b]$ and $f(a) = g(a)$;*
(ii) *f and g are differentiable on (a, b);*
(iii) *$f'(x) > g'(x)$, for all $x \in (a, b)$.*

Then, for all $x \in (a, b)$, we have $f(x) > g(x)$.

Exercise 13.1 Let $x, y \ge 0$ be real numbers such that $x + y = 2$. Prove the inequality

$$x^2y^2(x^2 + y^2) \le 2.$$

Solution We homogenize as follows

$$x^2y^2(x^2 + y^2) \le 2\left(\frac{x+y}{2}\right)^6 \quad \Leftrightarrow \quad (x + y)^6 \ge 32x^2y^2(x^2 + y^2). \tag{13.1}$$

If $xy = 0$ then the given inequality clearly holds.

Therefore let us assume that $xy \ne 0$.

Since (13.1) is homogenous, we may normalize with $xy = 1$.

So $y = \frac{1}{x}$, and inequality (13.1) becomes

$$\left(x + \frac{1}{x}\right)^6 \ge 32\left(x^2 + \frac{1}{x^2}\right). \tag{13.2}$$

Let $t = (x + \frac{1}{x})^2$, then clearly $x^2 + \frac{1}{x^2} = t - 2$.

Therefore (13.2) is equivalent to

$$t^3 \ge 32(t - 2).$$

Clearly $t = (x + \frac{1}{x})^2 \ge 2^2 = 4$.

Let us consider the function $f(t) = t^3 - 32(t - 2)$ on the interval $[4, \infty)$.

Since $f'(t) = 3t^2 - 32$ we have that $f'(t) \ge 0$ for all $t \ge \sqrt{\frac{32}{3}} > 4$, i.e. it follows that f is increasing on $[4, \infty)$, which implies that

$$f(t) \ge f(4) = 0$$

$$\begin{aligned} &\Leftrightarrow \quad t^3 - 32(t-2) \geq 0 \\ &\Leftrightarrow \quad t^3 \geq 32(t-2), \quad \text{for all } t \in [4, \infty), \end{aligned}$$

as required.

Exercise 13.2 Let x, y, z be non-negative real numbers such that $x + y + z = 1$. Prove the inequality

$$0 \leq xy + yz + zx - 2xyz \leq \frac{7}{27}.$$

Solution Let $f(x, y, z) = xy + yz + zx - 2xyz$.

Without loss of generality we may assume that $0 \leq x \leq y \leq z \leq 1$.

Since $x + y + z = 1$ we have

$$3x \leq x + y + z = 1, \quad \text{i.e.} \quad x \leq \frac{1}{3}. \tag{13.3}$$

Furthermore we have

$$f(x, y, z) = (1 - 3x)yz + xy + zx + xyz \overset{(1)}{\geq} 0,$$

and we are done with the left inequality.

It remains to prove the right inequality.

Since $AM \geq GM$ we obtain

$$yz \leq \left(\frac{y+z}{2}\right)^2 = \left(\frac{1-x}{2}\right)^2.$$

Since $1 - 2x > 0$ we get

$$\begin{aligned} f(x, y, z) = x(y+z) + yz(1-2x) &\leq x(1-x) + \left(\frac{1-x}{2}\right)^2 (1-2x) \\ &= \frac{-2x^3 + x^2 + 1}{4}. \end{aligned}$$

We'll show that

$$f(x) = \frac{-2x^3 + x^2 + 1}{4} \leq \frac{7}{27}, \quad \text{for all } x \in \left[0, \frac{1}{3}\right].$$

We have

$$f'(x) = \frac{-6x^2 + 2x}{4} = \frac{3x}{2}\left(\frac{1}{3} - x\right) \geq 0, \quad \text{for all } x \in \left[0, \frac{1}{3}\right].$$

Thus f is an increasing function on $[0, \frac{1}{3}]$, so it follows that

$$f(x) \leq f\left(\frac{1}{3}\right) = \frac{7}{27}, \quad x \in \left[0, \frac{1}{3}\right],$$

as required.

Exercise 13.3 Let $x > 0$ be a real number. Prove that $x - \frac{x^2}{2} < \ln(x + 1)$.

Solution Let us consider the functions

$$f(x) = \ln(x+1) \quad \text{and} \quad g(x) = x - \frac{x^2}{2} \quad \text{on the interval } [0, \alpha), \text{ where } \alpha \in \mathbb{R}.$$

We have

$$f(0) = 0 = g(0) \quad \text{and} \quad f'(x) = \frac{1}{1+x}, \qquad g'(x) = 1 - x.$$

For $x \in (0, \alpha)$ it follows that $\frac{1}{1+x} > 1 - x$, i.e.

$$f'(x) > g'(x), \quad \text{for all } x \in (0, \alpha).$$

According to Theorem 13.2 we have $f(x) > g(x)$, for all $x \in (0, \alpha)$ i.e.

$$\ln(x+1) > x - \frac{x^2}{2}, \quad x \in (0, \alpha).$$

Since α is arbitrary we conclude that $\ln(x + 1) > x - \frac{x^2}{2}$, for all $x \in (0, \infty)$.

Exercise 13.4 Prove that, for all $0 < x < \frac{\pi}{2}$, we have $\tan x > x$.

Solution Let $f(x) = \tan x$, $g(x) = x$ where $x \in (0, \frac{\pi}{2})$.
We have

$$f(0) = 0 = g(0) \quad \text{and} \quad f'(x) = \frac{1}{\cos^2 x} > 1 = g'(x), \quad \text{for all } x \in \left(0, \frac{\pi}{2}\right).$$

According to Theorem 13.2, we have $f(x) > g(x)$, i.e. $\tan x > x$ for all $x \in (0, \frac{\pi}{2})$.

Exercise 13.5 Prove that, for all $0 < x < \frac{\pi}{2}$ we have $\tan x > x + \frac{x^3}{3}$.

Solution Let $f(x) = \tan x$, $g(x) = x + \frac{x^3}{3}$, $x \in (0, \frac{\pi}{2})$.
Then $f(0) = 0 = g(0)$ and we have

$$f'(x) = \frac{1}{\cos^2 x} = 1 + \tan^2 x > 1 + x^2 = g'(x), \quad \text{for all } x \in \left(0, \frac{\pi}{2}\right).$$

Thus, due to Theorem 13.2, we get $f(x) > g(x)$, i.e. $\tan x > x + \frac{x^3}{3}$ for all $x \in (0, \frac{\pi}{2})$.

Chapter 14
One Method of Proving Symmetric Inequalities with Three Variables

In this section we'll give a wonderful method that will be used in proving symmetrical inequalities with three variables. I must emphasize that this method is a powerful instrument which can be used for proving inequalities of varying difficulty which can't be proved with previous methods and techniques. Also I must say that I respect this method so much, because it can be very valuable and workable for all symmetric inequalities.

Let $x, y, z \in \mathbb{R}^+$, and $p = x + y + z, q = xy + yz + zx, r = xyz$. Clearly $p, q, r \in \mathbb{R}^+$.

Using these notations we can easily prove the following identities:

I_1: $x^2 + y^2 + z^2 = p^2 - 2q$
I_2: $x^3 + y^3 + z^3 = p(p^2 - 3q) + 3r$
I_3: $x^2y^2 + y^2z^2 + z^2x^2 = q^2 - 2pr$
I_4: $x^4 + y^4 + z^4 = (p^2 - 2q)^2 - 2(q^2 - 2pr)$
I_5: $(x + y)(y + z)(z + x) = pq - r$
I_6: $(x + y)(y + z) + (y + z)(z + x) + (z + x)(x + y) = p^2 + q$
I_7: $(x + y)^2(y + z)^2 + (y + z)^2(z + x)^2 + (z + x)^2(x + y)^2 = (p^2 + q)^2 - 4p(pq - r)$
I_8: $xy(x + y) + yz(y + z) + zx(z + x) = pq - 3r$
I_9: $(1 + x)(1 + y)(1 + z) = 1 + p + q + r$
I_{10}: $(1 + x)(1 + y) + (1 + y)(1 + z) + (1 + z)(1 + x) = 3 + 2p + q$
I_{11}: $(1 + x)^2(1 + y)^2 + (1 + y)^2(1 + z)^2 + (1 + z)^2(1 + x)^2 = (3 + 2p + q)^2 - 2(3 + p)(1 + p + q + r)$
I_{12}: $x^2(y + z) + y^2(z + x) + z^2(x + y) = pq - 3r$
I_{13}: $x^3y^3 + y^3z^3 + z^3x^3 = q^3 - 3pqr - 3r^2$
I_{14}: $xy(x^2 + y^2) + yz(y^2 + z^2) + zx(z^2 + x^2) = p^2q - 2q^2 - pr$
I_{15}: $(1 + x^2)(1 + y^2)(1 + z^2) = p^2 + q^2 + r^2 - 2pr - 2q + 1$
I_{16}: $(1 + x^3)(1 + y^3)(1 + z^3) = p^3 + q^3 + r^3 - 3pqr - 3pq - 3r^2 + 3r + 1$.

The proofs, as mentioned, are quite simple, and are therefore left to the reader. Also, we will give some inequalities which will be used later, and which should be well-known.

Z. Cvetkovski, *Inequalities*,
DOI 10.1007/978-3-642-23792-8_14, © Springer-Verlag Berlin Heidelberg 2012

Some of them follow by the *mean inequalities* but some of them are direct consequences of *Schur's* and *Muirhead's inequalities*.

We will prove some of them, and some are left to the reader.

Theorem 14.1 *Let $x, y, z \geq 0$ and $p = x+y+z, q = xy+yz+zx, r = xyz$. Then we have*:

N_1: $p^3 - 4pq + 9r \geq 0$, N_2: $p^4 - 5p^2q + 4q^2 + 6pr \geq 0$.

Proof According to *Schur's inequality* we have: For any real numbers $x, y, z \geq 0, t \in \mathbb{R}$ we have $x^t(x-y)(x-z) + y^t(y-z)(y-x) + z^t(z-x)(z-y) \geq 0$.

For $t = 1$ and $t = 2$, we obtain the required inequalities N_1 and N_2, respectively. □

Theorem 14.2 *Let $x, y, z \geq 0$, and $p = x+y+z, q = xy+yz+zx, r = xyz$. Then we have the following inequalities*:

N_3: $pq - 9r \geq 0$,
N_4: $p^2 \geq 3q$,
N_5: $p^3 \geq 27r$,
N_6: $q^3 \geq 27r^2$,
N_7: $q^2 \geq 3pr$,
N_8: $2p^3 + 9r \geq 7pq$,
N_9: $p^4 + 3q^2 \geq 4p^2q$,
N_{10}: $2p^3 + 9r^2 \geq 7pqr$,
N_{11}: $p^2q + 3pr \geq 4q^2$,
N_{12}: $q^3 + 9r^2 \geq 4pqr$,
N_{13}: $pq^2 \geq 2p^2r + 3qr$.

Proof We have

$$N_3\colon\ pq = (x+y+z)(xy+yz+zx) \geq 3\sqrt[3]{xyz} \cdot 3\sqrt[3]{x^2y^2z^2} = 9r$$

$$\Leftrightarrow \quad pq - 9r \geq 0,$$

$$N_4\colon\ p^2 \geq 3q \quad \Leftrightarrow \quad (x+y+z)^2 \geq 3(xy+yz+zx)$$

$$\Leftrightarrow \quad x^2+y^2+z^2 \geq xy+yz+zx,$$

which clearly holds.

$$N_5\colon\ p = x+y+z \geq 3\sqrt[3]{xyz} = 3\sqrt[3]{r} \quad \Leftrightarrow \quad p^3 \geq 27r,$$

$$N_6\colon\ q = xy+yz+zx \geq 3\sqrt[3]{x^2y^2z^2} = 3\sqrt[3]{r^2} \quad \Leftrightarrow \quad q^3 \geq 27r^2,$$

$$\begin{aligned} N_7\colon\ q^2 &= (xy+yz+zx)^2 = x^2y^2+y^2z^2+z^2x^2+2xyz(x+y+z) \\ &\geq (xy)(yz)+(yz)(zx)+(zx)(xy)+2xyz(x+y+z) \\ &= 3xyz(x+y+z) = 3pr, \end{aligned}$$

$$
\begin{aligned}
N_8\colon 2p^3+9r\geq 7pq & \\
\Leftrightarrow\quad & 2(x+y+z)^3+9xyz\geq 7(x+y+z)(xy+yz+zx)\\
\Leftrightarrow\quad & 2(x^3+y^3+z^3)\geq x^2y+x^2z+y^2z+y^2x+z^2x+z^2y\\
\Leftrightarrow\quad & T[3,0,0]\geq T[2,1,0],
\end{aligned}
$$

which is true due to *Muirhead's theorem.* □

Exercise 14.1 Let $x, y, z > 0$ such that $x+y+z=1$. Prove the inequality

$$\left(1+\frac{1}{x}\right)\left(1+\frac{1}{y}\right)\left(1+\frac{1}{z}\right)\geq 64.$$

Solution Let $p=x+y+z=1, q=xy+yz+zx, r=xyz$.

Then the given inequality becomes

$$(1+x)(1+y)(1+z)\geq 64xyz. \tag{14.1}$$

Using I_9: $(1+x)(1+y)(1+z)=1+p+q+r$ we deduce

$$(1+x)(1+y)(1+z)=2+q+r.$$

So (14.1) is equivalent to

$$2+q+r\geq 64r \quad \text{i.e.} \quad 2+q\geq 63r. \tag{14.2}$$

By N_5: $p^3\geq 27r$ we get

$$r\leq\frac{1}{27}. \tag{14.3}$$

By N_3: $pq-9r\geq 0$ we get

$$pq\geq 9r, \quad \text{i.e.} \quad q\geq 9r. \tag{14.4}$$

Now using (14.4) we deduce that $2+q\geq 2+9r$.

So it suffices to show that $2+9r\geq 63r$, which is $2\geq 54r\Leftrightarrow r\leq\frac{1}{27}$, which clearly holds, by (14.3).

We have proved (14.2), and we are done.

Exercise 14.2 Let $x, y, z > 0$ be real numbers. Prove the inequality

$$(xy+yz+zx)\left(\frac{1}{(x+y)^2}+\frac{1}{(y+z)^2}+\frac{1}{(z+x)^2}\right)\geq\frac{9}{4}.$$

Solution The given inequality is equivalent to

$$4(xy+yz+zx)((z+x)^2(y+z)^2+(x+y)^2(z+x)^2+(x+y)^2(y+z)^2)$$
$$\geq 9(x+y)^2(y+z)^2(z+x)^2. \tag{14.5}$$

Let us denote $p=x+y+z, q=xy+yz+zx, r=xyz$.
By I_5 and I_7 we have

$$(x+y)^2(y+z)^2(z+x)^2=(pq-r)^2$$

and

$$(x+y)^2(y+z)^2+(y+z)^2(z+x)^2+(z+x)^2(x+y)^2=(p^2+q)^2-4p(pq-r).$$

So we can rewrite inequality (14.5) as follows

$$4q((p^2+q)^2-4p(pq-r))\geq 9(pq-r)^2$$
$$\Leftrightarrow \quad 4p^4q-17p^2q^2+4q^3+34pqr-9r^2\geq 0$$
$$\Leftrightarrow \quad 3pq(p^3-4pq+9r)+q(p^4-5p^2q+4q^2+6pr)+r(pq-9r)\geq 0.$$

The last inequality follows from N_1, N_2 and N_3, and the fact that $p,q,r>0$. Equality occurs if and only if $x=y=z$.

Exercise 14.3 Let $x,y,z\in\mathbb{R}^+$ such that $x+y+z=1$. Prove the inequality

$$\frac{1}{1-xy}+\frac{1}{1-yz}+\frac{1}{1-zx}\leq\frac{27}{8}.$$

Solution Let $p=x+y+z=1, q=xy+yz+zx, r=xyz$.
It can easily be shown that

$$(1-xy)(1-yz)(1-zx)=1-q+pr-r^2$$

and

$$(1-xy)(1-yz)+(1-yz)(1-zx)+(1-zx)(1-xy)=3-2q+pr.$$

So the given inequality becomes

$$8(3-2q+pr)\leq 27(1-q+pr-r^2)$$
$$\Leftrightarrow \quad 3-11q+19pr-27r^2\geq 0.$$

Since $p=1$, we need to show that

$$3-11q+19r-27r^2\geq 0.$$

By N_5: $p^3\geq 27r$ we have $1\geq 27r$, i.e. $r\geq 27r^2$.

Therefore

$$3-11q+19r-27r^2 \geq 3-11q+19r-r=3-11q+18r.$$

So it suffices to prove that

$$3-11q+18r \geq 0.$$

We have

$$\begin{aligned} &3-11q+18r \geq 0 \\ \Leftrightarrow\quad &3-11(xy+yz+zx)+18xyz \geq 0 \\ \Leftrightarrow\quad &11(xy+yz+zx)-18xyz \leq 3. \end{aligned}$$

Applying $AM \geq GM$ we deduce

$$\begin{aligned} 11(xy+yz+zx)-18xyz &= xy(11-18z)+11z(x+y) \\ &\leq \frac{(x+y)^2}{4}(11-18z)+11z(x+y) \\ &= \frac{(1-z)^2}{4}(11-18z)+11z(1-z) \\ &= \frac{(1-z)((1-z)(11-18z)+44z)}{4} \\ &= \frac{4z+3z^2-18z^3+11}{4}. \end{aligned}$$

So it remains to show that

$$\begin{aligned} &\frac{4z+3z^2-18z^3+11}{4} \leq 3 \\ \Leftrightarrow\quad &4z+3z^2-18z^3 \leq 1 \\ \Leftrightarrow\quad &18z^3-3z^2-4z+1 \geq 0 \\ \Leftrightarrow\quad &(3z-1)^2(2z+1) \geq 0, \end{aligned}$$

which is obvious.

Exercise 14.4 Let $a, b, c \in \mathbb{R}^+$ such that $\frac{1}{a+1}+\frac{1}{b+1}+\frac{1}{c+1}=2$. Prove the inequality

$$\frac{1}{8ab+1}+\frac{1}{8bc+1}+\frac{1}{8ca+1} \geq 1. \tag{14.6}$$

Solution Let $p=a+b+c, q=ab+bc+ca, r=abc$.

Since $\frac{1}{a+1}+\frac{1}{b+1}+\frac{1}{c+1}=2$ we have

$$(a+1)(b+1)+(b+1)(c+1)+(c+1)(a+1)=2(a+1)(b+1)(c+1). \quad (14.7)$$

Using the identities I_9 and I_{10}, identity (14.7) becomes $3+2p+q=2(1+p+q+r)$, from which it follows that

$$q+2r=1. \quad (14.8)$$

It can easily be shown that

$$(8ab+1)(8bc+1)+(8bc+1)(8ca+1)+(8ca+1)(8ab+1)=64pr+16q+3$$

and

$$(8ab+1)(8bc+1)(8ca+1)=512r^2+64pr+8q+1.$$

We need to prove that

$$64pr+16q+3\geq 512r^2+64pr+8q+1,$$

which is equivalent to

$$8q+2\geq 512r^2. \quad (14.9)$$

By $q^3\geq 27r^2$ and since $q=1-2r$ we obtain

$$\begin{aligned}&(1-2r)^3\geq 27r^2\\ \Leftrightarrow\quad & 8r^3+15r^2+6r-1\leq 0\\ \Leftrightarrow\quad & (8r-1)(r^2+2r+1)\leq 0.\end{aligned}$$

Thus

$$8r-1\leq 0, \quad \text{i.e.} \quad r\leq\frac{1}{8}. \quad (14.10)$$

Now since $q+2r=1$, inequality (14.9) becomes

$$\begin{aligned}&8(1-2r)+2\geq 512r^2\\ \Leftrightarrow\quad & 512r^2+16r-10\leq 0\\ \Leftrightarrow\quad & (8r-1)(64r+10)\leq 0,\end{aligned}$$

which follows due to (14.10).

Exercise 14.5 Let x, y, z be positive real numbers such that $x+y+z=1$. Prove the inequality

$$\frac{z-xy}{x^2+xy+y^2}+\frac{y-zx}{x^2+xz+z^2}+\frac{x-yz}{y^2+yz+z^2}\geq 2.$$

Solution Let $p = x + y + z = 1, q = xy + yz + zx, r = xyz$.
We have

$$x^2 + xy + y^2 = (x+y)^2 - xy = (1-z)^2 - xy = 1 - 2z + z^2 - xy$$
$$= 1 - z - z(1-z) - xy = 1 - z - z(x+y) - xy = 1 - z - q.$$

Similarly we deduce that

$$x^2 + xz + z^2 = 1 - y - q \quad \text{and} \quad y^2 + yz + z^2 = 1 - x - q.$$

According to the previous identities, I_3 and I_{12}, by using elementary algebraic transformations the given inequality becomes

$$q^3 + q^2 - 4q + 3qr + 4r + 1 \geq 0,$$

i.e.

$$27q^3 + 27q^2 - 108q + 27r(3q+4) + 27 \geq 0. \tag{14.11}$$

By N_1: $p^3 - 4pq + 9r \geq 0$, since $p = 1$ we get

$$9r \geq 4q - 1. \tag{14.12}$$

According to inequality (14.12) we obtain

$$27q^3 + 27q^2 - 108q + 27r(3q+4) + 27$$
$$\geq 27q^3 + 27q^2 - 108q + 3(4q-1)(3q+4) + 27$$
$$= (3q-1)(9q^2 + 24q - 15). \tag{14.13}$$

Since $p = 1$ due to N_1: $p^2 \geq 3q$ it follows that

$$q \leq \frac{1}{3}. \tag{14.14}$$

Finally by (14.13) and (14.14) we obtain

$$27q^3 + 27q^2 - 108q + 27r(3q+4) + 27 \geq (3q-1)(9q^2 + 24q - 15) \geq 0,$$

since $3q - 1 \leq 0$ and $9q^2 + 24q - 15 \leq 9 \cdot \frac{1}{9} + 24 \cdot \frac{1}{3} - 15 = -6$, as required.

Exercise 14.6 Let a, b, c be non-negative real numbers such that $a + b + c = 1$. Prove the inequality

$$7(ab + bc + ca) \leq 2 + 9abc.$$

Solution Let $p = a + b + c = 1, q = ab + bc + ca, r = abc$.
Then according to N_8: $2p^3 + 9r \geq 7pq$ we have

$$2 + 9r \geq 7q \quad \text{i.e.} \quad 2 + 9abc \geq 7(ab + bc + ca),$$

as required.

Exercise 14.7 Let $x, y, z \geq 0$ be real numbers such that $x + y + z = 1$. Prove the inequality

$$12(x^2y^2 + y^2z^2 + z^2x^2)(x^3 + y^3 + z^3) \leq xy + yz + zx.$$

Solution Let $p = x + y + z = 1, q = xy + yz + zx, r = xyz$.

By I_2 and I_3 we have

$$x^3 + y^3 + z^3 = p(p^2 - 3q) + 3r = 1 - 3q + 3r$$

and

$$x^2y^2 + y^2z^2 + z^2x^2 = q^2 - 2pr = q^2 - 2r.$$

Clearly $q \leq \frac{1}{3}$.

So the given inequality becomes

$$12(1 - 3q + 3r)(q^2 - 2r) \leq q. \tag{14.15}$$

Suppose that $q \geq \frac{1}{4}$.

By N_3: $pq - 9r \geq 0$ it follows that $r \leq \frac{q}{9}$, i.e.

$$0 \leq r \leq \frac{q}{9}. \tag{14.16}$$

Since $q \leq \frac{1}{3}$ we have

$$(1 - 3q + 3r)r \geq 0. \tag{14.17}$$

We'll prove that

$$12\left(1 - 3q + 3\frac{q}{9}\right)q^2 \leq q, \tag{14.18}$$

from which, together with (14.16) and (14.17), we'll have

$$12(1 - 3q + 3r)(q^2 - 2r) \leq 12(1 - 3q + 3r)q^2 \leq 12\left(1 - 3q + 3\frac{q}{9}\right)q^2 \leq q.$$

Hence

$$\begin{aligned} q &\geq 12\left(1 - 3q + 3\frac{q}{9}\right)q^2 \\ \Leftrightarrow \quad 1 &\geq 12\left(1 - 3q + \frac{q}{3}\right)q \\ \Leftrightarrow \quad 1 &\geq 12q - 32q^2. \end{aligned} \tag{14.19}$$

Let $f(q) = 12q - 32q^2$. Then $f'(q) = 12 - 64q$.

Since $q \geq \frac{1}{4}$ we deduce that $f'(q) = 12 - 64q \leq 12 - \frac{64}{4} = -4 < 0$, so it follows that f decreases on the interval $[1/4, 1/3]$, i.e. we have

$$f(q) \leq f\left(\frac{1}{4}\right) = 12\frac{1}{4} - 32\frac{1}{16} = 3 - 2 = 1,$$

and inequality (14.18) follows.

Now let us suppose that $0 \leq q \leq \frac{1}{4}$.

Let's rewrite inequality (14.15) as follows

$$q \geq 12q^2(1-3q) + 12r(3q^2 + 6q - 2) - 72r^2. \tag{14.20}$$

Since

$$12q(1-3q) = 4 \cdot 3q(1-3q) \leq 4\left(\frac{3q + (1-3q)}{2}\right)^2 = 1,$$

it follows that

$$12q^2(1-3q) \leq q. \tag{14.21}$$

Since $0 \leq q \leq \frac{1}{4}$ we get

$$3q^2 + 6q - 2 \leq 3\frac{1}{16} + 6\frac{1}{4} - 2 < 0. \tag{14.22}$$

By (14.21) and (14.22) we obtain

$$12q^2(1-3q) + 12r(3q^2 + 6q - 2) - 72r^2 \leq 12q^2(1-3q) \leq q,$$

as required.

Chapter 15
Method for Proving Symmetric Inequalities with Three Variables Defined on the Set of Real Numbers

This section will consider one method that is similar to the previous method of Chap. 14, for proving symmetrical inequalities with three variables that will be solvable only by elementary transformations and without major knowledge of inequalities (in the sense that for some of them the student has no need to know the powerful *Cauchy–Schwarz*, *Chebishev*, *Minkowski* and *Hölder* inequalities).

We must note that this method is suitable for proving inequalities that are defined on the set of real numbers, not just on the set of positive real numbers. For this purpose we will first state (without proof) two theorems from differential calculus.

Theorem *Let $f : I \to \mathbb{R}$ be a differentiable function on I. Then f is an increasing function on I if and only if $f'(x) \geq 0$ for all $x \in I$, and f is a decreasing function on I if and only if $f'(x) \leq 0$ for all $x \in I$.*

Theorem *Let $f(x)$ be a continuous function and twice differentiable on some interval that contains the point x_0.*
Suppose that $f'(x_0) = 0$. Then:

(1) *If $f''(x_0) < 0$, then f has a local maximum at x_0.*
(2) *If $f''(x_0) > 0$, then f has a local minimum at x_0.*

Let a, b, c be real numbers such that $a + b + c = 1$.

According to the obvious inequality $a^2 + b^2 + c^2 \geq ab + bc + ca$ (equality occurs iff $a = b = c$) it follows that

$$1 = (a + b + c)^2 = a^2 + b^2 + c^2 + 2(ab + bc + ca) \geq 3(ab + bc + ca),$$

i.e.

$$ab + bc + ca \leq \frac{1}{3}.$$

Z. Cvetkovski, *Inequalities*,
DOI 10.1007/978-3-642-23792-8_15, © Springer-Verlag Berlin Heidelberg 2012

Let $ab+bc+ca=\frac{1-q^2}{3}$, $(q\geq 0)$. We will find the maximum and minimum values of abc in terms of q.

If $q=0$ then $ab+bc+ca=\frac{1}{3}$, i.e. $a=b=c=\frac{1}{3}$.

Thus

$$abc=\frac{1}{27}.$$

If $q\neq 0$ then

$$ab+bc+ca=\frac{1-q^2}{3}<\frac{1}{3}=\frac{(a+b+c)^2}{3} \quad \Leftrightarrow \quad a^2+b^2+c^2>ab+bc+ca$$

$$\Leftrightarrow \quad (a-b)^2+(b-c)^2+(c-a)^2>0,$$

i.e. at least two of the numbers a, b, c are different.

Consider the function

$$f(x)=(x-a)(x-b)(x-c)=x^3-x^2+\frac{1-q^2}{3}x-abc.$$

We have

$$f'(x)=3x^2-2x+\frac{1-q^2}{3}, \quad \text{with zeros } x_1=\frac{1+q}{3} \quad \text{and} \quad x_2=\frac{1-q}{3}.$$

Hence $f'(x)<0$ for $x_2<x<x_1$, and $f'(x)>0$ for $x<x_2$ or $x>x_1$.

For $f''(x)$ we have

$$f''(x)=6x-2, \quad \text{i.e.} \quad f''(x_1)=6\left(\frac{1+q}{3}\right)-2=6q>0,$$

so it follows that $f(x)$ at x_1 has a local minimum.

Similarly $f''(x_1)=6(\frac{1-q}{3})-2=-6q<0$, i.e. $f(x)$ at x_2 has a local maximum.

Furthermore $f(x)$ has three zeros: a, b, c.

Then it follows that

$$f\left(\frac{1+q}{3}\right)=\frac{(1+q)^2(1-2q)}{27}-abc\leq 0 \quad \text{and}$$

$$f\left(\frac{1-q}{3}\right)=\frac{(1-q)^2(1+2q)}{27}-abc\geq 0.$$

Hence

$$\frac{(1+q)^2(1-2q)}{27}\leq abc\leq\frac{(1-q)^2(1+2q)}{27}.$$

Therefore we have the following theorem.

Theorem 15.1 *Let a, b, c be real numbers such that $a+b+c=1$ and let*

$$ab+bc+ca=\frac{1-q^2}{3} \quad (q \geq 0).$$

Then we have the following inequalities

$$\frac{(1+q)^2(1-2q)}{27} \leq abc \leq \frac{(1-q)^2(1+2q)}{27}.$$

Theorem 15.2 (Generalized) *Let a, b, c be real numbers such that $a+b+c=p$.*
Let $ab+bc+ca=\frac{p^2-q^2}{3}$, $(q \geq 0)$ and $abc=r$.
Then

$$\frac{(p+q)^2(p-2q)}{27} \leq r \leq \frac{(p-q)^2(p+2q)}{27}.$$

Equality occurs if and only if $(a-b)(b-c)(c-a)=0$.

If $a+b+c=p$ and $ab+bc+ca=\frac{p^2-q^2}{3}$ then we can easily show the following identities.

1° $a^2+b^2+c^2=\frac{p^2+2q^2}{3}$
2° $a^3+b^3+c^3=pq^2+3r$
3° $ab(a+b)+bc(b+c)+ca(c+a)=\frac{p(p^2-q^2)}{3}-3r$
4° $(a+b)(b+c)(c+a)=\frac{p(p^2-q^2)}{3}-r$
5° $a^2b^2+b^2c^2+c^2a^2=\frac{(p^2-q^2)^2}{9}-2pr$
6° $ab(a^2+b^2)+bc(b^2+c^2)+ca(c^2+a^2)=\frac{(p^2+2q^2)(p^2-q^2)}{9}-pr$
7° $a^4+b^4+c^4=\frac{-p^4+8p^2q^2+2q^4}{9}+4pr$.

Exercise 15.1 Let a, b, c be real numbers. Prove the inequality

$$a^4+b^4+c^4 \geq abc(a+b+c).$$

Solution Since the given inequality is homogenous, we may assume that $a+b+c=1$.

Then it becomes

$$\frac{-1+8q^2+2q^4}{9}+4r \geq r \quad \Leftrightarrow \quad -1+8q^2+2q^4+27r \geq 0.$$

According to Theorem 15.1, it follows that it suffices to show that

$$-1+8q^2+2q^4+27\frac{(1+q)^2(1-2q)}{27}\geq 0.$$

We have

$$\begin{aligned}
&-1+8q^2+2q^4+27\frac{(1+q)^2(1-2q)}{27}\\
&=-1+8q^2+2q^4+(1+q)^2(1-2q)\\
&=-1+8q^2+2q^4+(1+2q+q^2)(1-2q)\\
&=-1+8q^2+2q^4+(1-3q^2-2q^3)\\
&=2q^4+5q^2-2q^3=q^2(2q^2-2q+5)\\
&=q^2\frac{(4q^2-4q+10)}{2}=q^2\frac{((2q-1)^2+9)}{2}\geq 0,
\end{aligned}$$

as required. Equality occurs iff $a=b=c$.

Exercise 15.2 Let $a,b,c\in\mathbb{R}$. Prove the inequality

$$(a+b)^4+(b+c)^4+(c+a)^4\geq\frac{4}{7}(a^4+b^4+c^4).$$

Solution Since

$$\begin{aligned}
&(a+b)^4+(b+c)^4+(c+a)^4\\
&\quad=2(a^4+b^4+c^4)+4(a^3b+b^3a+b^3c+c^3b+c^3a+a^3c)\\
&\qquad+6(a^2b^2+b^2c^2+c^2a^2),
\end{aligned}$$

the given inequality becomes

$$\begin{aligned}
&5(a^4+b^4+c^4)+14(a^3b+b^3a+b^3c+c^3b+c^3a+a^3c)\\
&\quad+21(a^2b^2+b^2c^2+c^2a^2)\geq 0.
\end{aligned}$$

After setting $a+b+c=p, ab+bc+ca=\frac{p^2-q^2}{3}, r=abc$, due to 5°, 6° and 7° we deduce that the previous inequality is equivalent to

$$\begin{aligned}
&5\left(\frac{-p^4+8p^2q^2+2q^4}{9}+4pr\right)+14\left(\frac{(p^2+2q^2)(p^2-q^2)}{9}-pr\right)\\
&\quad+21\left(\frac{(p^2-q^2)^2}{9}-2pr\right)\geq 0,
\end{aligned}$$

i.e.

$$5(-p^4+8p^2q^2+2q^4+36pr)+14((p^2+2q^2)(p^2-q^2)-9pr)$$
$$+21((p^2-q^2)^2-18pr)\geq 0.$$

If $p=0$ then $10q^4-28q^4+21q^4\geq 0$, i.e. $3q^4\geq 0$, which is obvious.

Let $p\neq 0$.

Without loss of generality we may assume that $p=1$.

So we need to prove that

$$5(-1+8q^2+2q^4+36r)+14((1+2q^2)(1-q^2)-9r)+21((1-q^2)^2-18r)\geq 0,$$

i.e.

$$3q^4+4q^2+10-108r\geq 0.$$

Using Theorem 15.1, we obtain

$$\begin{aligned}3q^4+4q^2+10-108r &\geq 3q^4+4q^2+10-108\frac{(1-q)^2(1+2q)}{27}\\ &=3q^4+4q^2+10-4(1-q)^2(1+2q)\\ &=q^2(q-4)^2+2q^4+6\geq 0,\end{aligned}$$

which clearly holds.

Equality occurs if and only if $a=b=c=0$.

Exercise 15.3 Let a,b,c be real numbers such that $a^2+b^2+c^2=9$. Prove the inequality

$$2(a+b+c)-abc\leq 10.$$

Solution Let $a+b+c=p$, $ab+bc+ca=\frac{1-q^2}{3}$, $abc=r$.

Then using identity 1°, the condition can be rewritten as

$$9=a^2+b^2+c^2=\frac{p^2+2q^2}{3},$$

i.e.

$$p^2+2q^2=27. \tag{15.1}$$

By Theorem 15.2 we deduce

$$\begin{aligned}2(a+b+c)-abc=2p-r &\leq 2p-\frac{(p+q)^2(p-2q)}{27}\\ &=\frac{54p-p^3+3pq^2+2q^3}{27}\end{aligned}$$

$$= \frac{54p - p(p^2 + 2q^2) + 5pq^2 + 2q^3}{27}$$

$$\stackrel{(15.1)}{=} \frac{54p - 27p + 5pq^2 + 2q^3}{27}$$

$$= \frac{27p + 5pq^2 + 2q^3}{27} = \frac{p(27 + 5q^2) + 2q^3}{27}.$$

So it remains to prove that

$$\frac{p(27 + 5q^2) + 2q^3}{27} \le 10 \quad \text{or} \quad p(27 + 5q^2) \le 270 - 2q^3.$$

We have

$$(270 - 2q^3)^2 \ge (p(27 + 5q^2))^2$$

$$\Leftrightarrow \quad 27(q - 3)^2(2q^4 + 12q^3 + 49q^2 + 146q + 219) \ge 0$$

as required.

Equality occurs if and only if $(a, b, c) = (2, 2, -1)$ (up to permutation).

Exercise 15.4 Let a, b, c be positive real numbers. Prove the inequality

$$a^2 + b^2 + c^2 + 2abc + 1 \ge 2(ab + bc + ca).$$

Solution The given inequality is equivalent to

$$\frac{p^2 + 2q^2}{3} + 2r + 1 \ge 2\frac{1 - q^2}{3}$$

i.e.

$$6r + 3 + 4q^2 - p^2 \ge 0.$$

If $2q \ge p$ then we are done.

Therefore suppose that $p \ge 2q$.

By Theorem 15.2, it suffices to prove that

$$6r + 3 + 4q^2 - p^2 \ge 6\frac{(p + q)^2(p - 2q)}{27} + 3 + 4q^2 - p^2 \ge 0,$$

i.e.

$$\frac{2(p + q)^2(p - 2q)}{9} + 3 + 4q^2 - p^2 \ge 0$$

$$\Leftrightarrow \quad (p - 3)^2(2p + 3) \ge 2q^2(2q + 3p - 18). \tag{15.2}$$

If $2p \le 9$ it follows that $2q + 3p \le 4p \le 18$, and we are done.

If $2p \geq 9$ we have

$$
\begin{aligned}
2q^2(2q+3p-18) &\leq 2q^2(p+3p-18) = 4q^2(2p-9) \\
&\leq p^2(2p-9) = (p-3)^2(2p+3) - 27 < (p-3)^2(2p+3),
\end{aligned}
$$

so inequality (15.2) is true, as required.

Equality occurs if and only if $a = b = c = 1$.

Exercise 15.5 (Schur's inequality) Prove that for any non-negative real numbers a, b, c we have

$$a^3 + b^3 + c^3 + 3abc \geq ab(a+b) + bc(b+c) + ca(c+a).$$

Solution Since the above inequality is homogenous, we may assume that $a + b + c = 1$.

Then clearly $q \in [0, 1]$ and the given inequality becomes

$$27r + 4q^2 - 1 \geq 0.$$

If $q \geq \frac{1}{2}$, then we are done.

If $q \leq \frac{1}{2}$, by Theorem 15.1, we have

$$27r + 4q^2 - 1 \geq 27\frac{(1+q)^2(1-2q)}{27} + 4q^2 - 1 = q^2(1-2q) \geq 0,$$

as required.

Equality occurs iff $(a, b, c) = (t, t, t)$ or $(a, b, c) = (t, t, 0)$, where $t \geq 0$ is an arbitrary real number (up to permutation).

Chapter 16
Abstract Concreteness Method (ABC Method)

In this section we will present three theorems without proofs (the proofs can be found in [27]) which are the basis of a very useful method, the *Abstract Concreteness Method* (*ABC method*).

For this purpose we'll consider the function $f(abc, ab+bc+ca, a+b+c)$, as a one-variable function with variable abc on $\mathbb{R}$, i.e. on $\mathbb{R}^+$.

16.1 ABC Theorem

Theorem 16.1 *If the function* $f(abc, ab+bc+ca, a+b+c)$ *is monotonic then* f *achieves it's maximum and minimum values on* $\mathbb{R}$ *when* $(a-b)(b-c)(c-a)=0$, *and on* $\mathbb{R}^+$ *when* $(a-b)(b-c)(c-a)=0$ *or* $abc=0$.

Theorem 16.2 *If the function* $f(abc, ab+bc+ca, a+b+c)$ *is a convex function then it achieves it's maximum and minimum values on* $\mathbb{R}$ *when* $(a-b)(b-c)(c-a)=0$, *and on* $\mathbb{R}^+$ *when* $(a-b)(b-c)(c-a)=0$ *or* $abc=0$.

Theorem 16.3 *If the function* $f(abc, ab+bc+ca, a+b+c)$ *is a concave function then it achieves it's maximum and minimum values on* $\mathbb{R}$ *when* $(a-b)(b-c)(c-a)=0$, *and on* $\mathbb{R}^+$ *when* $(a-b)(b-c)(c-a)=0$ *or* $abc=0$.

Z. Cvetkovski, *Inequalities*,
DOI 10.1007/978-3-642-23792-8_16,

Consequence 16.1 *Let $f(abc, ab+bc+ca, a+b+c)$ be a linear function with variable abc. Then f achieves it's maximum and minimum values on $\mathbb{R}$ if and only if $(a-b)(b-c)(c-a)=0$, and on $\mathbb{R}^+$ if and only if $(a-b)(b-c)(c-a)=0$ or $abc=0$.*

Consequence 16.2 *Let $f(abc, ab+bc+ca, a+b+c)$ be a quadratic trinomial with variable abc, then f achieves it's maximum on $\mathbb{R}$ if and only if $(a-b)(b-c)(c-a)=0$, and on $\mathbb{R}^+$ if and only if $(a-b)(b-c)(c-a)=0$ or $abc=0$.*

Consequence 16.3 *All symmetric three-variable polynomials of degree less than or equal to 5 achieves their maximum and minimum values on $\mathbb{R}$ if and only if $(a-b)(b-c)(c-a)=0$, and on $\mathbb{R}^+$ if and only if $(a-b)(b-c)(c-a)=0$ or $abc=0$.*

Consequence 16.4 *All symmetric three-variables polynomials of degree less than or equal to 8 with non-negative coefficient of $(abc)^2$ in the representation form $f(abc, ab+bc+ca, a+b+c)$, achieves their maximum on $\mathbb{R}$ if and only if $(a-b)(b-c)(c-a)=0$, and on $\mathbb{R}^+$ if and only if $(a-b)(b-c)(c-a)=0$ or $abc=0$.*

Also we'll introduce some additional identities which will be very useful for the correct presentation of this method.

For that purpose, let $a=x+y+z, b=xy+yz+zx, c=xyz$. Then we have.

I_1: $x^2+y^2+z^2=a^2-2b$
I_2: $x^3+y^3+z^3=a^3-3ab+3c$
I_3: $x^4+y^4+z^4=a^4-4a^2b+2b^2+4ac$
I_4: $x^5+y^5+z^5=a^5-5a^3b+5ab^2+5a^2c-5bc$
I_5: $x^6+y^6+z^6=a^6-6a^4b+6a^3c+9a^2b^2-12abc+3c^2-2b^3$
I_6: $(xy)^2+(yz)^2+(zx)^2=b^2-2ac$
I_7: $(xy)^3+(yz)^3+(zx)^3=b^3-3abc+3c^2$
I_8: $(xy)^4+(yz)^4+(zx)^4=b^4-4ab^2c+2a^2c^2+4bc^2$
I_9: $(xy)^5+(yz)^5+(zx)^5=b^5-5ab^3c+5a^2bc^2+5b^2c^2-5ac^3$
I_{10}: $xy(x+y)+yz(y+z)+zx(z+x)=ab-3c$
I_{11}: $xy(x^2+y^2)+yz(y^2+z^2)+zx(z^2+x^2)=a^2b-2b^2-ac$
I_{12}: $xy(x^3+y^3)+yz(y^3+z^3)+zx(z^3+x^3)=a^3b-3ab^2-a^2c+5bc$
I_{13}: $x^2y^2(x+y)+y^2z^2(y+z)+z^2x^2(z+x)=ab^2-2a^2c-bc$
I_{14}: $x^3y^3(x+y)+y^3z^3(y+z)+z^3x^3(z+x)=ab^3-3a^2bc+5ac^2-b^2c$

I_{15}: $(x^2y+y^2z+z^2x)(xy^2+yz^2+zx^2)=9c^2+(a^3-6ab)c+b^3$
I_{16}: $(x^3y+y^3z+z^3x)(xy^3+yz^3+zx^3)=7a^2c^2+(a^5-5a^3b+ab^2)c+b^4$.

Exercise 16.1 Let $a,b,c>0$ be real numbers. Prove the inequality

$$\frac{abc}{a^3+b^3+c^3}+\frac{2}{3}\geq\frac{ab+bc+ca}{a^2+b^2+c^2}.$$

Solution The given inequality is equivalent to the following one

$$F=abc(a^2+b^2+c^2)+\frac{2}{3}(a^3+b^3+c^3)(a^2+b^2+c^2)$$
$$-(a^3+b^3+c^3)(ab+bc+ca)\geq 0.$$

The polynomial F is of third degree so it will achieves it's minimum when

$$(a-b)(b-c)(c-a)=0\quad\text{or } abc=0.$$

If $(a-b)(b-c)(c-a)=0$, then without loss of generality we may assume that $a=c$ and the given inequality becomes

$$\frac{a^2b}{2a^3+b^3}+\frac{2}{3}\geq\frac{a^2+2ab}{2a^2+b^2}\quad\Leftrightarrow\quad(a-b)^2\left(\frac{1}{2a^2+b^2}-\frac{2a+b}{3(2a^3+b^3)}\right)\geq 0$$
$$\Leftrightarrow\quad(a-b)^4(a+b)\geq 0,$$

which is obvious.

If $abc=0$ then without loss of generality we may assume that $c=0$ and the given inequality becomes

$$\frac{2}{3}\geq\frac{ab}{a^2+b^2}\quad\Leftrightarrow\quad a^2+b^2+3(a-b)^2\geq 0,$$

which is true. And we are done.

Exercise 16.2 Let $a,b,c>0$ be real numbers. Prove the inequality

$$\frac{a^3+b^3+c^3}{4abc}+\frac{1}{4}\geq\left(\frac{a^2+b^2+c^2}{ab+bc+ca}\right)^2.$$

Solution Observe that by applying the previous identities the given inequality can be rewritten as a seventh-degree symmetric polynomial with variables a,b,c, but it's only a first-degree polynomial with variable abc.

Therefore by Consequence 16.1, we need to consider only the following two cases.

First case: If $(a-b)(b-c)(c-a)=0$, then without lose of generality we may assume that $a=c$ and the given inequality becomes

$$\frac{2a^3+b^3}{4a^2b}+\frac{1}{4}\geq\left(\frac{2a^2+b^2}{a^2+2ab}\right)^2 \quad\Leftrightarrow\quad \frac{2a^3+b^3}{4a^2b}-\frac{3}{4}\geq\left(\frac{2a^2+b^2}{a^2+2ab}\right)^2-1$$

$$\Leftrightarrow\quad \frac{(a-b)^2(2a+b)}{4a^2b}\geq\frac{(a-b)^2(3a^2+b^2+2ab)}{(a^2+2ab)^2}$$

$$\Leftrightarrow\quad (a-b)^2((2b-a)^2+a^2)\geq 0,$$

which is obvious.

Second case: If $abc=0$ then the given inequality is trivially correct.

Exercise 16.3 Let $a,b,c>0$ be real numbers. Prove the inequality

$$(ab+bc+ca)\left(\frac{1}{(a+b)^2}+\frac{1}{(b+c)^2}+\frac{1}{(c+a)^2}\right)\geq\frac{9}{4}.$$

Solution We can rewrite the given inequality in the following form

$$\begin{aligned}
&f(a+b+c,ab+bc+ca,abc)\\
&\quad=9((a+b)(b+c)(c+a))^2\\
&\qquad-4(ab+bc+ca)((a+b)^2(b+c)^2+(b+c)^2(c+a)^2+(c+a)^2(a+b)^2)\\
&\quad=k(abc)^2+mabc+n,
\end{aligned}$$

where $k\geq 0$ and k,m,n are quantities containing constants or $a+b+c, ab+bc+ca, abc$, which we also consider as constants, i.e. in the form as a sixth-degree symmetric polynomial with variables a,b,c and a second-degree polynomial with variable abc and positive coefficients.

Let us explain this:

The expression $(a+b)(b+c)(c+a)$ has the form $kabc+m$ so it follows that $9((a+b)(b+c)(c+a))^2$ has the form $k^2(abc)^2+mabc+n$.

Furthermore

$$4(ab+bc+ca)((a+b)^2(b+c)^2+(b+c)^2(c+a)^2+(c+a)^2(a+b)^2)=4kA,$$

where $k=ab+bc+ca$, and A is a fourth-degree polynomial and also has the form $kabc+m$.

Therefore the expression of the left side of $f(a+b+c,ab+bc+ca,abc)$ has the form $k(abc)^2+mabc+n$.

Then the function achieves it's minimum value when $(a-b)(b-c)(c-a)=0$ or when $abc=0$.

If $(a-b)(b-c)(c-a)=0$, then without loss of generality we may assume that $a=c$, and the given inequality is equivalent to

$$\begin{aligned}
&(a^2+2ab)\left(\frac{1}{4a^2}+\frac{2}{(a+b)^2}\right)\ge\frac{9}{4}\\
\Leftrightarrow\quad &(a-b)^2\left(\frac{2a+b}{2a(a+b)^2}-\frac{1}{(a+b)^2}\right)\ge 0\\
\Leftrightarrow\quad &b(a-b)^2\ge 0,
\end{aligned}$$

as required.

If $abc=0$, we may assume that $c=0$ and the given inequality becomes

$$ab\left(\frac{1}{(a+b)^2}+\frac{1}{a^2}+\frac{1}{b^2}\right)\ge\frac{9}{4}\quad\Leftrightarrow\quad(a-b)^2\left(\frac{1}{ab}-\frac{1}{4(a+b)^2}\right)\ge 0$$

$$\Leftrightarrow\quad(a-b)^2(4a^2+4b^2+7ab)\ge 0,$$

and the problem is solved.

Exercise 16.4 Let $a,b,c>0$ be real numbers such that $a^2+b^2+c^2=1$. Prove the inequality

$$\frac{a}{a^3+bc}+\frac{b}{b^3+ca}+\frac{c}{c^3+ab}\ge 3.$$

Solution If we transform the given inequality as a symmetric polynomial we obtain a ninth-degree polynomial with variables a,b,c, and a third-degree polynomial with variable abc. But, as we know, this case is not in the previously mentioned consequences, so the problem cannot be solved with ABC (for now).

Therefore we'll make some algebraic transformations.

If we take

$$x=\frac{bc}{a},\qquad y=\frac{ac}{b},\qquad z=\frac{ab}{c},$$

then clearly $xy+yz+zx=a^2+b^2+c^2=1$, and the given inequality becomes

$$\frac{1}{xy+z}+\frac{1}{yz+x}+\frac{1}{zx+y}\ge 3. \tag{16.1}$$

If we transform the inequality (16.1) we'll get a second-degree polynomial with variable xyz, with a non-negative coefficient in front of $(xyz)^2$.

So we need to consider just the following cases:

If $x=z$ then inequality (16.1) becomes

$$\frac{2}{xy+x}+\frac{1}{x^2+y}\ge 3.$$

Since $2xy + x^2 = 1$ it follows that $y = \frac{1-x^2}{2x}$, and after using these, the previous inequality easily follows.

If $z = 0$ then inequality (16.1) becomes

$$\frac{1}{xy} + \frac{1}{x} + \frac{1}{y} \geq 3, \quad \text{with } xy = 1.$$

We have $\frac{1}{xy} + \frac{1}{x} + \frac{1}{y} \geq 1 + \frac{2}{\sqrt{xy}} = 3$, as required.

Exercise 16.5 Let a, b, c be positive real numbers such that $ab+bc+ca+abc = 4$. Prove the inequality

$$\frac{1}{a} + \frac{1}{b} + \frac{1}{c} \geq a + b + c.$$

Solution Since $ab + bc + ca + abc = 4$ there exist real numbers x, y, z such that

$$a = \frac{2x}{y+z}, \qquad b = \frac{2y}{z+x}, \qquad c = \frac{2z}{x+y},$$

and the given inequality becomes

$$\frac{x+y}{z} + \frac{z+x}{y} + \frac{y+z}{x} \geq 4\left(\frac{x}{y+z} + \frac{y}{z+x} + \frac{z}{x+y}\right). \tag{16.2}$$

Inequality (16.2) is homogenous, so we may assume that $x + y + z = 1, xy + yz + zx = u, xyz = v$.

After some algebraic transformations we find that inequality (16.2) can be rewritten as follows

$$9v^2 + 4(1 - u)v - v^2 \geq 0.$$

So, according to the *ABC theorem*, we need to consider just two cases:

If $z = 0$ then inequality (16.2) is trivially correct.

If $y = z = 1$ (we can do this because of the homogenous property) inequality (16.2) becomes

$$2(x+1) + \frac{2}{x} \geq 4\left(\frac{x}{2} + \frac{2}{x+1}\right) \quad \text{i.e.} \quad 2(x-1)^2 \geq 0,$$

which is obvious.

Chapter 17
Sum of Squares (SOS Method)

One of the basic procedures for proving inequalities is to rewrite them as a sum of squares (*SOS*) and then, according to the most elementary property that the square of a real number is non-negative, to prove a certain inequality. This property is the basis of the SOS method.

The advantage of the *method of squares* is that it requires knowledge only of basic inequalities, which we met earlier, and basic skills in elementary operations.

Let's start with one well-known inequality.

Example 17.1 Let $a, b, c \geq 0$. Prove the inequality

$$a^3 + b^3 + c^3 \geq 3abc.$$

Solution We have

$$a^3 + b^3 + c^3 - 3abc = \frac{a+b+c}{2}((a-b)^2 + (b-c)^2 + (c-a)^2) \geq 0,$$

which is obviously true.

The whole idea is to rewrite the given inequality in the form

$$S_a(b-c)^2 + S_b(a-c)^2 + S_c(a-b)^2,$$

where S_a, S_b, S_c are functions of a, b, c.

We must mention that this method works well for proving symmetrical inequalities where we can assume that $a \geq b \geq c$, while if we work with cyclic inequalities we need to consider the additional case $c \geq b \geq a$.

We will discuss symmetrical inequalities with three variables, and for that purpose firstly we'll give three properties that we will use for the proof of the main theorem.

Proposition 17.1 *Let $a, b, c \in \mathbb{R}$. Then $(a-c)^2 \leq 2(a-b)^2 + 2(b-c)^2$.*

Z. Cvetkovski, *Inequalities*,
DOI 10.1007/978-3-642-23792-8_17, © Springer-Verlag Berlin Heidelberg 2012

Proof We have

$$
\begin{aligned}
&(a-c)^2 \le 2(a-b)^2 + 2(b-c)^2\\
&\quad\Leftrightarrow\quad a^2 - 2ac + c^2 \le 2(a^2 - 2ab + b^2) + 2(b^2 - 2bc + c^2)\\
&\quad\Leftrightarrow\quad a^2 + 4b^2 + c^2 - 4ab - 4bc + 2ac \ge 0\\
&\quad\Leftrightarrow\quad (a+c-2b)^2 \ge 0,
\end{aligned}
$$

which clearly holds. □

Proposition 17.2 *Let $a \ge b \ge c$. Then $(a-c)^2 \ge (a-b)^2 + (b-c)^2$.*

Proof We have

$$
\begin{aligned}
&(a-c)^2 \ge (a-b)^2 + (b-c)^2\\
&\quad\Leftrightarrow\quad a^2 - 2ac + c^2 \ge (a^2 - 2ab + b^2) + (b^2 - 2bc + c^2)\\
&\quad\Leftrightarrow\quad b^2 + ac - ab - bc \le 0\\
&\quad\Leftrightarrow\quad (b-a)(b-c) \le 0,
\end{aligned}
$$

which is true since $a \ge b \ge c$. □

Proposition 17.3 *Let $a \ge b \ge c$. Then $\frac{a-c}{b-c} \ge \frac{a}{b}$.*

Proof We have

$$
\frac{a-c}{b-c} \ge \frac{a}{b} \quad\Leftrightarrow\quad b(a-c) \ge a(b-c) \quad\Leftrightarrow\quad ac \ge bc \quad\Leftrightarrow\quad a \ge b.
$$
□

Theorem 17.1 (SOS method) *Consider the expression $S = S_a(b-c)^2 + S_b(a-c)^2 + S_c(a-b)^2$, where S_a, S_b, S_c are functions of a, b, c.*

1° *If $S_a, S_b, S_c \ge 0$ then $S \ge 0$.*
2° *If $a \ge b \ge c$ or $a \le b \le c$ and $S_b, S_b + S_a, S_b + S_c \ge 0$ then $S \ge 0$.*
3° *If $a \ge b \ge c$ or $a \le b \le c$ and $S_a, S_c, S_a + 2S_b, S_c + 2S_b \ge 0$ then $S \ge 0$.*
4° *If $a \ge b \ge c$ and $S_b, S_c, a^2S_b + b^2S_a \ge 0$ then $S \ge 0$.*
5° *If $S_a + S_b \ge 0$ or $S_b + S_c \ge 0$ or $S_c + S_a \ge 0$ ($S_a + S_b + S_c \ge 0$) and $S_aS_b + S_bS_c + S_cS_a \ge 0$ then $S \ge 0$.*

Proof 1° If $S_a, S_b, S_c \ge 0$ then clearly $S \ge 0$.

2° Let us assume that $a \geq b \geq c$ and $S_b, S_b + S_a, S_b + S_c \geq 0$.
By Proposition 17.2, it follows that $(a-c)^2 \geq (a-b)^2 + (b-c)^2$, so we have

$$\begin{aligned} S &= S_a(b-c)^2 + S_b(a-c)^2 + S_c(a-b)^2 \\ &\geq S_a(b-c)^2 + S_b((a-b)^2 + (b-c)^2) + S_c(a-b)^2 \\ &= (b-c)^2(S_a + S_b) + (a-b)^2(S_b + S_c). \end{aligned}$$

Now since $S_b + S_a, S_b + S_c \geq 0$ it follows that $S \geq 0$.
3° Let $a \geq b \geq c$ and $S_a, S_c, S_a + 2S_b, S_c + 2S_b \geq 0$.
Then if $S_b \geq 0$ clearly $S \geq 0$.
Suppose that $S_b \leq 0$.
By Proposition 17.1, we have that $(a-c)^2 \leq 2(a-b)^2 + 2(b-c)^2$.
Therefore

$$\begin{aligned} S &= S_a(b-c)^2 + S_b(a-c)^2 + S_c(a-b)^2 \\ &\geq S_a(b-c)^2 + S_b(2(a-b)^2 + 2(b-c)^2) + S_c(a-b)^2 \\ &= (b-c)^2(S_a + 2S_b) + (a-b)^2(S_c + 2S_b), \end{aligned}$$

and since $S_a + 2S_b, S_c + 2S_b \geq 0$ it follows that $S \geq 0$.
4° Let $a \geq b \geq c$ and suppose that $S_b, S_c, a^2S_b + b^2S_a \geq 0$.
By Proposition 17.3, it follows that $\frac{a-c}{b-c} \geq \frac{a}{b}$.
Therefore

$$\begin{aligned} S &= S_a(b-c)^2 + S_b(a-c)^2 + S_c(a-b)^2 \geq S_a(b-c)^2 + S_b(a-c)^2 \\ &= (b-c)^2\left(S_a + S_b\left(\frac{a-c}{b-c}\right)^2\right) \geq (b-c)^2\left(S_a + S_b\left(\frac{a}{b}\right)^2\right) \\ &= (b-c)^2\left(\frac{b^2S_a + a^2S_b}{b^2}\right), \end{aligned}$$

since $a^2S_b + b^2S_a \geq 0$ we obtain $S \geq 0$.
5° Assume that $S_b + S_c \geq 0$.
We have

$$\begin{aligned} S &= S_a(b-c)^2 + S_b(a-c)^2 + S_c(a-b)^2 \\ &= S_a(b-c)^2 + S_b\big((c-b) + (b-a)\big)^2 + S_c(a-b)^2 \\ &= (S_b + S_c)(a-b)^2 + 2S_b(c-b)(b-a) + (S_a + S_b)(b-c)^2 \\ &= (S_b + S_c)\left(b - a + \frac{S_b}{S_b + S_c}(c-b)\right)^2 + \frac{S_aS_b + S_bS_c + S_cS_a}{S_b + S_c}(c-b)^2 \\ &\geq 0. \end{aligned}$$

□

The main difficulty with using the S.O.S. method is the transformation of the given inequality into mentioned (S.O.S.) form.

Every difference $\sum_{cyc} x_1^{\alpha_1} x_2^{\alpha_2} \cdots x_n^{\alpha_n} - \sum_{cyc} x_1^{\beta_1} x_2^{\beta_2} \cdots x_n^{\beta_n}$ where $\alpha_1 + \alpha_2 + \cdots + \alpha_n = \beta_1 + \beta_2 + \cdots + \beta_n$ can be written in S.O.S. form, so almost all symmetrical or permutation homogeneous inequalities can be written in S.O.S. form. In fact there is a huge class of algebraic expressions which can be written in S.O.S form (the algorithm which helps to transform algebraic expressions into S.O.S. form is explicitly explained for example in [27]).

Here we will introduce the reader to the simplest and most often used forms which are as follows:

1° $a^2 + b^2 + c^2 - ab - bc - ca = \frac{(a-b)^2+(b-c)^2+(c-a)^2}{2}$

2° $a^3 + b^3 + c^3 - 3abc = (a+b+c) \cdot (\frac{(a-b)^2+(b-c)^2+(c-a)^2}{2})$

3° $a^2b + b^2c + c^2a - ab^2 - bc^2 - ca^2 = \frac{(a-b)^3+(b-c)^3+(c-a)^3}{3}$

4° $a^3 + b^3 + c^3 - a^2b - b^2c - c^2a = \frac{(2a+b)(a-b)^2+(2b+c)(b-c)^2+(2c+a)(c-a)^2}{3}$

5° $a^3b + b^3c + c^3a - ab^3 - bc^3 - ca^3 = (a+b+c)(\frac{(b-a)^3+(c-b)^3+(a-c)^3}{3})$

6° $a^4 + b^4 + c^4 - a^2b^2 - b^2c^2 - c^2a^2 = \frac{(a+b)^2(a-b)^2+(b+c)^2(b-c)^2+(c+a)^2(c-a)^2}{2}$

Exercise 17.1 Let $x, y, z \in \mathbb{R}$ such that $xyz \geq 1$. Prove the inequality

$$\frac{x^5 - x^2}{x^5 + y^2 + z^2} + \frac{y^5 - y^2}{y^5 + z^2 + x^2} + \frac{z^5 - z^2}{z^5 + x^2 + y^2} \geq 0.$$

Solution We'll homogenize as follows

$$\frac{x^5 - x^2}{x^5 + y^2 + z^2} \geq \frac{x^5 - x^2 \cdot xyz}{x^5 + xyz(y^2 + z^2)} = \frac{x^4 - x^2yz}{x^4 + yz(y^2 + z^2)}$$

$$\geq \frac{x^4 - x^2(\frac{y^2+z^2}{2})}{x^4 + (\frac{y^2+z^2}{2})(y^2 + z^2)} = \frac{2x^4 - x^2(y^2 + z^2)}{2x^4 + (y^2 + z^2)^2}.$$

Similarly we get

$$\frac{y^5 - y^2}{y^5 + z^2 + x^2} \geq \frac{2y^4 - y^2(z^2 + x^2)}{2y^4 + (z^2 + x^2)^2} \quad \text{and} \quad \frac{z^5 - z^2}{z^5 + x^2 + y^2} \geq \frac{2z^4 - z^2(x^2 + y^2)}{2z^4 + (x^2 + y^2)^2}.$$

So it suffices to show that

$$\frac{2x^4 - x^2(y^2 + z^2)}{2x^4 + (y^2 + z^2)^2} + \frac{2y^4 - y^2(z^2 + x^2)}{2y^4 + (z^2 + x^2)^2} + \frac{2z^4 - z^2(x^2 + y^2)}{2z^4 + (x^2 + y^2)^2} \geq 0. \quad (17.1)$$

Let $x^2 = a, y^2 = b, z^2 = c$. Then inequality (17.1) becomes

$$\frac{2a^2 - a(b+c)}{2a^2 + (b+c)^2} + \frac{2b^2 - b(c+a)}{2b^2 + (c+a)^2} + \frac{2c^2 - c(a+b)}{2c^2 + (a+b)^2} \geq 0. \quad (17.2)$$

After some algebraic operations we can rewrite inequality (17.2) as follows

$$(b-c)^2\frac{a^2+a(b+c)+b^2-bc+c^2}{(2b^2+(c+a)^2)(2c^2+(a+b)^2)}$$
$$+(c-a)^2\frac{b^2+b(a+c)+c^2-ca+a^2}{(2a^2+(b+c)^2)(2c^2+(a+b)^2)}$$
$$+(a-b)^2\frac{c^2+c(a+b)+a^2-ab+b^2}{(2a^2+(b+c)^2)(2b^2+(c+a)^2)}\geq 0,$$

which is true due to the obvious inequality: if $x, y\in\mathbb{R}$ then $x^2-xy+y^2\geq 0$.

Exercise 17.2 Let a, b, c be positive real numbers. Prove the inequality

$$\frac{a^2+b^2+c^2}{ab+bc+ca}+\frac{8abc}{(a+b)(b+c)(c+a)}\geq 2. \tag{17.3}$$

Solution Observe that

$$a^2+b^2+c^2-(ab+bc+ca)=\frac{1}{2}((a-b)^2+(b-c)^2+(c-a)^2)$$

and

$$(a+b)(b+c)(c+a)-8abc=a(b-c)^2+b(c-a)^2+c(a-b)^2.$$

Inequality (17.3) becomes

$$\frac{a^2+b^2+c^2}{ab+bc+ca}-1\geq 1-\frac{8abc}{(a+b)(b+c)(c+a)}$$
$$\Leftrightarrow\quad \frac{a^2+b^2+c^2-(ab+bc+ca)}{ab+bc+ca}\geq\frac{(a+b)(b+c)(c+a)-8abc}{(a+b)(b+c)(c+a)}$$
$$\Leftrightarrow\quad \frac{(a-b)^2+(b-c)^2+(c-a)^2}{ab+bc+ca}\geq\frac{2a(b-c)^2+2b(c-a)^2+2c(a-b)^2}{(a+b)(b+c)(c+a)}$$
$$\Leftrightarrow\quad (b-c)^2\left(\frac{(a+b)(b+c)(c+a)}{ab+bc+ca}-2a\right)$$
$$+(c-a)^2\left(\frac{(a+b)(b+c)(c+a)}{ab+bc+ca}-2b\right)$$
$$+(a-b)^2\left(\frac{(a+b)(b+c)(c+a)}{ab+bc+ca}-2c\right)\geq 0.$$

Let

$$S_a=\frac{(a+b)(b+c)(c+a)}{ab+bc+ca}-2a=b+c-a-\frac{abc}{ab+bc+ca},$$

$$S_b = \frac{(a+b)(b+c)(c+a)}{ab+bc+ca} - 2b = a+c-b - \frac{abc}{ab+bc+ca},$$

$$S_c = \frac{(a+b)(b+c)(c+a)}{ab+bc+ca} - 2c = a+b-c - \frac{abc}{ab+bc+ca}.$$

Since inequality (17.3) is symmetric, we may assume that $a \geq b \geq c$.

Then clearly

$$S_b, S_c \geq 0, \quad \text{i.e.} \quad S_b + S_c \geq 0.$$

According to 2° from Theorem 17.1, it suffices to show that $S_b + S_a \geq 0$.

We have

$$S_b + S_a = 2c - 2\frac{abc}{ab+bc+ca} = \frac{2c^2(a+b)}{ab+bc+ca} \geq 0,$$

as required.

Exercise 17.3 Let a, b, c be positive real numbers such that $ab+bc+ac = 1$. Prove the inequality

$$\frac{1+a^2b^2}{(a+b)^2} + \frac{1+b^2c^2}{(b+c)^2} + \frac{1+c^2a^2}{(c+a)^2} \geq \frac{5}{2}.$$

Solution The given inequality is equivalent to

$$\sum_{cyc} \frac{(ab+bc+ac)^2 + a^2b^2}{(a+b)^2} \geq \frac{5}{2}(ab+bc+ac)$$

$$\Leftrightarrow \quad 2\sum_{cyc} \frac{2ab(ab+bc+ac) + (bc+ca)^2}{(a+b)^2} \geq 5(ab+bc+ac)$$

$$\Leftrightarrow \quad 4(ab+bc+ca)\left(\frac{ab}{(a+b)^2} + \frac{bc}{(b+c)^2} + \frac{ca}{(c+a)^2}\right)$$
$$+ 2(a^2+b^2+c^2) \geq 5(ab+bc+ac)$$

$$\Leftrightarrow \quad (ab+bc+ca)\left(\frac{4ab}{(a+b)^2} + \frac{4bc}{(b+c)^2} + \frac{4ca}{(c+a)^2} - 3\right)$$
$$+ 2(a^2+b^2+c^2-ab-bc-ca) \geq 0$$

$$\Leftrightarrow -(ab+bc+ca)\left(\frac{(a-b)^2}{(a+b)^2} + \frac{(b-c)^2}{(b+c)^2} + \frac{(c-a)^2}{(c+a)^2}\right)$$
$$+ ((a-b)^2 + (b-c)^2 + (c-a)^2) \geq 0$$

$$\Leftrightarrow \quad \left(1-\frac{ab+bc+ca}{(a+b)^2}\right)(a-b)^2+\left(1-\frac{ab+bc+ca}{(b+c)^2}\right)(b-c)^2$$
$$+\left(1-\frac{ab+bc+ca}{(c+a)^2}\right)(c-a)^2 \geq 0.$$

Let

$$S_a = 1-\frac{ab+bc+ca}{(b+c)^2}, \qquad S_b = 1-\frac{ab+bc+ca}{(c+a)^2} \quad \text{and}$$
$$S_c = 1-\frac{ab+bc+ca}{(a+b)^2}.$$

Without loss of generality we may assume that $a \geq b \geq c$, and then clearly $S_a \leq S_b \leq S_c$.

We have

$$S_c = 1-\frac{ab+bc+ca}{(a+b)^2} = \frac{a^2+(a+b)(b-c)}{(a+b)^2} > 0,$$

and it follows that $S_b \geq S_c > 0$.

Also we have

$$\begin{aligned}
a^2S_b + b^2S_a &= a^2\left(1-\frac{ab+bc+ca}{(c+a)^2}\right)+b^2\left(1-\frac{ab+bc+ca}{(b+c)^2}\right)\\
&= a^2\frac{c^2+(c+a)(a-b)}{(c+a)^2}+b^2\frac{c^2+(b+c)(b-a)}{(b+c)^2}\\
&= c^2\left(\frac{a^2}{(c+a)^2}+\frac{b^2}{(b+c)^2}\right)+(a-b)\left(\frac{a^2}{c+a}-\frac{b^2}{b+c}\right)\\
&= c^2\left(\frac{a^2}{(c+a)^2}+\frac{b^2}{(b+c)^2}\right)+(a-b)^2\frac{ab+bc+ca}{(c+a)(b+c)} > 0,
\end{aligned}$$

and according to 4° from Theorem 17.1 we are done.

Equality occurs iff $a=b=c=\frac{1}{\sqrt{3}}$.

Chapter 18
Strong Mixing Variables Method (SMV Theorem)

This method is very useful in proving symmetric inequalities with more than two variables. The *SMV method* (strong mixing variables method) is a simple and concise method that "works" in proving inequalities that have either a too complicated or a too long proof. In order to better describe the given method, first we will give a *lemma* (without proof) and then we will introduce the reader to the *SMV theorem* and its applications through exercises. We should point out that this theorem is part of a more comprehensive method, the *Mixing Variable method* (MV method), which can be found in [27].

Lemma 18.1 *Let* $(x_1, x_2, \ldots, x_n)$ *be an arbitrary real sequence.*

1° *Choose* $i, j \in \{1, 2, \ldots, n\}$, *such that* $x_i = \min\{x_1, x_2, \ldots, x_n\}, x_j = \max\{x_1, x_2, \ldots, x_n\}$.

2° *Replace* x_i *and* x_j *by it's average* $\frac{x_i + x_j}{2}$ *(their orders don't change).*

After infinitely many of the above transformations, each number $x_i, i = 1, 2, \ldots, n$, *tends to the same limit* $x = \frac{x_1 + x_2 + \cdots + x_n}{n}$.

Theorem 18.1 (SMV theorem) *Let* $F : I \subset \mathbb{R}^n \to \mathbb{R}$ *be a symmetric, continuous, function satisfying* $F(a_1, a_2, \ldots, a_n) \geq F(b_1, b_2, \ldots, b_n)$, *where the sequence* $(b_1, b_2, \ldots, b_n)$ *is a sequence obtained from the sequence* $(a_1, a_2, \ldots, a_n)$ *by some predefined transformation (a* Δ*-transformation). Then we have* $F(x_1, x_2, \ldots, x_n) \geq F(x, x, \ldots, x)$, *with* $x = \frac{x_1 + x_2 + \cdots + x_n}{n}$.

Lets us note that the transformation Δ can be different, i.e. Δ can be defined according to the current problem; for example it can be defined as $\frac{a+b}{2}, \sqrt{ab}, \sqrt{\frac{a^2+b^2}{2}}$, etc.

Z. Cvetkovski, *Inequalities*,
DOI 10.1007/978-3-642-23792-8_18,

Exercise 18.1 Let $a, b, c > 0$ be real numbers. Prove the inequality

$$\frac{a}{b+c}+\frac{b}{c+a}+\frac{c}{a+b}\geq\frac{3}{2}.$$

Solution Let $f(a,b,c)=\frac{a}{b+c}+\frac{b}{c+a}+\frac{c}{a+b}$.

We have

$$\begin{aligned}
&f(a,b,c)-f\left(\frac{a+b}{2},\frac{a+b}{2},c\right)\\
&=\frac{a}{b+c}+\frac{b}{c+a}+\frac{c}{a+b}-\left(\frac{a+b}{a+b+2c}+\frac{a+b}{a+b+2c}+\frac{c}{a+b}\right)\\
&=\frac{a}{b+c}+\frac{b}{c+a}-\frac{2(a+b)}{a+b+2c}=\frac{a^3+ca^2+cb^2+b^3-2abc-ab^2-a^2b}{(b+c)(a+c)(a+b+2c)}.
\end{aligned} \tag{18.1}$$

Since $AM\geq GM$ we obtain

$$\begin{aligned}
a^3+ca^2+cb^2+b^3&=\frac{a^3+a^3+b^3}{3}+\frac{a^3+b^3+b^3}{3}\\
&\quad+ca^2+cb^2\geq a^2b+ab^2+2abc.
\end{aligned} \tag{18.2}$$

From (18.1) and (18.2) it follows that

$$f(a,b,c)-f\left(\frac{a+b}{2},\frac{a+b}{2},c\right)\geq 0,$$

i.e.

$$f(a,b,c)\geq f\left(\frac{a+b}{2},\frac{a+b}{2},c\right).$$

Therefore by the *SMV theorem* it suffices to prove that $f(t,t,c)\geq\frac{3}{2}$.

We have

$$f(t,t,c)\geq\frac{3}{2}\quad\Leftrightarrow\quad\frac{t}{t+c}+\frac{t}{t+c}+\frac{c}{2t}\geq\frac{3}{2}\quad\Leftrightarrow\quad 2(t-c)^2\geq 0,$$

which is obviously true.

Equality occurs if and only if $a=b=c$.

Exercise 18.2 (Turkevicius inequality) Let a, b, c, d be non-negative real numbers. Prove the inequality

$$a^4+b^4+c^4+d^4+2abcd\geq a^2b^2+b^2c^2+c^2d^2+d^2a^2+a^2c^2+b^2d^2.$$

Solution Without loss of generality we may assume that $a\geq b\geq c\geq d$.

Let us denote

$$\begin{aligned}f(a,b,c,d) &= a^4+b^4+c^4+d^4+2abcd-a^2b^2-b^2c^2-c^2d^2\\&\quad -d^2a^2-a^2c^2-b^2d^2\\&= a^4+b^4+c^4+d^4+2abcd-a^2c^2-b^2d^2-(a^2+c^2)(b^2+d^2).\end{aligned}$$

We have

$$\begin{aligned}&f(a,b,c,d)-f(\sqrt{ac},b,\sqrt{ac},d)\\&\quad = a^4+b^4+c^4+d^4+2abcd-a^2c^2-b^2d^2-(a^2+c^2)(b^2+d^2)\\&\qquad -(a^2c^2+b^4+a^2c^2+d^4+2abcd-a^2c^2-b^2d^2-2ac(b^2+d^2))\\&\quad = a^4+c^4-2a^2c^2-(b^2+d^2)(a^2+c^2+2ac)\\&\quad = (a^2-c^2)^2-(b^2+d^2)(a-c)^2=(a-c)^2((a+c)^2-(b^2+d^2))\ge 0.\end{aligned}$$

Thus

$$f(a,b,c,d)\ge f(\sqrt{ac},b,\sqrt{ac},d).$$

By the *SMV theorem* we only need to prove that $f(a,b,c,d)\ge 0$, in the case when $a=b=c=t\ge d$.

We have

$$f(t,t,t,d)\ge 0 \quad\Leftrightarrow\quad 3t^4+d^4+2t^3d\ge 3t^4+3t^2d^2 \quad\Leftrightarrow\quad d^4+2t^3d\ge 3t^2d^2,$$

which immediately follows from $AM\ge GM$.

Equality occurs iff $a=b=c=d$ or $a=b=c, d=0$ (up to permutation).

Exercise 18.3 Let a,b,c,d be non-negative real numbers such that $a+b+c+d=4$. Prove the inequality

$$(1+3a)(1+3b)(1+3c)(1+3d)\le 125+131abcd.$$

Solution Let us denote

$$f(a,b,c,d)=(1+3a)(1+3b)(1+3c)(1+3d)-131abcd.$$

Without loss of generality we may assume that $a\ge b\ge c\ge d$.

We have

$$\begin{aligned}&f(a,b,c,d)-f\left(\frac{a+c}{2},b,\frac{a+c}{2},d\right)\\&\quad =9(1+3b)(1+3d)\left(ac-\frac{(a+c)^2}{4}\right)-131bd\left(ac-\frac{(a+c)^2}{4}\right)\\&\quad =\frac{(a-c)^2}{4}(131bd-9(1+3b)(1+3d)).\end{aligned}\tag{18.3}$$

Note that

$$b+d \le \frac{1}{2}(a+b+c+d)=2,$$

and clearly

$$bd \le \frac{(b+d)^2}{4}=1, \tag{18.4}$$

therefore

$$\begin{aligned}
&131bd-9(1+3b)(1+3d)\\
&\quad=131bd-9-27(b+d)-81bd\\
&\quad=50bd-27(b+d)-9=50bd-27(b+d)-9 \overset{A\ge G}{\le} 50bd-54\sqrt{bd}\\
&\quad\overset{(18.4)}{\le} 50bd-54bd=-4bd\le 0.
\end{aligned}$$

By (18.3) and the last inequality we deduce that

$$f(a,b,c,d)-f\left(\frac{a+c}{2},b,\frac{a+c}{2},d\right)\le 0,$$

i.e.

$$f(a,b,c,d)\le f\left(\frac{a+c}{2},b,\frac{a+c}{2},d\right).$$

According to the *SMV theorem* it follows that it's enough to prove that

$$f(a,b,c,d)\le 125,$$

when $a=b=c=t\ge d$, i.e.

$$f(t,t,t,d)\le 125, \quad \text{when } 3t+d=4.$$

Clearly $3t\le 4$.

We have

$$\begin{aligned}
&f(t,t,t,d)\le 125\\
&\quad\Leftrightarrow\quad (1+3t)^3(1+3(4-3t))-131t^3(4-3t)\le 125\\
&\quad\Leftrightarrow\quad 150t^4-416t^3+270t^2+108t-112\le 0\\
&\quad\Leftrightarrow\quad (t-1)^2(3t-4)(50t+28)\le 0, \text{ which is true.}
\end{aligned}$$

Equality occurs iff $a=b=c=d=1$ or $a=b=c=\frac{4}{3}, d=0$ (up to permutation).

Exercise 18.4 Let a, b, c, d be non-negative real numbers such that $a + b + c + d = 4$. Prove the inequality

$$16 + 2abcd \geq 3(ab + ac + ad + bc + bd + cd).$$

Solution Without loss of generality we may assume that $a \geq b \geq c \geq d$.

Let us denote

$$f(a, b, c, d) = 3(ab + ac + ad + bc + bd + cd) - 2abcd.$$

We have

$$\begin{aligned}
&f\left(\frac{a+c}{2}, b, \frac{a+c}{2}, d\right) - f(a, b, c, d) \\
&= 3\left(\left(\frac{a+c}{2}\right)b + \left(\frac{a+c}{2}\right)^2 + \left(\frac{a+c}{2}\right)d + \left(\frac{a+c}{2}\right)b + bd + \left(\frac{a+c}{2}\right)d\right) \\
&\quad - 2bd\left(\frac{a+c}{2}\right)^2 - (3(ab + ac + ad + bc + bd + cd) - 2abcd) \\
&= 3\left(\left(\frac{a+c}{2}\right)^2 - ac\right) - 2bd\left(\left(\frac{a+c}{2}\right)^2 - ac\right) \\
&= \left(\frac{a-c}{2}\right)^2 (3 - 2bd). \qquad (18.5)
\end{aligned}$$

Also $2\sqrt{bd} \leq b + d \leq \frac{1}{2}(a + b + c + d) = 2$, from which it follows that $bd \leq 1$.

By (18.5) and the last conclusion we get

$$\begin{aligned}
f\left(\frac{a+c}{2}, b, \frac{a+c}{2}, d\right) - f(a, b, c, d) &= \left(\frac{a-c}{2}\right)^2 (3 - 2bd) \\
&\geq \left(\frac{a-c}{2}\right)^2 (3 - 2) \geq 0,
\end{aligned}$$

i.e. it follows that

$$f\left(\frac{a+c}{2}, b, \frac{a+c}{2}, d\right) \geq f(a, b, c, d).$$

By the *SMV theorem* it follows that we only need to prove the inequality $f(a, b, c, d) \leq 16$, in the case when $a = b = c = t \geq d$, i.e. we need to prove that $f(t, t, t, d) \leq 16$, when $3t + d = 4$.

Clearly $3t \leq 4$.

Thus we have

$$\begin{aligned}
&f(t, t, t, d) \leq 16 \\
&\Leftrightarrow \quad 9(t^2 + dt) - 2t^3 d - 16 \leq 0
\end{aligned}$$

$$\Leftrightarrow \quad 9t^2+9t(4-3t)-2t^3(4-3t)-16\le 0$$

$$\Leftrightarrow \quad 2(3t-4)(t-1)^2(t+2)\le 0, \quad \text{which is true.}$$

Equality occurs if and only if $a=b=c=d=1$ or $a=b=c=4/3, d=0$ (up to permutation).

Exercise 18.5 Let a,b,c,d be non-negative real numbers such that $a+b+c+d=1$. Prove the inequality

$$abc+bcd+cda+dab\le \frac{1}{27}+\frac{176}{27}abcd.$$

Solution Without loss of generality we may assume that $a\le b\le c\le d$.
Let $f(a,b,c,d)=abc+bcd+cda+dab-\frac{176}{27}abcd$ i.e.

$$f(a,b,c,d)=ac(b+d)+bd\left(a+c-\frac{176}{27}ac\right).$$

Since $a\le b\le c\le d$ we have

$$a+c\le \frac{1}{2}(a+b+c+d)=\frac{1}{2},$$

from which it follows that

$$\frac{1}{a}+\frac{1}{c}\ge \frac{4}{a+c}\ge 8>\frac{176}{27}. \tag{18.6}$$

We have

$$\begin{aligned}
&f(a,b,c,d)-f\left(a,\frac{b+d}{2},c,\frac{b+d}{2}\right)\\
&=ac(b+d)+bd\left(a+c-\frac{176}{27}ac\right)\\
&\quad -ac(b+d)-\left(\frac{b+d}{2}\right)^2\left(a+c-\frac{176}{27}ac\right)\\
&=\left(a+c-\frac{176}{27}ac\right)\left(bd-\left(\frac{b+d}{2}\right)^2\right)\\
&=-\left(a+c-\frac{176}{27}ac\right)\frac{(b-d)^2}{4}\overset{(18.6)}{\le} 0.
\end{aligned}$$

Therefore

$$f(a,b,c,d)\le f\left(a,\frac{b+d}{2},c,\frac{b+d}{2}\right).$$

By the *SMV theorem* we have

$$f(a,b,c,d) \le f(a,t,t,t), \quad \text{when } t = \frac{b+c+d}{3}.$$

Now we need to prove only the inequality

$$f(a,t,t,t) \le \frac{1}{27}, \quad \text{with } a+3t=1.$$

Let us note that $3t \le a+3t=1$.

The inequality $f(a,t,t,t) \le \frac{1}{27}$ is equivalent to

$$3at^2 + t^3 \le \frac{1}{27} + \frac{176}{27}at^3. \tag{18.7}$$

After putting $a = 1-3t$ by (18.7) we get $(1-3t)(4t-1)^2(11t+1) \ge 0$, which is obviously true (since $3t \le 1$), and the problem is solved.

Equality occurs if and only if $a=b=c=d=1/4$ or $a=b=c=1/3, d=0$ (up to permutation).

Chapter 19
Method of Lagrange Multipliers

This method is intended for conditional inequalities. It requires elementary skills of differential calculus but it is very easy to apply. We'll give the main theorem, without proof, and we'll introduce some exercises to see how this method works.

Theorem 19.1 (Lagrange multipliers theorem) *Let $f(x_1, x_2, \ldots, x_m)$ be a continuous and differentiable function on $I \subseteq \mathbb{R}^m$, and let $g_i(x_1, x_2, \ldots, x_m) = 0, i = 1, 2, \ldots, k$, where $(k < m)$ are the conditions that must be satisfied. Then the maximum or minimum values of f with the conditions $g_i(x_1, x_2, \ldots, x_m) = 0, i = 1, 2, \ldots, k$, occur at the bounds of the interval I or occur at the points at which the partial derivatives (according to the variables $x_1, x_2, \ldots, x_m$) of the function $L = f - \sum_{i=1}^{k} \lambda_i g_i$, are all zero.*

Exercise 19.1 Let $x_1, x_2, \ldots, x_n$ be positive real numbers such that $x_1 + x_2 + \cdots + x_n = a$. Find the maximal value of the expression $A = \sqrt[n]{x_1 x_2 \cdots x_n}$.

Solution Let $g = x_1 + x_2 + \cdots + x_n - a$. Then *Lagrange's function* is

$$F = A - \lambda g = \sqrt[n]{x_1 x_2 \cdots x_n} - \lambda(x_1 + x_2 + \cdots + x_n - a).$$

For the first partial derivatives we have

$$\begin{cases} F'_{x_1} = \frac{\sqrt[n]{x_1 x_2 \cdots x_n}}{x_1} - \lambda, \\ F'_{x_2} = \frac{\sqrt[n]{x_1 x_2 \cdots x_n}}{x_2} - \lambda, \\ \vdots \\ F'_{x_n} = \frac{\sqrt[n]{x_1 x_2 \cdots x_n}}{x_n} - \lambda \end{cases}$$

from which easily we deduce that we must have $x_1 = x_2 = \cdots = x_n = \frac{a}{n}$.

Hence $\max A = \frac{a}{n}$, i.e. $\sqrt[n]{x_1 x_2 \cdots x_n} \leq \frac{x_1 + x_2 + \cdots + x_n}{n}$ which is the well-known inequality $AM \geq GM$.

Z. Cvetkovski, *Inequalities*,
DOI 10.1007/978-3-642-23792-8_19, © Springer-Verlag Berlin Heidelberg 2012

Exercise 19.2 Let $a, b, c \in \mathbb{R}^+$ such that $a+b+c=1$. Prove the inequality

$$7(ab+bc+ca) \leq 9abc+2.$$

Solution Let

$$f(a,b,c)=7(ab+bc+ca)-9abc-2, \qquad g(a,b,c)=a+b+c-1$$

and

$$L=f-\lambda g=7(ab+bc+ca)-9abc-2-\lambda(a+b+c-1).$$

We have

$$\begin{aligned}
\frac{\partial L}{\partial a}&=7(b+c)-9bc-\lambda=0 \quad \Rightarrow \quad \lambda=7(b+c)-9bc,\\
\frac{\partial L}{\partial b}&=7(c+a)-9ca-\lambda=0 \quad \Rightarrow \quad \lambda=7(c+a)-9ca,\\
\frac{\partial L}{\partial c}&=7(a+b)-9ab-\lambda=0 \quad \Rightarrow \quad \lambda=7(a+b)-9ab.
\end{aligned}$$

So

$$7(b+c)-9bc=\lambda=7(c+a)-9ca \quad \Leftrightarrow \quad (b-a)(7-9c)=0. \tag{19.1}$$

In the same way we obtain

$$(c-b)(7-9a)=0 \tag{19.2}$$

and

$$(a-c)(7-9b)=0. \tag{19.3}$$

Let us consider the identity (19.1).

If $a=b$ then if $b=c$ we get $a=b=c=1/3$, and then

$$f(a,b,c)=7(ab+bc+ca)-9abc-2=\frac{21}{9}-\frac{9}{27}-2=0.$$

If $a=b$ and $b \neq c$ then by (19.2) we must have $a=\frac{7}{9}=b$ and then $a+b=\frac{14}{9}>1$, a contradiction, since $a+b<a+b+c=1$.

If $7-9c=0$ then we can't have $7-9a=0$ or $7-9b=0$ for the same reasons as before, so according to (19.2) and (19.3) we must have $b=c$ and $a=c$, i.e. $a=b=c=7/9$, which is impossible.

Therefore $\min L=0$, i.e. $7(ab+bc+ca) \leq 9abc+2$.

Exercise 19.3 Let $a, b, c \in \mathbb{R}$ such that $a^2+b^2+c^2+abc=4$. Find the minimal value of the expression $a+b+c$.

Solution Let

$$f(a,b,c) = a+b+c, g(a,b,c) = a^2+b^2+c^2+abc-4$$

and

$$L = f - \lambda g = a+b+c-\lambda(a^2+b^2+c^2+abc-4).$$

We have

$$\frac{\partial L}{\partial a} = 1-\lambda a-\lambda bc = 0 \quad \Rightarrow \quad \lambda = \frac{1}{2a+bc},$$
$$\frac{\partial L}{\partial b} = 1-\lambda b-\lambda ac = 0 \quad \Rightarrow \quad \lambda = \frac{1}{2b+ac},$$
$$\frac{\partial L}{\partial c} = 1-\lambda c-\lambda ab = 0 \quad \Rightarrow \quad \lambda = \frac{1}{2c+ab}.$$

So

$$\frac{1}{2a+bc} = \frac{1}{2b+ac} \quad \Leftrightarrow \quad (a-b)(2-c) = 0. \tag{19.4}$$

In the same way we obtain

$$(b-c)(2-a) = 0 \tag{19.5}$$

and

$$(c-a)(2-b) = 0. \tag{19.6}$$

If $a = b = 2$ then since $a^2+b^2+c^2+abc = 4$ we get $c = -2$, and therefore $a+b+c = 2$.

If $a = b = c \neq 2$ then from the given condition we deduce that

$$3a^2+a^3 = 4 \quad \Leftrightarrow \quad (a-1)(a+2)^2 = 0,$$

and therefore $a = b = c = 1$ or $a = b = c = -2$, i.e. $a+b+c = 3$ or $a+b+c = -6$.

Thus $\min\{a+b+c\} = -6$.

Exercise 19.4 Let $a,b,c,d \in \mathbb{R}^+$ such that $a+b+c+d = 1$. Prove the inequality

$$abc+bcd+cda+dab \leq \frac{1}{27} + \frac{176}{27}abcd.$$

Solution Let $f = abc+bcd+cda+dab - \frac{176}{27}abcd$.

We'll prove that

$$f \leq \frac{1}{27}.$$

Define $g = a + b + c + d - 1$ and

$$L = f - \lambda g = abc + bcd + cda + dab - \frac{176}{27}abcd - \lambda(a + b + c + d - 1).$$

For the first partial derivatives we have

$$\frac{\partial L}{\partial a} = bc + cd + db - \frac{176}{27}bcd - \lambda = 0,$$
$$\frac{\partial L}{\partial b} = ac + cd + da - \frac{176}{27}acd - \lambda = 0,$$
$$\frac{\partial L}{\partial c} = ab + bd + da - \frac{176}{27}abd - \lambda = 0,$$
$$\frac{\partial L}{\partial d} = bc + ac + ab - \frac{176}{27}abc - \lambda = 0.$$

Therefore

$$\lambda = bc + cd + db - \frac{176}{27}bcd = ac + cd + da - \frac{176}{27}acd$$
$$= ab + bd + da - \frac{176}{27}abd = bc + ac + ab - \frac{176}{27}abc.$$

Since

$$bc + cd + db - \frac{176}{27}bcd = ac + cd + da - \frac{176}{27}acd,$$

we deduce that

$$(b - a)\left(c + d - \frac{176}{27}cd\right) = 0.$$

Similarly we get

$$(b - c)\left(a + d - \frac{176}{27}ad\right) = 0,$$
$$(b - d)\left(a + c - \frac{176}{27}ac\right) = 0,$$
$$(a - c)\left(b + d - \frac{176}{27}bd\right) = 0,$$
$$(a - d)\left(c + b - \frac{176}{27}cb\right) = 0,$$
$$(c - d)\left(a + b - \frac{176}{27}ab\right) = 0.$$

By solving these equations we must have $a = b = c = d$, and since $a+b+c+d = 1$ it follows that $a = b = c = d = 1/4$.

Then

$$f(1/4, 1/4, 1/4, 1/4) = 1/27,$$

and we are done.

Exercise 19.5 Let $a, b, c \in \mathbb{R}$ be real numbers such that $a + b + c > 0$. Prove the inequality

$$a^3 + b^3 + c^3 \le (a^2 + b^2 + c^2)^{3/2} + 3abc.$$

Solution If we define

$$x = \frac{a}{\sqrt{a^2 + b^2 + c^2}}, \qquad y = \frac{b}{\sqrt{a^2 + b^2 + c^2}}, \qquad z = \frac{c}{\sqrt{a^2 + b^2 + c^2}},$$

then the given inequality becomes

$$x^3 + y^3 + z^3 \le (x^2 + y^2 + z^2)^{3/2} + 3xyz, \quad \text{with } x^2 + y^2 + z^2 = 1.$$

So it suffices to prove that

$$a^3 + b^3 + c^3 \le (a^2 + b^2 + c^2)^{3/2} + 3abc, \quad \text{with condition } a^2 + b^2 + c^2 = 1,$$

i.e.

$$a^3 + b^3 + c^3 \le 1 + 3abc, \quad \text{with } a^2 + b^2 + c^2 = 1.$$

Let us define

$$f = a^3 + b^3 + c^3 - 3abc, \qquad g = a^2 + b^2 + c^2 - 1$$

and

$$L = f - \lambda g = a^3 + b^3 + c^3 - 3abc - \lambda(a^2 + b^2 + c^2 - 1).$$

We obtain

$$\frac{\partial L}{\partial a} = 3a^2 - 3bc - 2\lambda a = 0,$$
$$\frac{\partial L}{\partial b} = 3b^2 - 3ac - 2\lambda b = 0,$$
$$\frac{\partial L}{\partial c} = 3c^2 - 3ab - 2\lambda c = 0$$

i.e.

$$\lambda = \frac{3(a^2 - bc)}{2a} = \frac{3(b^2 - ac)}{2b} = \frac{3(c^2 - ab)}{2c}.$$

Thus

$$\frac{3(a^2 - bc)}{2a} = \frac{3(b^2 - ac)}{2b} \quad \Leftrightarrow \quad (a - b)(ab + bc + ca) = 0.$$

Similarly we deduce

$$(b-c)(ab+bc+ca)=0 \quad \text{and} \quad (c-a)(ab+bc+ca)=0.$$

By solving these equations we deduce that we must have $a=b=c$ or $ab+bc+ca=0$.

If $a=b=c$ then $f(a,a,a)=0<1$.

If $ab+bc+ca=0$ then

$$(a+b+c)^2=a^2+b^2+c^2+2(ab+bc+ca)=1,$$

and since $a+b+c>0$ we obtain $a+b+c=1$.

Therefore

$$f(a,b,c)=a^3+b^3+c^3-3abc=(a+b+c)(a^2+b^2+c^2-ab-bc-ca)=1,$$

and the problem is solved.

Chapter 20
Problems

1 Let n be a positive integer. Prove that

$$1+\frac{1}{2^2}+\frac{1}{3^2}+\cdots+\frac{1}{n^2}<2.$$

2 Let $a_n=1+\frac{1}{2}+\frac{1}{3}+\cdots+\frac{1}{n}$. Prove that for any $n\in\mathbb{N}$ we have

$$\frac{1}{a_1^2}+\frac{1}{2a_2^2}+\frac{1}{3a_3^2}+\cdots+\frac{1}{na_n^2}<2.$$

3 Let x, y, z be real numbers. Prove the inequality

$$x^4+y^4+z^4\geq 4xyz-1.$$

4 Prove that for any real number x, the following inequality holds

$$x^{2002}-x^{1999}+x^{1996}-x^{1995}+1>0.$$

5 Let x, y be real numbers. Prove the inequality

$$3(x+y+1)^2+1\geq 3xy.$$

6 Let a, b, c be positive real numbers such that $a+b+c\geq abc$. Prove that at least two of the following inequalities

$$\frac{2}{a}+\frac{3}{b}+\frac{6}{c}\geq 6,\qquad \frac{2}{b}+\frac{3}{c}+\frac{6}{a}\geq 6,\qquad \frac{2}{c}+\frac{3}{a}+\frac{6}{b}\geq 6$$

are true.

7 Let $a, b, c, x, y, z>0$. Prove the inequality

$$\frac{ax}{a+x}+\frac{by}{b+y}+\frac{cz}{c+z}\leq\frac{(a+b+c)(x+y+z)}{a+b+c+x+y+z}.$$

Z. Cvetkovski, *Inequalities*,
DOI 10.1007/978-3-642-23792-8_20, © Springer-Verlag Berlin Heidelberg 2012

8 Let $a, b, c \in \mathbb{R}^+$. Prove the inequality

$$\frac{2a}{a^2+bc}+\frac{2b}{b^2+ac}+\frac{2c}{c^2+ab}\le\frac{a}{bc}+\frac{b}{ac}+\frac{c}{ab}.$$

9 Let $a, b, c, x, y, z \in \mathbb{R}^+$ such that $a+x=b+y=c+z=1$. Prove the inequality

$$(abc+xyz)\left(\frac{1}{ay}+\frac{1}{bz}+\frac{1}{cx}\right)\ge 3.$$

10 Let $a_1, a_2, \ldots, a_n$ be positive real numbers and let $b_1, b_2, \ldots, b_n$ be their permutation. Prove the inequality

$$\frac{a_1^2}{b_1}+\frac{a_2^2}{b_2}+\cdots+\frac{a_n^2}{b_n}\ge a_1+a_2+\cdots+a_n.$$

11 Let $x \in \mathbb{R}^+$. Find the minimum value of the expression $\frac{x^2+1}{x+1}$.

12 Let $a, b, c \in \mathbb{R}^+$ such that $abc=1$. Prove the inequality

$$\frac{a}{(a+1)(b+1)}+\frac{b}{(b+1)(c+1)}+\frac{c}{(c+1)(a+1)}\ge\frac{3}{4}.$$

13 Let $x, y \ge 0$ be real numbers such that $y(y+1)\le(x+1)^2$. Prove the inequality

$$y(y-1)\le x^2.$$

14 Let $x, y \in \mathbb{R}^+$ such that $x^3+y^3\le x-y$. Prove that

$$x^2+y^2\le 1.$$

15 Let $a, b, x, y \in \mathbb{R}$ such that $ay-bx=1$. Prove that

$$a^2+b^2+x^2+y^2+ax+by\ge\sqrt{3}.$$

16 Let a, b, c, d be non-negative real numbers such that $a^2+b^2+c^2+d^2=1$. Prove the inequality

$$(1-a)(1-b)(1-c)(1-d)\ge abcd.$$

17 Let x, y be non-negative real numbers. Prove the inequality

$$4(x^9+y^9)\ge(x^2+y^2)(x^3+y^3)(x^4+y^4).$$

18 Let $x, y, z \in \mathbb{R}^+$ such that $xyz = 1$ and $\frac{1}{x} + \frac{1}{y} + \frac{1}{z} \geq x + y + z$. Prove that for any natural number n the inequality

$$\frac{1}{x^n} + \frac{1}{y^n} + \frac{1}{z^n} \geq x^n + y^n + z^n$$

is true.

19 Let x, y, z be real numbers different from 1, such that $xyz = 1$. Prove the inequality

$$\left(\frac{3-x}{1-x}\right)^2 + \left(\frac{3-y}{1-y}\right)^2 + \left(\frac{3-z}{1-z}\right)^2 > 7.$$

20 Let $x, y, z \leq 1$ be real numbers such that $x + y + z = 1$. Prove the inequality

$$\frac{1}{1+x^2} + \frac{1}{1+y^2} + \frac{1}{1+z^2} \leq \frac{27}{10}.$$

21 Let $a, b, c \in \mathbb{R}^+$. Prove the inequality

$$\frac{1}{a(1+b)} + \frac{1}{b(1+c)} + \frac{1}{c(1+a)} \geq \frac{3}{1+abc}.$$

22 Let x, y, z be positive real numbers. Prove the inequality

$$9(a+b)(b+c)(c+a) \geq 8(a+b+c)(ab+bc+ca).$$

23 Let a, b, c be real numbers. Prove the inequality

$$(a^2+b^2+c^2)^2 \geq 3(a^3b+b^3c+c^3a).$$

24 Let a, b, c be positive real numbers such that $a^2 + b^2 + c^2 = 3$. Prove the inequality

$$a^3(b+c) + b^3(c+a) + c^3(a+b) \leq 6.$$

25 Let a, b, c be positive real numbers. Prove the inequality

$$\sqrt{\frac{a}{b+c}} + \sqrt{\frac{b}{c+a}} + \sqrt{\frac{c}{a+b}} > 2.$$

26 Let a, b, c be positive real numbers such that $a^2 + b^2 + c^2 = 3$. Prove the inequality

$$\frac{a}{b+2} + \frac{b}{c+2} + \frac{c}{a+2} \leq 1.$$

27 Let x, y, z be distinct nonnegative real numbers. Prove the inequality

$$\frac{1}{(x-y)^2}+\frac{1}{(y-z)^2}+\frac{1}{(z-x)^2}\geq\frac{4}{xy+yz+zx}.$$

28 Let a, b, c be non-negative real numbers. Prove the inequality

$$3(a^2-a+1)(b^2-b+1)(c^2-c+1)\geq 1+abc+(abc)^2.$$

29 Let $a, b \in \mathbb{R}, a \neq 0$. Prove the inequality

$$a^2+b^2+\frac{1}{a^2}+\frac{b}{a}\geq\sqrt{3}.$$

30 Let $a, b, c \in \mathbb{R}^+$. Prove the inequality

$$\frac{a^2+1}{b+c}+\frac{b^2+1}{c+a}+\frac{c^2+1}{a+b}\geq 3.$$

31 Let x, y, z be positive real numbers such that $xy+yz+zx=5$. Prove the inequality

$$3x^2+3y^2+z^2\geq 10.$$

32 Let a, b, c be positive real numbers such that $ab+bc+ca>a+b+c$. Prove the inequality

$$a+b+c>3.$$

33 Let a, b be real numbers such that $9a^2+8ab+7b^2\leq 6$. Prove that

$$7a+5b+12ab\leq 9.$$

34 Let $x, y, z \in \mathbb{R}^+$, such that $xyz\geq xy+yz+zx$. Prove the inequality

$$xyz\geq 3(x+y+z).$$

35 Let $a, b, c \in \mathbb{R}^+$ with $a^2+b^2+c^2=3$. Prove the inequality

$$\frac{ab}{c}+\frac{bc}{a}+\frac{ca}{b}\geq 3.$$

36 Let a, b, c be positive real numbers such that $a+b+c=\sqrt{abc}$. Prove the inequality

$$ab+bc+ca\geq 9(a+b+c).$$

37 Let a, b, c be positive real numbers such that $abc \geq 1$. Prove the inequality

$$\left(a+\frac{1}{a+1}\right)\left(b+\frac{1}{b+1}\right)\left(c+\frac{1}{c+1}\right) \geq \frac{27}{8}.$$

38 Let $a, b, c, d \in \mathbb{R}^+$ such that $a^2+b^2+c^2+d^2=4$. Prove the inequality

$$a+b+c+d \geq ab+bc+cd+da.$$

39 Let $a, b, c \in (-3, 3)$ such that $\frac{1}{3+a}+\frac{1}{3+b}+\frac{1}{3+c}=\frac{1}{3-a}+\frac{1}{3-b}+\frac{1}{3-c}$.
Prove the inequality

$$\frac{1}{3+a}+\frac{1}{3+b}+\frac{1}{3+c} \geq 1.$$

40 Let $a, b, c \in \mathbb{R}^+$ such that $a^2+b^2+c^2=3$. Prove the inequality

$$\frac{1}{a+bc+abc}+\frac{1}{b+ca+bca}+\frac{1}{c+ab+cab} \geq 1.$$

41 Let $a, b, c \in \mathbb{R}^+$ such that $a+b+c=3$. Prove the inequality

$$\frac{a^2b^2+a^2+b^2}{ab+1}+\frac{b^2c^2+b^2+c^2}{bc+1}+\frac{c^2a^2+c^2+a^2}{ca+1} \geq \frac{9}{2}.$$

42 Let a, b, c, d be positive real numbers such that $a^2+b^2+c^2+d^2=4$. Prove the inequality

$$\frac{a^2+b^2+3}{a+b}+\frac{b^2+c^2+3}{b+c}+\frac{c^2+d^2+3}{c+d}+\frac{d^2+a^2+3}{d+a} \geq 10.$$

43 Let a, b, c be positive real numbers. Prove the inequality

$$\frac{1}{ab(a+b)}+\frac{1}{bc(b+c)}+\frac{1}{ca(c+a)} \geq \frac{9}{2(a^3+b^3+c^3)}.$$

44 Let $a, b, c \in \mathbb{R}^+$ such that $a\sqrt{bc}+b\sqrt{ca}+c\sqrt{ab} \geq 1$. Prove the inequality

$$a+b+c \geq \sqrt{3}.$$

45 Let a, b, c be positive real numbers such that $abc=1$. Prove the inequality

$$\frac{b+c}{\sqrt{a}}+\frac{c+a}{\sqrt{b}}+\frac{a+b}{\sqrt{c}} \geq \sqrt{a}+\sqrt{b}+\sqrt{c}+3.$$

46 Let x, y, z be positive real numbers such that $x+y+z=4$. Prove the inequality

$$\frac{1}{2xy+xz+yz}+\frac{1}{xy+2xz+yz}+\frac{1}{xy+xz+2yz} \leq \frac{1}{xyz}.$$

47 Let $a, b, c \in \mathbb{R}^+$. Prove the inequality

$$abc \geq (a+b-c)(b+c-a)(c+a-b).$$

48 Let a, b, c be positive real numbers such that $a+b+c=3$. Prove the inequality

$$abc + \frac{12}{ab+bc+ac} \geq 5.$$

49 Let a, b, c be positive real numbers such that $abc=1$. Prove that

$$\left(a-1+\frac{1}{b}\right)\left(b-1+\frac{1}{c}\right)\left(c-1+\frac{1}{a}\right) \leq 1.$$

50 Let a, b, c be positive real numbers such that $abc=1$. Prove the inequality

$$\frac{1}{1+a+b} + \frac{1}{1+b+c} + \frac{1}{1+c+a} \leq \frac{1}{2+a} + \frac{1}{2+b} + \frac{1}{2+c}.$$

51 Let $a, b, c > 0$. Prove the inequality

$$(a+b)^2 + (a+b+4c)^2 \geq \frac{100abc}{a+b+c}.$$

52 Let $a, b, c > 0$ such that $abc=1$. Prove the inequality

$$\frac{1+ab}{1+a} + \frac{1+bc}{1+b} + \frac{1+ac}{1+a} \geq 3.$$

53 Let a, b, c be real numbers such that $ab+bc+ca=1$. Prove the inequality

$$\left(a+\frac{1}{b}\right)^2 + \left(b+\frac{1}{c}\right)^2 + \left(c+\frac{1}{a}\right)^2 \geq 16.$$

54 Let a, b, c be positive real numbers such that $abc \geq 1$. Prove the inequality

$$a+b+c \geq \frac{1+a}{1+b} + \frac{1+b}{1+c} + \frac{1+c}{1+a}.$$

55 Let $a, b \in \mathbb{R}^+$. Prove the inequality

$$\left(a^2+b+\frac{3}{4}\right)\left(b^2+a+\frac{3}{4}\right) \geq \left(2a+\frac{1}{2}\right)\left(2b+\frac{1}{2}\right).$$

56 Let $a, b, c \in \mathbb{R}^+$ such that $abc=1$. Prove the inequality

$$\frac{a}{a^2+2} + \frac{b}{b^2+2} + \frac{c}{c^2+2} \leq 1.$$

57 Let $x, y, z > 0$ be real numbers such that $x + y + z = xyz$. Prove the inequality

$$(x-1)(y-1)(z-1) \le 6\sqrt{3} - 10.$$

58 Let $a, b, c \in (1, 2)$ be real numbers. Prove the inequality

$$\frac{b\sqrt{a}}{4b\sqrt{c} - c\sqrt{a}} + \frac{c\sqrt{b}}{4c\sqrt{a} - a\sqrt{b}} + \frac{a\sqrt{c}}{4a\sqrt{b} - b\sqrt{c}} \ge 1.$$

59 Let $a, b, c \in \mathbb{R}^+$ such that $a + b + c = 3$. Prove the inequality

$$\sqrt{a(b+c)} + \sqrt{b(c+a)} + \sqrt{c(a+b)} \ge 3\sqrt{2abc}.$$

60 Let a, b, c be positive real numbers such that $a + b + c = 1$. Prove the inequality

$$\sqrt{a+bc} + \sqrt{b+ca} + \sqrt{c+ab} \le 2.$$

61 Let a, b, c be positive real numbers such that $a + b + c + 1 = 4abc$. Prove that

$$\frac{b^2+c^2}{a} + \frac{c^2+a^2}{b} + \frac{a^2+b^2}{c} \ge 2(ab+bc+ca).$$

62 Let $a, b, c \in (-1, 1)$ be real numbers such that $ab + bc + ac = 1$. Prove the inequality

$$6\sqrt[3]{(1-a^2)(1-b^2)(1-c^2)} \le 1 + (a+b+c)^2.$$

63 Let a, b, c, d be positive real numbers such that $a^2 + b^2 + c^2 + d^2 = 1$. Prove the inequality

$$\sqrt{1-a} + \sqrt{1-b} + \sqrt{1-c} + \sqrt{1-d} \ge \sqrt{a} + \sqrt{b} + \sqrt{c} + \sqrt{d}.$$

64 Let x, y, z be positive real numbers such that $xyz = 1$. Prove the inequality

$$\frac{1}{(x+1)^2+y^2+1} + \frac{1}{(y+1)^2+z^2+1} + \frac{1}{(z+1)^2+x^2+1} \le \frac{1}{2}.$$

65 Let $a, b, c \in \mathbb{R}^+$. Prove the inequality

$$\sqrt{\frac{a^3}{a^3+(b+c)^3}} + \sqrt{\frac{a^3}{a^3+(b+c)^3}} + \sqrt{\frac{a^3}{a^3+(b+c)^3}} \ge 1.$$

66 Let $x, y, z \in \mathbb{R}^+$. Prove the inequality

$$(x+y+z)^2(xy+yz+zx)^2 \le 3(x^2+xy+y^2)(y^2+yz+z^2)(z^2+zx+x^2).$$

67 Let a, b, c be real numbers such that $a+b+c=3$. Prove the inequality

$$2(a^2b^2+b^2c^2+c^2a^2)+3 \leq 3(a^2+b^2+c^2).$$

68 Let a, b, c, d be positive real numbers. Prove the inequality

$$\frac{a-b}{b+c}+\frac{b-c}{c+d}+\frac{c-d}{d+a}+\frac{d-a}{a+b} \geq 0.$$

69 Let $a, b, c \in \mathbb{R}^+$ such that $a+b+c=1$. Prove the inequality

$$\frac{a}{(b+c)^2}+\frac{b}{(c+a)^2}+\frac{c}{(a+b)^2} \geq \frac{9}{4}.$$

70 Let $a, b, c \in \mathbb{R}^+$ such that $abc=1$. Prove the inequality

$$\frac{a^3c}{(b+c)(c+a)}+\frac{b^3a}{(c+a)(a+b)}+\frac{c^3b}{(a+b)(b+c)} \geq \frac{3}{4}.$$

71 Let $a, b, c > 0$ be real numbers such that $abc=1$. Prove that

$$(a+b)(b+c)(c+a) \geq 4(a+b+c-1).$$

72 Let a, b, c be positive real numbers such that $abc=1$. Prove the inequality

$$1+\frac{3}{a+b+c} \geq \frac{6}{ab+bc+ca}.$$

73 Let x, y, z be positive real numbers such that $x^2+y^2+z^2=xyz$. Prove the following inequalities:

$1°\ xyz \geq 27$ $\quad$ $2°\ xy+yz+zx \geq 27$
$3°\ x+y+z \geq 9$ $\quad$ $4°\ xy+yz+zx \geq 2(x+y+z)+9$.

74 Let a, b, c be real numbers such that $a^3+b^3+c^3-3abc=1$. Prove the inequality

$$a^2+b^2+c^2 \geq 1.$$

75 Let $a, b, c, d \in \mathbb{R}^+$ such that $\frac{1}{1+a^4}+\frac{1}{1+b^4}+\frac{1}{1+c^4}+\frac{1}{1+d^4}=1$. Prove that

$$abcd \geq 3.$$

76 Let a, b, c be non-negative real numbers. Prove the inequality

$$\sqrt{\frac{ab+bc+ca}{3}} \leq \sqrt[3]{\frac{(a+b)(b+c)(c+a)}{8}}.$$

77 Let a, b, c, d be positive real numbers such that $a+b+c+d=1$. Prove that

$$16(abc+bcd+cda+dab) \leq 1.$$

78 Let a, b, c, d, e be positive real numbers such that $a+b+c+d+e=5$. Prove the inequality

$$abc+bcd+cde+dea+eab \leq 5.$$

79 Let $a, b, c > 0$ be real numbers. Prove the inequality

$$\frac{a}{b}+\frac{b}{c}+\frac{c}{a} \geq \frac{a+b}{b+c}+\frac{b+c}{c+a}+1.$$

80 Let $a, b, c > 0$ be real numbers such that $abc=1$. Prove the inequality

$$\left(1+\frac{a}{b}\right)\left(1+\frac{b}{c}\right)\left(1+\frac{c}{a}\right) \geq 2(1+a+b+c).$$

81 Let a, b, c be positive real numbers such that $a+b+c \geq \frac{1}{a}+\frac{1}{b}+\frac{1}{c}$. Prove the inequality

$$a+b+c \geq \frac{3}{a+b+c}+\frac{2}{abc}.$$

82 Let a, b, c, d be positive real numbers such that $abcd=1$. Prove the inequality

$$\frac{1+ab}{1+a}+\frac{1+bc}{1+b}+\frac{1+cd}{1+c}+\frac{1+da}{1+d} \geq 4.$$

83 Let $a, b, c \in \mathbb{R}^{+}$. Prove the inequality

$$\frac{1}{b(a+b)}+\frac{1}{c(b+c)}+\frac{1}{a(c+a)} \geq \frac{27}{2(a+b+c)^{2}}.$$

84 Let a, b, c be positive real numbers such that $a+b+c=3$. Prove the inequality

$$\frac{a^{2}}{b^{2}-2b+3}+\frac{b^{2}}{c^{2}-2c+3}+\frac{c^{2}}{a^{2}-2a+3} \geq \frac{3}{2}.$$

85 Let a, b, c be positive real numbers such that $ab+bc+ca=3$. Prove the inequality

$$\frac{1}{1+a^{2}(b+c)}+\frac{1}{1+b^{2}(c+a)}+\frac{1}{1+c^{2}(a+b)} \leq \frac{1}{abc}.$$

86 Let a, b, c be positive real numbers such that $a+b+c=1$. Prove the inequality

$$a\sqrt[3]{1+b-c}+b\sqrt[3]{1+c-a}+a\sqrt[3]{1+a-b} \leq 1.$$

87 Let $a, b, c \in \mathbb{R}^+$ such that $a+b+c=1$. Prove the inequality

$$\frac{1-2ab}{c}+\frac{1-2bc}{a}+\frac{1-2ca}{b} \geq 7.$$

88 Let a, b, c be non negative real numbers such that $a^2+b^2+c^2=1$. Prove the inequality

$$\frac{1-ab}{7-3ac}+\frac{1-ab}{7-3ac}+\frac{1-ab}{7-3ac} \geq \frac{1}{3}.$$

89 Let $x, y, z \in \mathbb{R}^+$ such that $x+y+z=1$. Prove the inequality

$$\frac{xy}{\sqrt{\frac{1}{3}+z^2}}+\frac{zx}{\sqrt{\frac{1}{3}+y^2}}+\frac{yz}{\sqrt{\frac{1}{3}+x^2}} \leq \frac{1}{2}.$$

90 Let a, b, c be positive real numbers such that $a+b+c=1$. Prove the inequality

$$\frac{a-bc}{a+bc}+\frac{b-ca}{b+ca}+\frac{c-ab}{c+ab} \leq \frac{3}{2}.$$

91 Let a, b, c be positive real numbers such that $abc=1$. Prove the inequality

$$\sqrt{\frac{a+b}{a+1}}+\sqrt{\frac{b+c}{c+1}}+\sqrt{\frac{c+a}{a+1}} \geq 3.$$

92 Let $x, y, z \geq 0$ be real numbers such that $xy+yz+zx=1$. Prove the inequality

$$\frac{x}{1+x^2}+\frac{y}{1+y^2}+\frac{z}{1+z^2} \leq \frac{3\sqrt{3}}{4}.$$

93 Let a, b, c be non-negative real numbers such that $ab+bc+ca=1$. Prove the inequality

$$\frac{1}{1+a}+\frac{1}{1+b}+\frac{1}{1+c} \geq \frac{3\sqrt{3}}{\sqrt{3}+1}.$$

94 Let a, b, c be non-negative real numbers such that $ab+bc+ca=1$. Prove the inequality

$$\frac{a^2}{1+a}+\frac{b^2}{1+b}+\frac{c^2}{1+c} \geq \frac{\sqrt{3}}{\sqrt{3}+1}.$$

95 Let $a, b, c \in \mathbb{R}^+$ such that $(a+b)(b+c)(c+a)=8$. Prove the inequality

$$\frac{a+b+c}{3} \geq \sqrt[27]{\frac{a^3+b^3+c^3}{3}}.$$

96 Find the maximum value of $\frac{x^4-x^2}{x^6+2x^3-1}$, where $x \in \mathbb{R}, x > 1$.

97 Let a, b, c be positive real numbers. Prove the inequality

$$\frac{a+\sqrt{ab}+\sqrt[3]{abc}}{3} \leq \sqrt[3]{a \cdot \frac{a+b}{2} \cdot \frac{a+b+c}{3}}.$$

98 Let a, b, c be positive real numbers such that $abc(a+b+c)=3$. Prove the inequality

$$(a+b)(b+c)(c+a) \geq 8.$$

99 Let a, b, c be positive real numbers. Prove the inequality

$$\sqrt{\frac{2a}{b+c}}+\sqrt{\frac{2b}{c+a}}+\sqrt{\frac{2c}{a+b}} \leq 3.$$

100 Let $a, b, c \in \mathbb{R}^+$ such that $ab+bc+ca=1$. Prove the inequality

$$\frac{1}{a(a+b)}+\frac{1}{b(b+c)}+\frac{1}{c(c+a)} \geq \frac{9}{2}.$$

101 Let $0 \leq a \leq b \leq c \leq 1$ be real numbers. Prove that

$$a^2(b-c)+b^2(c-b)+c^2(1-c) \leq \frac{108}{529}.$$

102 Let $a, b, c \in \mathbb{R}^+$ such that $a+b+c=1$. Prove the inequality

$$S=a^4b+b^4c+c^4a \leq \frac{256}{3125}.$$

103 Let $a, b, c > 0$ be real numbers. Prove the inequality

$$\frac{a^2}{b^2}+\frac{b^2}{c^2}+\frac{c^2}{a^2} \geq \frac{a}{b}+\frac{b}{c}+\frac{c}{a}.$$

104 Prove that for all positive real numbers a, b, c we have

$$\frac{a^3}{b^2}+\frac{b^3}{c^2}+\frac{c^3}{a^2} \geq a+b+c.$$

105 Prove that for all positive real numbers a, b, c we have

$$\frac{a^3}{b^2}+\frac{b^3}{c^2}+\frac{c^3}{a^2} \geq \frac{a^2}{b}+\frac{b^2}{c}+\frac{c^2}{a}.$$

106 Prove that for all positive real numbers a, b, c we have

$$\frac{a^3}{b}+\frac{b^3}{c}+\frac{c^3}{a}\geq ab+bc+ca.$$

107 Prove that for all positive real numbers a, b, c we have

$$\frac{a^5}{b^3}+\frac{b^5}{c^3}+\frac{c^5}{a^3}\geq a^2+b^2+c^2.$$

108 Let $a, b, c \in \mathbb{R}^+$ such that $a+b+c=3$. Prove the inequality

$$\frac{a^3}{b(2c+a)}+\frac{b^3}{c(2a+b)}+\frac{c^3}{a(2b+c)}\geq 1.$$

109 Let $a, b, c \in \mathbb{R}^+$ and $a^2+b^2+c^2=3$. Prove the inequality

$$\frac{a^3}{b+2c}+\frac{b^3}{c+2a}+\frac{c^3}{a+2b}\geq 1.$$

110 Let a, b, c be positive real numbers such that $a^2+b^2+c^2=3$. Prove the inequality

$$\frac{1}{a^3+2}+\frac{1}{b^3+2}+\frac{1}{c^3+2}\geq 1.$$

111 Let $a, b, c \in \mathbb{R}^+$ such that $a+b+c=1$. Prove the inequality

$$\frac{a^3}{a^2+b^2}+\frac{b^3}{b^2+c^2}+\frac{c^3}{c^2+a^2}\geq \frac{1}{2}.$$

112 Let a, b, c be positive real numbers such that $a+b+c=3$. Prove the inequality

$$\frac{1}{1+2a^2b}+\frac{1}{1+2b^2c}+\frac{1}{1+2c^2a}\geq 1.$$

113 Let a, b, c, d be positive real numbers such that $a+b+c+d=4$. Prove the inequality

$$\frac{a}{1+b^2c}+\frac{b}{1+c^2d}+\frac{c}{1+d^2a}+\frac{d}{1+a^2b}\geq 2.$$

114 Let a, b, c, d be positive real numbers. Prove the inequality

$$\frac{a^3}{a^2+b^2}+\frac{b^3}{b^2+c^2}+\frac{c^3}{c^2+d^2}+\frac{d^3}{d^2+a^2}\geq \frac{a+b+c+d}{2}.$$

115 Let a, b, c be positive real numbers such that $a+b+c=3$. Prove the inequality

$$\frac{a^2}{a+2b^2}+\frac{b^2}{b+2c^2}+\frac{c^2}{c+2a^2}\geq 1.$$

116 Let a, b, c be positive real numbers such that $a+b+c=3$. Prove the inequality

$$\frac{a^2}{a+2b^3}+\frac{b^2}{b+2c^3}+\frac{c^2}{c+2a^3}\geq 1.$$

117 Let a, b, c be positive real numbers such that $a^2+b^2+c^2=3$. Find the minimum value of the expression

$$a+b+c+\frac{16}{a+b+c}.$$

118 Let $a, b, c \geq 0$ be real numbers such that $a^2+b^2+c^2=1$. Find the minimal value of the expression

$$A=a+b+c+\frac{1}{abc}.$$

119 Let a, b, c be positive real numbers such that $a+b+c=6$. Prove the inequality

$$\sqrt[3]{ab+bc}+\sqrt[3]{bc+ca}+\sqrt[3]{ca+ab}+\sqrt[3]{\frac{9}{4}(a^2+b^2+c^2)}\leq 9.$$

120 Let $a, b, c \in \mathbb{R}^+$ such that $a+2b+3c\geq 20$. Prove the inequality

$$S=a+b+c+\frac{3}{a}+\frac{9}{2b}+\frac{4}{c}\geq 13.$$

121 Let $a, b, c \in \mathbb{R}^+$. Prove the inequality

$$S=30a+3b^2+\frac{2c^3}{9}+36\left(\frac{1}{ab}+\frac{1}{bc}+\frac{1}{ca}\right)\geq 84.$$

122 Let $a, b, c \in \mathbb{R}^+$ such that $ac\geq 12$ and $bc\geq 8$. Prove the inequality

$$S=a+b+c+2\left(\frac{1}{ab}+\frac{1}{bc}+\frac{1}{ca}\right)+\frac{8}{abc}\geq\frac{121}{12}.$$

123 Let $a, b, c, d > 0$ be real numbers. Determine the minimal value of the expression

$$A=\left(1+\frac{2a}{3b}\right)\left(1+\frac{2b}{3c}\right)\left(1+\frac{2c}{3d}\right)\left(1+\frac{2d}{3a}\right).$$

124 Let $a, b, c > 0$ be real numbers such that $a^2 + b^2 + c^2 = 12$. Determine the maximal value of the expression

$$A = a\sqrt[3]{b^2 + c^2} + b\sqrt[3]{c^2 + a^2} + c\sqrt[3]{a^2 + b^2}.$$

125 Let $a, b, c \geq 0$ such that $a + b + c = 3$. Prove the inequality

$$(a^2 - ab + b^2)(b^2 - bc + c^2)(c^2 - ca + a^2) \leq 12.$$

126 Let a, b, c be positive real numbers. Prove the inequality

$$(a^5 - a^2 + 3)(b^5 - b^2 + 3)(c^5 - c^2 + 3) \geq (a + b + c)^3.$$

127 Let $x, y, z \in \mathbb{R}^+$ such that $x + y + z = 1$. Prove the inequality

$$\frac{xy}{\sqrt{1 + z^2}} + \frac{zx}{\sqrt{1 + y^2}} + \frac{yz}{\sqrt{1 + x^2}} \leq \frac{1}{\sqrt{10}}.$$

128 Let $a, b, c \in \mathbb{R}^+$. Prove the inequality

$$(a + b + c)^6 \geq 27(a^2 + b^2 + c^2)(ab + bc + ca)^2.$$

129 Let $a, b, c \in [1, 2]$ be real numbers. Prove the inequality

$$a^3 + b^3 + c^3 \leq 5abc.$$

130 Let a, b, c be positive real numbers such that $ab + bc + ca = 3$. Prove the inequality

$$(a^7 - a^4 + 3)(b^5 - b^2 + 3)(c^4 - c + 3) \geq 27.$$

131 Let $a, b, c \in [1, 2]$ be real numbers. Prove the inequality

$$(a + b + c)\left(\frac{1}{a} + \frac{1}{b} + \frac{1}{c}\right) \leq 10.$$

132 Let $a, b, c \in \mathbb{R}^+$ such that $a + b + c = 1$. Prove the inequality

$$10(a^3 + b^3 + c^3) - 9(a^5 + b^5 + c^5) \geq 1.$$

133 Let $n \in \mathbb{N}$ and $x_1, x_2, \ldots, x_n \in (0, \pi)$. Find the maximum value of the expression

$$\sin x_1 \cos x_2 + \sin x_2 \cos x_3 + \cdots + \sin x_n \cos x_1.$$

134 Let $\alpha_i \in [\frac{\pi}{4}, \frac{5\pi}{4}]$, for $i = 1, 2, \ldots, n$. Prove the inequality

$$\left(\sin\alpha_1 + \sin\alpha_2 + \cdots + \sin\alpha_n + \frac{1}{4}\right)^2 \geq (\cos\alpha_1 + \cos\alpha_2 + \cdots + \cos\alpha_n).$$

135 Let $a_1, a_2, \ldots, a_n$; $a_{n+1} = a_1, a_{n+2} = a_2$ be positive real numbers. Prove the inequality

$$\sum_{i=1}^{n} \frac{a_i - a_{i+2}}{a_{i+1} + a_{i+2}} \geq 0.$$

136 Let $n \geq 2, n \in \mathbb{N}$ and $x_1, x_2, \ldots, x_n$ be positive real numbers such that

$$\frac{1}{x_1 + 1998} + \frac{1}{x_2 + 1998} + \cdots + \frac{1}{x_n + 1998} = \frac{1}{1998}.$$

Prove the inequality

$$\sqrt[n]{x_1 x_2 \cdots x_n} \geq 1998(n-1).$$

137 Let $a_1, a_2, \ldots, a_n \in \mathbb{R}^+$. Prove the inequality

$$\sum_{k=1}^{n} k a_k \leq \binom{n}{2} + \sum_{k=1}^{n} a_k^k.$$

138 Let $a_1, a_2, \ldots, a_n$ be positive real numbers such that $a_1 + a_2 + \cdots + a_n = n$. Prove that for every natural number k the following inequality holds

$$a_1^k + a_2^k + \cdots + a_n^k \geq a_1^{k-1} + a_2^{k-1} + \cdots + a_n^{k-1}.$$

139 Let a, b, c, d be positive real numbers. Prove the inequality

$$\left(\frac{a}{a+b}\right)^5 + \left(\frac{b}{b+c}\right)^5 + \left(\frac{c}{c+d}\right)^5 + \left(\frac{d}{d+a}\right)^5 \geq \frac{1}{8}.$$

140 Let $x_1, x_2, \ldots, x_n$ be positive real numbers not greater then 1. Prove the inequality

$$(1+x_1)^{\frac{1}{x_2}} (1+x_2)^{\frac{1}{x_3}} \cdots (1+x_n)^{\frac{1}{x_1}} \geq 2^n.$$

141 Let $x_1, x_2, \ldots, x_n$ be non-negative real numbers such that $x_1 + x_2 + \cdots + x_n \leq \frac{1}{2}$. Prove the inequality

$$(1-x_1)(1-x_2)\cdots(1-x_n) \geq \frac{1}{2}.$$

142 Let $a, b, c \in \mathbb{R}^+$ such that $abc = 1$. Prove the inequality

$$\frac{1}{a^3 + b^3 + 1} + \frac{1}{b^3 + c^3 + 1} + \frac{1}{c^3 + a^3 + 1} \leq 1.$$

143 Let $0 \le a, b, c \le 1$. Prove the inequality

$$\frac{c}{7+a^3+b^3}+\frac{b}{7+c^3+a^3}+\frac{a}{7+b^3+c^3}\le\frac{1}{3}.$$

144 Let $a, b, c \in \mathbb{R}^+$ such that $abc = 1$. Prove the inequality

$$\frac{ab}{a^5+ab+b^5}+\frac{bc}{b^5+bc+c^5}+\frac{ca}{c^5+ca+a^5}\le 1.$$

145 Let $a, b, c \in \mathbb{R}^+$ such that $a+b+c=3$. Prove the inequality

$$\frac{a^3}{a^2+ab+b^2}+\frac{b^3}{b^2+bc+c^2}+\frac{c^3}{c^2+ca+a^2}\ge 1.$$

146 Let a, b, c be positive real numbers such that $a^2+b^2+c^2=3abc$. Prove the inequality

$$\frac{a}{b^2c^2}+\frac{b}{c^2a^2}+\frac{c}{a^2b^2}\ge\frac{9}{a+b+c}.$$

147 Let a, b, c, x, y, z be positive real number, and let $a+b=3$. Prove the inequality

$$\frac{x}{ay+bz}+\frac{y}{az+bx}+\frac{z}{ax+by}\ge 1.$$

148 Let $x, y, z > 0$ be real numbers. Prove the inequality

$$\frac{x}{x+2y+3z}+\frac{y}{y+2z+3x}+\frac{z}{z+2x+3y}\ge\frac{1}{2}.$$

149 Let $a, b, c, d \in \mathbb{R}^+$. Prove the inequality

$$\frac{c}{a+3b}+\frac{d}{b+3c}+\frac{a}{c+3d}+\frac{b}{d+3a}\ge 1.$$

150 Let a, b, c, d, e be positive real numbers. Prove the inequality

$$\frac{a}{b+c}+\frac{b}{c+d}+\frac{c}{d+e}+\frac{d}{e+a}+\frac{e}{a+b}\ge\frac{5}{2}.$$

151 Prove that for all positive real numbers a, b, c the following inequality holds

$$\frac{a^3}{a^2+ab+b^2}+\frac{b^3}{b^2+bc+c^2}+\frac{c^3}{c^2+ca+a^2}\ge\frac{a^2+b^2+c^2}{a+b+c}.$$

152 Let a, b, c be positive real numbers such that $ab + bc + ca = 1$. Prove the inequality

$$\frac{1}{4a^2 - bc + 1} + \frac{1}{4b^2 - ca + 1} + \frac{1}{4c^2 - ab + 1} \geq \frac{3}{2}.$$

153 Let a, b, c be positive real numbers such that

$$\frac{1}{a^2 + b^2 + 1} + \frac{1}{b^2 + c^2 + 1} + \frac{1}{c^2 + a^2 + 1} \geq 1.$$

Prove the inequality

$$ab + bc + ca \leq 3.$$

154 Let a, b, c be positive real numbers such that $ab + bc + ca = 1/3$. Prove the inequality

$$\frac{a}{a^2 - bc + 1} + \frac{b}{b^2 - ca + 1} + \frac{c}{c^2 - ab + 1} \geq \frac{1}{a + b + c}.$$

155 Let a, b, c be positive real numbers. Prove the inequality

$$\frac{a^3}{a^3 + b^3 + abc} + \frac{b^3}{b^3 + c^3 + abc} + \frac{c^3}{c^3 + a^3 + abc} \geq 1.$$

156 Let a, b, c be positive real numbers such that $a^2 + b^2 + c^2 = 3$. Prove the inequality

$$\frac{a}{a^2 + 2b + 3} + \frac{b}{b^2 + 2c + 3} + \frac{c}{c^2 + 2a + 3} \leq \frac{1}{2}.$$

157 Let $a, b, c, d > 1$ be real numbers. Prove the inequality

$$\sqrt{a - 1} + \sqrt{b - 1} + \sqrt{c - 1} + \sqrt{d - 1} \leq \sqrt{(ab + 1)(cd + 1)}.$$

158 Let $a_1, a_2, \ldots, a_n \in \mathbb{R}^+$ such that $a_1 a_2 \cdots a_n = 1$. Prove the inequality

$$\sqrt{a_1} + \sqrt{a_2} + \cdots + \sqrt{a_n} \leq a_1 + a_2 + \cdots + a_n.$$

159 Let a, b, c be positive real numbers such that $a + b + c = 1$. Prove the inequality

$$a\sqrt{b} + b\sqrt{c} + c\sqrt{a} \leq \frac{1}{\sqrt{3}}.$$

160 Let $a, b, c \in (0, 1)$ be real numbers. Prove the inequality

$$\sqrt{abc} + \sqrt{(1 - a)(1 - b)(1 - c)} < 1.$$

161 Let a, b, c be positive real numbers such that $a+b+c=3$. Prove the inequality

$$\frac{a^3+2}{b+2}+\frac{b^3+2}{c+2}+\frac{c^3+2}{a+2}\geq 3.$$

162 Let a, b, c be positive real numbers such that $a^2+b^2+c^2=3$. Prove the inequality

$$\frac{1}{2-a}+\frac{1}{2-b}+\frac{1}{2-c}\geq 3.$$

163 Let a, b, c be positive real numbers such that $abc=8$. Prove the inequality

$$\frac{a-2}{a+1}+\frac{b-2}{b+1}+\frac{c-2}{c+1}\leq 0.$$

164 Let $a, b, c \in \mathbb{R}^+$ such that $a^2+b^2+c^2=1$. Prove the inequality

$$a+b+c-2abc\leq\sqrt{2}.$$

165 Let $x, y, z \in \mathbb{R}^+$ such that $x^2+y^2+z^2=2$. Prove the inequality

$$x+y+z\leq 2+xyz.$$

166 Let $x, y, z > -1$ be real numbers. Prove the inequality

$$\frac{1+x^2}{1+y+z^2}+\frac{1+y^2}{1+z+x^2}+\frac{1+z^2}{1+x+y^2}\geq 2.$$

167 Let a, b, c, d be positive real numbers such that $abcd=1$. Prove the inequality

$$(1+a^2)(1+b^2)(1+c^2)(1+d^2)\geq(a+b+c+d)^2.$$

168 Let $a, b, c, d \in \mathbb{R}^+$ such that $\frac{1}{a}+\frac{1}{b}+\frac{1}{c}+\frac{1}{d}=4$. Prove the inequality

$$\sqrt[3]{\frac{a^3+b^3}{2}}+\sqrt[3]{\frac{b^3+c^3}{2}}+\sqrt[3]{\frac{c^3+d^3}{2}}+\sqrt[3]{\frac{d^3+a^3}{2}}\leq 2(a+b+c+d)-4.$$

169 Let $x, y, z \in [-1, 1]$ be real numbers such that $x+y+z+xyz=0$. Prove the inequality

$$\sqrt{x+1}+\sqrt{y+1}+\sqrt{z+1}\leq 3.$$

170 Let $a, b, c > 0$ be positive real numbers such that $a+b+c=abc$. Prove the inequality

$$ab+bc+ca\geq 3+\sqrt{a^2+1}+\sqrt{b^2+1}+\sqrt{c^2+1}.$$

171 Let a, b, c, x, y, z be positive real numbers such that $ax+by+cz=xyz$. Prove the inequality

$$\sqrt{a+b}+\sqrt{b+c}+\sqrt{c+a}<x+y+z.$$

172 Let a, b, c be non-negative real numbers such that $a^2+b^2+c^2=1$. Prove the inequality

$$\frac{a}{b^2+1}+\frac{b}{c^2+1}+\frac{c}{a^2+1}\geq\frac{3}{4}(a\sqrt{a}+b\sqrt{b}+c\sqrt{c})^2.$$

173 Let a, b, c be positive real numbers. Prove the inequality

$$\frac{a}{(b+c)^2}+\frac{b}{(c+a)^2}+\frac{c}{(a+b)^2}\geq\frac{9}{4(a+b+c)}.$$

174 Let $x\geq y\geq z>0$ be real numbers. Prove the inequality

$$\frac{x^2y}{z}+\frac{y^2z}{x}+\frac{z^2x}{y}\geq x^2+y^2+z^2.$$

175 Let a, b, c be positive real numbers such that $abc=1$. Prove the inequality

$$\frac{1}{2+a}+\frac{1}{2+b}+\frac{1}{2+c}\leq 1.$$

176 Let a, b, c be positive real numbers such that $abc\geq 1$. Prove the inequality

$$\frac{1}{a^4+b^3+c^2}+\frac{1}{b^4+c^3+a^2}+\frac{1}{c^4+a^3+b^2}\leq 1.$$

177 Let a, b, c, d be positive real numbers such that $abcd=1$. Prove the inequality

$$\frac{1}{a(1+b)}+\frac{1}{b(1+c)}+\frac{1}{c(1+d)}+\frac{1}{d(1+a)}\geq 2.$$

178 Let a, b, c be non-negative real numbers such that $a+b+c=1$. Prove the inequality

$$\frac{ab}{c+1}+\frac{bc}{a+1}+\frac{ca}{b+1}\leq\frac{1}{4}.$$

179 Let a, b, c be positive real numbers such that $abc=1$. Prove the inequality

$$\frac{1}{(a+1)^2(b+c)}+\frac{1}{(b+1)^2(c+a)}+\frac{1}{(c+1)^2(a+b)}\leq\frac{3}{8}.$$

180 Let x, y, z be positive real numbers. Prove the inequality

$$xy(x+y-z)+yz(y+z-x)+zx(z+x-y)\geq\sqrt{3(x^3y^3+y^3z^3+z^3x^3)}.$$

181 Let a, b, c be positive real numbers. Prove the inequality

$$\frac{ab(a^3+b^3)}{a^2+b^2}+\frac{bc(b^3+c^3)}{b^2+c^2}+\frac{ca(c^3+a^3)}{c^2+a^2} \geq \sqrt{3abc(a^3+b^3+c^3)}.$$

182 Let a, b, c be positive real numbers. Prove the inequality.

$$ab\frac{a+c}{b+c}+bc\frac{b+a}{c+a}+ca\frac{c+b}{a+b} \geq \sqrt{3abc(a+b+c)}.$$

183 Let a, b, c and x, y, z be positive real numbers. Prove the inequality

$$a(y+z)+b(z+x)+c(x+y) \geq 2\sqrt{(xy+yz+zx)(ab+bc+ca)}.$$

184 Let a, b, c be positive real numbers such that $abc \geq 1$. Prove the inequality

$$a^3+b^3+c^3 \geq ab+bc+ca.$$

185 Let $a, b, c > 0$ be real numbers such that $a^{2/3}+b^{2/3}+c^{2/3}=3$. Prove the inequality

$$a^2+b^2+c^2 \geq a^{4/3}+b^{4/3}+c^{4/3}.$$

186 Let a, b, c be positive real numbers such that $a+b+c=3$. Prove the inequality

$$\frac{1}{c^2+a+b}+\frac{1}{a^2+b+c}+\frac{1}{b^2+c+a} \leq 1.$$

187 Let $a, b, c \in \mathbb{R}^+$. Prove the inequality

$$\frac{2a^2}{b+c}+\frac{2b^2}{c+a}+\frac{2c^2}{a+b} \geq a+b+c.$$

188 Let a, b, c be positive real numbers such that $abc=2$. Prove the inequality

$$a^3+b^3+c^3 \geq a\sqrt{b+c}+b\sqrt{c+a}+c\sqrt{a+b}.$$

189 Let $a_1, a_2, \ldots, a_n$ be positive real numbers. Prove the inequality

$$\frac{1}{\frac{1}{1+a_1}+\frac{1}{1+a_2}+\cdots+\frac{1}{1+a_n}}-\frac{1}{\frac{1}{a_1}+\frac{1}{a_2}+\cdots+\frac{1}{a_n}} \geq \frac{1}{n}.$$

190 Let $a, b, c, d \in \mathbb{R}^+$ such that $ab+bc+cd+da=1$. Prove the inequality

$$\frac{a^3}{b+c+d}+\frac{b^3}{a+c+d}+\frac{c^3}{b+d+a}+\frac{d^3}{b+c+a} \geq \frac{1}{3}.$$

191 Let α, x, y, z be positive real numbers such that $xyz = 1$ and $\alpha \geq 1$. Prove the inequality

$$\frac{x^\alpha}{y+z} + \frac{y^\alpha}{z+x} + \frac{z^\alpha}{x+y} \geq \frac{3}{2}.$$

192 Let $x_1, x_2, \ldots, x_n$ be positive real numbers such that

$$\frac{1}{1+x_1} + \frac{1}{1+x_2} + \cdots + \frac{1}{1+x_n} = 1.$$

Prove the inequality

$$\frac{\sqrt{x_1} + \sqrt{x_2} + \cdots + \sqrt{x_n}}{n-1} \geq \frac{1}{\sqrt{x_1}} + \frac{1}{\sqrt{x_2}} + \cdots + \frac{1}{\sqrt{x_n}}.$$

193 Let $x_1, x_2, \ldots, x_n > 0$ be real numbers. Prove the inequality

$$x_1^{x_1} x_2^{x_2} \cdots x_n^{x_n} \geq (x_1 x_2 \cdots x_n)^{\frac{x_1+x_2+\cdots+x_n}{n}}.$$

194 Let $a, b, c > 0$ be real numbers such that $a + b + c = 1$. Prove the inequality

$$\frac{a^2+b}{b+c} + \frac{b^2+c}{c+a} + \frac{c^2+a}{a+b} \geq 2.$$

195 Let $a, b, c > 1$ be positive real numbers such that $\frac{1}{a^2-1} + \frac{1}{b^2-1} + \frac{1}{c^2-1} = 1$. Prove the inequality

$$\frac{1}{a+1} + \frac{1}{b+1} + \frac{1}{c+1} \leq 1.$$

196 Let a, b, c, d be positive real numbers such that $a^2 + b^2 + c^2 + d^2 = 4$. Prove the inequality

$$\frac{1}{5-a} + \frac{1}{5-b} + \frac{1}{5-c} + \frac{1}{5-d} \leq 1.$$

197 Let $a, b, c, d \in \mathbb{R}$ such that $\frac{1}{4+a} + \frac{1}{4+b} + \frac{1}{4+c} + \frac{1}{4+d} + \frac{1}{4+e} = 1$. Prove the inequality

$$\frac{a}{4+a^2} + \frac{b}{4+b^2} + \frac{c}{4+c^2} + \frac{d}{4+d^2} + \frac{e}{4+e^2} \leq 1.$$

198 Let a, b, c be real numbers different from 1, such that $a + b + c = 1$. Prove the inequality

$$\frac{1+a^2}{1-a^2} + \frac{1+b^2}{1-b^2} + \frac{1+c^2}{1-c^2} \geq \frac{15}{4}.$$

199 Let $x, y, z > 0$, such that $xyz = 1$. Prove the inequality

$$\frac{x^3}{(1+y)(1+z)} + \frac{y^3}{(1+z)(1+x)} + \frac{z^3}{(1+x)(1+y)} \geq \frac{3}{4}.$$

200 Let $a, b, c, d > 0$ be real numbers. Prove the inequality

$$\frac{a}{b+2c+3d} + \frac{b}{c+2d+3a} + \frac{c}{d+2a+3b} + \frac{d}{a+2b+3c} \geq \frac{2}{3}.$$

201 Let a, b, c be positive real numbers. Prove the inequality

$$\frac{a^2+bc}{b+c} + \frac{b^2+ca}{c+a} + \frac{c^2+ab}{a+b} \geq a+b+c.$$

202 Let $a, b > 0, n \in \mathbb{N}$. Prove the inequality

$$\left(1+\frac{a}{b}\right)^n + \left(1+\frac{b}{a}\right)^n \geq 2^{n+1}.$$

203 Let $a, b, c > 0$ be real numbers such that $a+b+c = 1$. Prove the inequality

$$\left(a+\frac{1}{a}\right)^2 + \left(b+\frac{1}{b}\right)^2 + \left(c+\frac{1}{c}\right)^2 \geq \frac{100}{3}.$$

204 Let $x, y, z > 0$ be real numbers. Prove the inequality

$$\frac{x}{2x+y+z} + \frac{y}{x+2y+z} + \frac{z}{x+y+2z} \leq \frac{3}{4}.$$

205 Let $a, b, c, d > 0$ be real numbers such that $a \leq 1, a+b \leq 5, a+b+c \leq 14, a+b+c+d \leq 30$. Prove that

$$\sqrt{a}+\sqrt{b}+\sqrt{c}+\sqrt{d} \leq 10.$$

206 Let a, b, c, d be positive real numbers such that $a+b+c+d = 4$. Prove the inequality

$$\frac{a}{b^2+b} + \frac{b}{c^2+c} + \frac{c}{d^2+d} + \frac{d}{a^2+a} \geq \frac{8}{(a+c)(b+d)}.$$

207 Let $x_1, x_2, \ldots, x_n > 0$ and $n \in \mathbb{N}, n > 1$, such that $x_1+x_2+\cdots+x_n = 1$. Prove the inequality

$$\frac{x_1}{\sqrt{1-x_1}} + \frac{x_2}{\sqrt{1-x_2}} + \cdots + \frac{x_n}{\sqrt{1-x_n}} \geq \frac{\sqrt{x_1}+\sqrt{x_2}+\cdots+\sqrt{x_n}}{\sqrt{n-1}}.$$

208 Let $n \in \mathbb{N}, n \geq 2$. Determine the minimal value of

$$\frac{x_1^5}{x_2+x_3+\cdots+x_n}+\frac{x_2^5}{x_1+x_3+\cdots+x_n}+\cdots+\frac{x_n^5}{x_1+x_2+\cdots+x_{n-1}},$$

where $x_1, x_2, \ldots, x_n \in \mathbb{R}^+$ such that $x_1^2+x_2^2+\cdots+x_n^2=1$.

209 Let P, L, R denote the area, perimeter and circumradius of $\triangle ABC$, respectively. Determine the maximum value of the expression $\frac{LP}{R^3}$.

210 Let $a, b, c \in \mathbb{R}^+$ such that $a+b+c=abc$. Prove the inequality

$$\frac{1}{\sqrt{1+a^2}}+\frac{1}{\sqrt{1+b^2}}+\frac{1}{\sqrt{1+c^2}} \leq \frac{3}{2}.$$

211 Let $a, b, c \in \mathbb{R}$ such that $abc+a+c=b$. Prove the inequality

$$\frac{2}{a^2+1}-\frac{2}{b^2+1}+\frac{3}{c^2+1} \leq \frac{10}{3}.$$

212 Let $x, y, z > 1$ be real numbers such that $\frac{1}{x}+\frac{1}{y}+\frac{1}{z}=2$. Prove the inequality

$$\sqrt{x-1}+\sqrt{y-1}+\sqrt{z-1} \leq \sqrt{x+y+z}.$$

213 Let a, b, c be positive real numbers such that $a+b+c=1$. Prove the inequality

$$\sqrt{\frac{1}{a}-1}\sqrt{\frac{1}{b}-1}+\sqrt{\frac{1}{b}-1}\sqrt{\frac{1}{c}-1}+\sqrt{\frac{1}{c}-1}\sqrt{\frac{1}{a}-1} \geq 6.$$

214 Let a, b, c be positive real numbers such that $a+b+c+1=4abc$. Prove the inequalities

$$\frac{1}{a}+\frac{1}{b}+\frac{1}{c} \geq 3 \geq \frac{1}{\sqrt{ab}}+\frac{1}{\sqrt{bc}}+\frac{1}{\sqrt{ca}}.$$

215 Let a, b, c be non-negative real numbers such that $ab+bc+ca=1$. Prove the inequality

$$\frac{a}{1+a^2}+\frac{b}{1+b^2}+\frac{c}{1+c^2} \leq \frac{3\sqrt{3}}{4}.$$

216 Let a, b, c be positive real numbers such that $a+b+c=1$. Prove the inequality

$$\sqrt{\frac{ab}{c+ab}}+\sqrt{\frac{bc}{a+bc}}+\sqrt{\frac{ca}{b+ca}} \leq \frac{3}{2}.$$

217 Let $a, b, c > 0$ be real numbers such that $(a+b)(b+c)(c+a) = 1$. Prove the inequality

$$ab + bc + ca \leq \frac{3}{4}.$$

218 Let $a, b, c \geq 0$ be real numbers such that $a^2 + b^2 + c^2 + abc = 4$. Prove the inequality

$$0 \leq ab + bc + ca - abc \leq 2.$$

219 Let a, b, c be positive real numbers. Prove the inequality

$$a^2 + b^2 + c^2 + 2abc + 3 \geq (1+a)(1+b)(1+c).$$

220 Let a, b, c be real numbers. Prove the inequality

$$\sqrt{a^2 + (1-b)^2} + \sqrt{b^2 + (1-c)^2} + \sqrt{c^2 + (1-a)^2} \geq \frac{3\sqrt{2}}{2}.$$

221 Let $a_1, a_2, \ldots, a_n \in \mathbb{R}^+$ such that $\sum_{i=1}^{n} a_i^3 = 3$ and $\sum_{i=1}^{n} a_i^5 = 5$. Prove the inequality

$$\sum_{i=1}^{n} a_i > \frac{3}{2}.$$

222 Let a, b, c be positive real numbers such that $ab + bc + ca = 3$. Prove the inequality

$$(1+a^2)(1+b^2)(1+c^2) \geq 8.$$

223 Let a, b, c be positive real numbers such that $ab + bc + ca = 1$. Prove the inequality

$$(a^2 + ab + b^2)(b^2 + bc + c^2)(c^2 + ca + a^2) \geq 1.$$

224 Let a, b, c be positive real numbers such that $abc = 1$. Prove the inequality

$$\frac{a}{\sqrt{7 + b^2 + c^2}} + \frac{b}{\sqrt{7 + c^2 + a^2}} + \frac{c}{\sqrt{7 + a^2 + b^2}} \geq 1.$$

225 Let $a_1, a_2, \ldots, a_n$ be positive real numbers such that $a_1 + a_2 + \cdots + a_n = 1$. Prove the inequality

$$\frac{a_1}{\sqrt{1-a_1}} + \frac{a_2}{\sqrt{1-a_2}} + \cdots + \frac{a_n}{\sqrt{1-a_n}} \geq \sqrt{\frac{n}{n-1}}.$$

226 Let a, b, c be positive real numbers. Prove the inequality

$$\frac{a}{\sqrt{2b^2+2c^2-a^2}}+\frac{b}{\sqrt{2c^2+2a^2-b^2}}+\frac{c}{\sqrt{2a^2+2b^2-c^2}}\geq\sqrt{3}.$$

227 Let a, b, c be positive real numbers such that $ab+bc+ca\geq 3$. Prove the inequality

$$\frac{a}{\sqrt{a+b}}+\frac{b}{\sqrt{b+c}}+\frac{c}{\sqrt{c+a}}\geq\frac{3}{\sqrt{2}}.$$

228 Let $a, b, c\geq 1$ be real numbers such that $a+b+c=2abc$. Prove the inequality

$$\sqrt[3]{(a+b+c)^2}\geq\sqrt[3]{ab-1}+\sqrt[3]{bc-1}+\sqrt[3]{ca-1}.$$

229 Let t_a, t_b, t_c be the lengths of the medians, and a, b, c be the lengths of the sides of a given triangle. Prove the inequality

$$t_at_b+t_bt_c+t_ct_a<\frac{5}{4}(ab+bc+ca).$$

230 Let a, b, c and t_a, t_b, t_c be the lengths of the sides and lengths of the medians of an arbitrary triangle, respectively. Prove the inequality

$$at_a+bt_b+ct_c\leq\frac{\sqrt{3}}{2}(a^2+b^2+c^2).$$

231 Let a, b, c be the lengths of the sides of a triangle. Prove the inequality

$$\sqrt{a+b-c}+\sqrt{c+a-b}+\sqrt{b+c-a}\leq\sqrt{a}+\sqrt{b}+\sqrt{c}.$$

232 Let P be the area of the triangle with side lengths a, b and c, and T be the area of the triangle with side lengths $a+b, b+c$ and $c+a$. Prove that $T\geq 4P$.

233 Let a, b, c be the lengths of the sides of a triangle, such that $a+b+c=3$. Prove the inequality

$$a^2+b^2+c^2+\frac{4abc}{3}\geq\frac{13}{3}.$$

234 Let a, b, c be the lengths of the sides of a triangle. Prove that

$$\sqrt[3]{\frac{a^3+b^3+c^3+3abc}{2}}\geq\max\{a, b, c\}.$$

235 Let a, b, c be the lengths of the sides of a triangle. Prove the inequality

$$abc<a^2(s-a)+b^2(s-a)+c^2(s-a)\leq\frac{3}{2}abc.$$

236 Let a, b, c be the lengths of the sides of a triangle. Prove that

$$\frac{1}{\sqrt{a}+\sqrt{b}-\sqrt{c}}+\frac{1}{\sqrt{b}+\sqrt{c}-\sqrt{a}}+\frac{1}{\sqrt{c}+\sqrt{a}-\sqrt{b}} \geq \frac{3(\sqrt{a}+\sqrt{b}+\sqrt{c})}{a+b+c}.$$

237 Let a, b, c be the lengths of the sides of a triangle with area P. Prove that

$$a^2+b^2+c^2 \geq 4\sqrt{3}P.$$

238 (*Hadwinger–Finsler*) Let a, b, c be the lengths of the sides of a triangle. Prove the inequality

$$a^2+b^2+c^2 \geq 4\sqrt{3}P+(a-b)^2+(b-c)^2+(c-a)^2.$$

239 Let a, b, c be the lengths of the sides of a triangle. Prove that

$$\frac{1}{8abc+(a+b-c)^3}+\frac{1}{8abc+(b+c-a)^3}+\frac{1}{8abc+(c+a-b)^3} \leq \frac{1}{3abc}.$$

240 In the triangle ABC, $\overline{AC}^2$ is the arithmetic mean of $\overline{BC}^2$ and $\overline{AB}^2$. Prove that

$$\cot^2\beta \geq \cot\alpha \cdot \cot\gamma.$$

241 Let d_1, d_2 and d_3 be the distances from an arbitrary point to the sides BC, CA, AB, respectively, of the triangle ABC. Prove the inequality

$$\frac{9}{4}(d_1^2+d_2^2+d_3^2) \geq \left(\frac{P}{R}\right)^2.$$

242 Let a, b, c be the side lengths, and h_a, h_b, h_c be the lengths of the altitudes (respectively) of a given triangle. Prove the inequality

$$\frac{h_a+h_b+h_c}{a+b+c} \leq \frac{\sqrt{3}}{2}.$$

243 Let O be an arbitrary point in the interior of $\triangle ABC$. Let x, y and z be the distances from O to the sides BC, CA, AB, respectively, and let R be the circumradius of the triangle $\triangle ABC$. Prove the inequality

$$\sqrt{x}+\sqrt{y}+\sqrt{z} \leq 3\sqrt{\frac{R}{2}}.$$

244 Let D, E and F be the feet of the altitudes of the triangle ABC dropped from the vertices A, B and C, respectively. Prove the inequality

$$\left(\frac{\overline{EF}}{a}\right)^2+\left(\frac{\overline{FD}}{b}\right)^2+\left(\frac{\overline{DE}}{c}\right)^2 \geq \frac{3}{4}.$$

245 Let a, b, c be the side-lengths, h_a, h_b, h_c be the lengths of the respective altitudes, and s be the semi-perimeter of a given triangle. Prove the inequality

$$\frac{h_a}{a} + \frac{h_b}{b} + \frac{h_c}{c} \le \frac{s}{2r}.$$

246 Let a, b, c be the side lengths, h_a, h_b, h_c be the altitudes, respectively, of a triangle. Prove the inequality

$$\frac{a^2}{h_b^2 + h_c^2} + \frac{b^2}{h_a^2 + h_c^2} + \frac{c^2}{h_a^2 + h_b^2} \ge 2.$$

247 Let a, b, c be the side lengths, h_a, h_b, h_c be the altitudes, respectively, and r be the inradius of a triangle. Prove the inequality

$$\frac{1}{h_a - 2r} + \frac{1}{h_b - 2r} + \frac{1}{h_c - 2r} \ge \frac{3}{r}.$$

248 Let $a, b, c; l_\alpha, l_\beta, l_\gamma$ be the lengths of the sides and the bisectors of respective angles. Let s be the semi-perimeter and r denote the inradius of a given triangle. Prove the inequality

$$\frac{l_\alpha}{a} + \frac{l_\beta}{b} + \frac{l_\gamma}{c} \le \frac{s}{2r}.$$

249 Let $a, b, c; l_\alpha, l_\beta, l_\gamma$ be the lengths of the sides and of the bisectors of respective angles. Let R and r be the circumradius and inradius, respectively, of a given triangle. Prove the inequality

$$18r^2\sqrt{3} \le al_\alpha + bl_\beta + cl_\gamma < 9R^2.$$

250 Let a, b, c be the lengths of the sides of a triangle, with circumradius $r = 1/2$. Prove the inequality

$$\frac{a^4}{b+c-a} + \frac{b^4}{a+c-b} + \frac{c^4}{a+b-c} \ge 9\sqrt{3}.$$

251 Let a, b, c be the side-lengths of a triangle. Prove the inequality

$$\frac{a}{3a-b+c} + \frac{b}{3b-c+a} + \frac{c}{3c-a+b} \ge 1.$$

252 Let h_a, h_b and h_c be the lengths of the altitudes, and R and r be the circumradius and inradius, respectively, of a given triangle. Prove the inequality

$$h_a + h_b + h_c \le 2R + 5r.$$

253 Let a, b, c be the side-lengths, and α, β and γ be the angles of a given triangle, respectively. Prove the inequality

$$a\left(\frac{1}{\beta}+\frac{1}{\gamma}\right)+b\left(\frac{1}{\gamma}+\frac{1}{\alpha}\right)+c\left(\frac{1}{\alpha}+\frac{1}{\beta}\right)\geq 2\left(\frac{a}{\alpha}+\frac{b}{\beta}+\frac{c}{\gamma}\right).$$

254 Let a, b, c be the lengths of the sides of a given triangle, and α, β, γ be the respective angles (in radians). Prove the inequalities

1° $\frac{1}{\alpha}+\frac{1}{\beta}+\frac{1}{\gamma}\geq\frac{9}{\pi}$
2° $\frac{b+c-a}{\alpha}+\frac{c+a-b}{\beta}+\frac{a+b-c}{\gamma}\geq\frac{6s}{\pi}$, where $s=\frac{a+b+c}{2}$
3° $\frac{b+c-a}{a\alpha}+\frac{c+a-b}{b\beta}+\frac{a+b-c}{c\gamma}\geq\frac{9}{\pi}$.

255 Let X be an arbitrary interior point of a given regular n-gon with side-length a. Let $h_1, h_2, \ldots, h_n$ be the distances from X to the sides of the n-gon. Prove that

$$\frac{1}{h_1}+\frac{1}{h_2}+\cdots+\frac{1}{h_n}>\frac{2\pi}{a}.$$

256 Prove that among the lengths of the sides of an arbitrary n-gon ($n\geq 3$), there always exist two of them (let's denote them by b and c) such that $1\leq\frac{b}{c}<2$.

257 Let a_1, a_2, a_3, a_4 be the lengths of the sides, and s be the semi-perimeter of arbitrary quadrilateral. Prove that

$$\sum_{i=1}^{4}\frac{1}{s+a_i}\leq\frac{2}{9}\sum_{1\leq i<j\leq 4}\frac{1}{\sqrt{(s-a_i)(s-a_j)}}.$$

258 Let $n\in\mathbb{N}$, and α, β, γ be the angles of a given triangle. Prove the inequality

$$\cot^n\frac{\alpha}{2}+\cot^n\frac{\beta}{2}+\cot^n\frac{\gamma}{2}\geq 3^{\frac{n+2}{2}}.$$

259 Let α, β, γ be the angles of an arbitrary acute triangle. Prove that

$$2(\sin\alpha+\sin\beta+\sin\gamma)>3(\cos\alpha+\cos\beta+\cos\gamma).$$

260 Let α, β, γ be the angles of a triangle. Prove the inequality

$$\sin\alpha+\sin\beta+\sin\gamma\geq\sin 2\alpha+\sin 2\beta+\sin 2\gamma.$$

261 Let α, β, γ be the angles of a triangle. Prove the inequality

$$\cos\alpha+\sqrt{2}(\cos\beta+\cos\gamma)\leq 2.$$

262 Let α, β, γ be the angles of a triangle and let t be a real number. Prove the inequality

$$\cos\alpha + t(\cos\beta + \cos\gamma) \leq 1 + \frac{t^2}{2}.$$

263 Let $0 \leq \alpha, \beta, \gamma \leq 90^\circ$ such that $\sin\alpha + \sin\beta + \sin\gamma = 1$. Prove the inequality

$$\tan^2\alpha + \tan^2\beta + \tan^2\gamma \geq \frac{3}{8}.$$

264 Let a, b, c be positive real numbers such that $a+b+c = 3$. Prove the inequality

$$(1+a+a^2)(1+b+b^2)(1+c+c^2) \geq 9(ab+bc+ca).$$

265 Let $a, b, c > 0$ such that $a+b+c = 1$. Prove the inequality

$$6(a^3+b^3+c^3)+1 \geq 5(a^2+b^2+c^2).$$

266 Let $x, y, z \in \mathbb{R}^+$ such that $x+y+z = 1$. Prove the inequality

$$(1-x^2)^2 + (1-z^2)^2 + (1-z^2)^2 \leq (1+x)(1+y)(1+z).$$

267 Let x, y, z be non-negative real numbers such that $x^2+y^2+z^2 = 1$. Prove the inequality

$$(1-xy)(1-yz)(1-zx) \geq \frac{8}{27}.$$

268 Let $a, b, c \in \mathbb{R}^+$ such that $\frac{1}{a+1} + \frac{1}{b+1} + \frac{1}{c+1} = 2$. Prove the inequalities:

1° $\frac{1}{8a^2+1} + \frac{1}{8b^2+1} + \frac{1}{8c^2+1} \geq 1$
2° $\frac{1}{4ab+1} + \frac{1}{4bc+1} + \frac{1}{4ca+1} \geq \frac{3}{2}$.

269 Let $a, b, c > 0$ be real numbers such that $ab+bc+ca = 1$. Prove the inequality

$$\frac{1}{a+b} + \frac{1}{b+c} + \frac{1}{c+a} - \frac{1}{a+b+c} \geq 2.$$

270 Let $a, b, c \geq 0$ be real numbers. Prove the inequality

$$\frac{ab+4bc+ca}{a^2+bc} + \frac{bc+4ca+ab}{b^2+ca} + \frac{ca+4ab+bc}{c^2+ab} \geq 6.$$

271 Let a, b, c be positive real numbers such that $a+b+c+1 = 4abc$. Prove the inequality

$$\frac{1}{a^4+b+c} + \frac{1}{b^4+c+a} + \frac{1}{c^4+a+b} \leq \frac{3}{a+b+c}.$$

272 Let $x, y, z > 0$ be real numbers such that $x + y + z = 1$. Prove the inequality

$$(x^2 + y^2)(y^2 + z^2)(z^2 + x^2) \leq \frac{1}{32}.$$

273 Let $x, y, z \in \mathbb{R}^+$ such that $x + y + z = 1$. Prove the inequalities:

$$1 \leq \frac{x}{1 - yz} + \frac{y}{1 - zx} + \frac{z}{1 - xy} \leq \frac{9}{8}.$$

274 Let $x, y, z \in \mathbb{R}^+$, such that $xyz = 1$. Prove the inequality

$$\frac{1}{(1+x)^2} + \frac{1}{(1+y)^2} + \frac{1}{(1+z)^2} + \frac{2}{(1+x)(1+y)(1+z)} \geq 1.$$

275 Let $a, b, c \geq 0$ such that $a + b + c = 1$. Prove the inequalities:

1° $ab + bc + ca \leq a^3 + b^3 + c^3 + 6abc$

2° $a^3 + b^3 + c^3 + 6abc \leq a^2 + b^2 + c^2$

3° $a^2 + b^2 + c^2 \leq 2(a^3 + b^3 + c^3) + 3abc$.

276 Let $x, y, z \geq 0$ be real numbers such that $xy + yz + zx + xyz = 4$. Prove the inequality

$$3(x^2 + y^2 + z^2) + xyz \geq 10.$$

277 Let $a, b, c \in \mathbb{R}^+$. Prove the inequality

$$x^4(y + z) + y^4(z + x) + z^4(x + y) \leq \frac{1}{12}(x + y + z)^5.$$

278 Let $a, b, c \in \mathbb{R}^+$ such that $a + b + c = 1$. Prove the inequality

$$\frac{1}{a} + \frac{1}{b} + \frac{1}{c} + 48(ab + bc + ca) \geq 25.$$

279 Let a, b, c be non-negative real numbers such that $a + b + c = 2$. Prove the inequality

$$a^4 + b^4 + c^4 + abc \geq a^3 + b^3 + c^3.$$

280 Let a, b, c be non-negative real numbers. Prove the inequality

$$2(a^2 + b^2 + c^2) + abc + 8 \geq 5(a + b + c).$$

281 Let a, b, c be non-negative real numbers. Prove the inequality

$$a^3 + b^3 + c^3 + 4(a + b + c) + 9abc \geq 8(ab + bc + ca).$$

282 Let a, b, c be non-negative real numbers. Prove the inequality

$$\frac{a^3}{b^2 - bc + c^2} + \frac{b^3}{c^2 - ca + a^2} + \frac{c^3}{a^2 - ab + b^2} \geq a + b + c.$$

283 Let a, b, c be non-negative real numbers such that $a + b + c = 2$. Prove the inequality

$$a^3 + b^3 + c^3 + \frac{15abc}{4} \geq 2.$$

284 Let a, b, c be positive real numbers such that $abc = 1$. Prove the inequality

$$\frac{a^2 + bc}{a^2(b + c)} + \frac{b^2 + ca}{b^2(c + a)} + \frac{c^2 + ab}{c^2(a + b)} \geq ab + bc + ca.$$

285 Let a, b, c be positive real numbers such that $a^2 + b^2 + c^2 = 3$. Prove the inequality

$$\frac{a^3 + abc}{(b + c)^2} + \frac{b^3 + abc}{(c + a)^2} + \frac{c^3 + abc}{(a + b)^2} \geq \frac{3}{2}.$$

286 Let a, b, c be positive real numbers such that $a^4 + b^4 + c^4 = 3$. Prove the inequality

$$\frac{1}{4 - ab} + \frac{1}{4 - bc} + \frac{1}{4 - ca} \leq 1.$$

287 Let a, b, c be positive real numbers such that $ab + bc + ca = 3$. Prove the inequality

$$(a^3 - a + 5)(b^5 - b^3 + 5)(c^7 - c^5 + 5) \geq 125.$$

288 Let x, y, z be positive real numbers. Prove the inequality

$$\frac{1}{x^2 + xy + y^2} + \frac{1}{y^2 + yz + z^2} + \frac{1}{z^2 + zx + x^2} \geq \frac{9}{(x + y + z)^2}.$$

289 Let x, y, z be positive real numbers such that $xyz = x + y + z + 2$. Prove the inequalities

1° $xy + yz + zx \geq 2(x + y + z)$

2° $\sqrt{x} + \sqrt{y} + \sqrt{z} \leq \frac{3\sqrt{xyz}}{2}$.

290 Let x, y, z be positive real numbers. Prove the inequality

$$8(x^3 + y^3 + z^3) \geq (x+y)^3 + (y+z)^3 + (z+x)^3.$$

291 Let a, b, c be non-negative real numbers. Prove the inequality

$$a^3 + b^3 + c^3 + abc \geq \frac{1}{7}(a+b+c)^3.$$

292 Let a, b, c be positive real numbers such that $a+b+c=1$. Prove the inequality

$$a^2 + b^2 + c^2 + 3abc \geq \frac{4}{9}.$$

293 Let $a_1, a_2, \ldots, a_n$ be positive real numbers. Prove the inequality

$$(1+a_1)(1+a_2)\cdots(1+a_n) \leq \left(1+\frac{a_1^2}{a_2}\right)\left(1+\frac{a_2^2}{a_3}\right)\cdots\left(1+\frac{a_n^2}{a_1}\right).$$

294 Let a, b, c, d be positive real numbers such that $abcd = 1$. Prove the inequality

$$\frac{1}{(1+a)^2} + \frac{1}{(1+b)^2} + \frac{1}{(1+c)^2} + \frac{1}{(1+d)^2} \geq 1.$$

295 Let $a, b, c, d \geq 0$ be real numbers such that $a+b+c+d=4$. Prove the inequality

$$abc + bcd + cda + dab + (abc)^2 + (bcd)^2 + (cda)^2 + (dab)^2 \leq 8.$$

296 Let $a, b, c, d \geq 0$ such that $a+b+c+d=1$. Prove the inequality

$$a^4 + b^4 + c^4 + d^4 + \frac{148}{27}abcd \geq \frac{1}{27}.$$

297 Let a, b, c be positive real numbers such that $a^2 + b^2 + c^2 = 3$. Prove the inequality

$$a^2b^2 + b^2c^2 + c^2a^2 \leq a+b+c.$$

298 Let $a, b, c, d \geq 0$ be real numbers such that $a+b+c+d=4$. Prove the inequality

$$(1+a^2)(1+b^2)(1+c^2)(1+d^2) \geq (1+a)(1+b)(1+c)(1+d).$$

299 Let a, b, c be positive real numbers such that $abc = 1$. Prove the inequality

$$\frac{1}{a} + \frac{1}{b} + \frac{1}{c} + \frac{6}{a+b+c} \geq 5.$$

300 Let a, b, c be positive real numbers such that $a+b+c=3$. Prove the inequality

$$12\left(\frac{1}{a}+\frac{1}{b}+\frac{1}{c}\right) \geq 4(a^3+b^3+c^3)+21.$$

301 Let a, b, c, d be non-negative real numbers such that $a+b+c+d+e=5$. Prove the inequality

$$4(a^2+b^2+c^2+d^2+e^2)+5abcd \geq 25.$$

302 Let a, b, c be positive real numbers such that $a+b+c=3$. Prove the inequality

$$\frac{1}{2+a^2+b^2}+\frac{1}{2+b^2+c^2}+\frac{1}{2+c^2+a^2} \leq \frac{3}{4}.$$

303 Let a, b, c be positive real numbers such that $a^2+b^2+c^2=3$. Prove the inequality

$$ab+bc+ca \leq abc+2.$$

304 Let a, b, c be positive real numbers. Prove the inequality

$$\frac{a}{b}+\frac{b}{c}+\frac{c}{a} \geq \frac{a+b}{b+c}+\frac{b+c}{c+a}+\frac{a+c}{a+b}.$$

305 Let a, b, c be positive real numbers. Prove the inequality

$$\frac{a^2}{b^2+c^2}+\frac{b^2}{c^2+a^2}+\frac{c^2}{a^2+b^2} \geq \frac{a}{b+c}+\frac{b}{c+a}+\frac{c}{a+b}.$$

306 Let a, b, c be positive real numbers such that $a \geq b \geq c$. Prove the inequality

$$a^2b(a-b)+b^2c(b-c)+c^2a(c-a) \geq 0.$$

307 Let a, b, c be the lengths of the sides of a triangle. Prove the inequality

$$\frac{(b+c)^2}{a^2+bc}+\frac{(c+a)^2}{b^2+ca}+\frac{(a+b)^2}{c^2+ab} \geq 6.$$

308 Let a, b, c be positive real numbers. Prove the inequality

$$\frac{a+b}{b+c}+\frac{b+c}{c+a}+\frac{c+a}{a+b}+3\frac{ab+bc+ca}{(a+b+c)^2} \geq 4.$$

309 Let a, b, c be real numbers. Prove the inequality

$$3(a^2 - ab + b^2)(b^2 - bc + c^2)(c^2 - ca + a^2) \geq a^3b^3 + b^3c^3 + c^3a^3.$$

310 Let $a, b, c, d \in \mathbb{R}^+$ such that $a + b + c + d + abcd = 5$. Prove the inequality

$$\frac{1}{a} + \frac{1}{b} + \frac{1}{c} + \frac{1}{d} \geq 4.$$

Chapter 21
Solutions

1 Let n be a positive integer. Prove that

$$1+\frac{1}{2^2}+\frac{1}{3^2}+\cdots+\frac{1}{n^2}<2.$$

Solution For each $k \geq 2$ we have

$$\frac{1}{k^2}<\frac{1}{k(k-1)}=\frac{1}{k-1}-\frac{1}{k}.$$

So

$$\begin{aligned}1+\frac{1}{2^2}+\frac{1}{3^2}+\cdots+\frac{1}{n^2}&<1+\left(1-\frac{1}{2}\right)+\left(\frac{1}{2}-\frac{1}{3}\right)+\cdots+\left(\frac{1}{n-1}-\frac{1}{n}\right)\\&=2-\frac{1}{n}<2.\end{aligned}$$

■

2 Let $a_n=1+\frac{1}{2}+\frac{1}{3}+\cdots+\frac{1}{n}$. Prove that for any $n \in \mathbb{N}$ we have

$$\frac{1}{a_1^2}+\frac{1}{2a_2^2}+\frac{1}{3a_3^2}+\cdots+\frac{1}{na_n^2}<2.$$

Solution Note that for any $k \geq 2$ we have

$$\frac{1}{a_{k-1}}-\frac{1}{a_k}=\frac{a_k-a_{k-1}}{a_{k-1}a_k}=\frac{1}{ka_ka_{k-1}}>\frac{1}{ka_k^2}.$$

Adding these inequalities for $k=2,3,\ldots,n$ we get

$$\frac{1}{2a_2^2}+\frac{1}{3a_3^2}+\cdots+\frac{1}{na_n^2}<\frac{1}{a_1}-\frac{1}{a_n}<\frac{1}{a_1},$$

Z. Cvetkovski, *Inequalities*,
DOI 10.1007/978-3-642-23792-8_21, © Springer-Verlag Berlin Heidelberg 2012

and since $a_1 = 1$, we obtain

$$\frac{1}{a_1^2} + \frac{1}{2a_2^2} + \frac{1}{3a_3^2} + \cdots + \frac{1}{na_n^2} < \frac{2}{a_1} = 2. \qquad \blacksquare$$

3 Let x, y, z be real numbers. Prove the inequality

$$x^4 + y^4 + z^4 \geq 4xyz - 1.$$

Solution We have

$$\begin{aligned} &x^4 + y^4 + z^4 - 4xyz + 1 \\ &\quad = (x^4 - 2x^2 + 1) + (y^4 - 2y^2z^2 + z^4) + (2y^2z^2 - 4xyz + 2x^2) \\ &\quad = (x^2 - 1)^2 + (y^2 - z^2)^2 + 2(yz - x)^2 \geq 0, \end{aligned}$$

so it follows that

$$x^4 + y^4 + z^4 \geq 4xyz - 1.$$

When does equality occur? $\blacksquare$

4 Prove that for any real number x, the following inequality holds

$$x^{2002} - x^{1999} + x^{1996} - x^{1995} + 1 > 0.$$

Solution Denote

$$x^{2002} - x^{1999} + x^{1996} - x^{1995} + 1 > 0. \tag{1}$$

We will consider five cases:

1° If $x < 0$, then all summands on the left side of the inequality (1) are positive, so the inequality is true.

2° If $x = 0$, inequality (1) is equivalent to $1 > 0$, which is obviously true.

3° If $0 < x < 1$, then (1) is

$$x^{2002} + x^{1996}(1 - x^3) + (1 - x^{1995}) > 0.$$

Since

$$1 - x^3 = (1 - x)(1 + x + x^2) > 0 \quad \text{and}$$
$$1 - x^5 = (1 - x)(1 + x + x^2 + x^3 + x^4) > 0,$$

we deduce that the required inequality is true.

4° If $x = 1$, then (1) is equivalent to $1 > 0$, which is clearly true.

5° If $x > 1$, rewrite (1) in following way

$$x^{1999}(x^3-1)+x^{1995}(x-1)+1>0.$$

Since $x > 1$ we have $x^3 > 1$.

So $x^{1999}(x^3-1)+x^{1995}(x-1)+1>0$, and we are done. ∎

5 Let x, y be real numbers. Prove the inequality

$$3(x+y+1)^2+1 \geq 3xy.$$

Solution Observe that for any real numbers a and b we have

$$a^2+ab+b^2=\left(a+\frac{b}{2}\right)^2+\frac{3b^2}{4}\geq 0,$$

with equality if and only if $a=b=0$.

Let x, y be real numbers. Then according to the above inequality we have

$$\left(x+\frac{2}{3}\right)^2+\left(x+\frac{2}{3}\right)\left(y+\frac{2}{3}\right)+\left(y+\frac{2}{3}\right)^2\geq 0, \quad \text{i.e.}$$

$$3x^2+3y^2+3xy+6x+6y+4\geq 0,$$

which is equivalent to

$$3(x+y+1)^2+1\geq 3xy.$$

Equality occurs iff $x+\frac{2}{3}=y+\frac{2}{3}=0$, i.e. $x=y=-\frac{2}{3}$. ∎

6 Let a, b, c be positive real numbers such that $a+b+c\geq abc$. Prove that at least two of the following inequalities

$$\frac{2}{a}+\frac{3}{b}+\frac{6}{c}\geq 6, \qquad \frac{2}{b}+\frac{3}{c}+\frac{6}{a}\geq 6, \qquad \frac{2}{c}+\frac{3}{a}+\frac{6}{b}\geq 6$$

are true.

Solution Set $\frac{1}{a}=x$, $\frac{1}{b}=y$, $\frac{1}{c}=z$.

Then $x, y, z>0$ and the initial condition becomes $xy+yz+zx\geq 1$.

We need to prove that at least two of the following inequalities $2x+3y+6z\geq 6$, $2y+3z+6x\geq 6$, $2z+3x+6y\geq 6$, hold.

Assume the contrary, i.e. we may assume that $2x+3y+6z<6$ and $2z+3x+6y<6$.

Adding these inequalities we get $5x+9y+8z<12$.

But we have $x\geq \frac{1-yz}{y+z}$.

Thus, $12>\frac{5-5yz}{y+z}+9y+8z$, i.e.

$$12(y+z)>5+9y^2+8z^2+12yz \quad \Leftrightarrow \quad (2z-1)^2+(3y+2z-2)^2<0,$$

which is impossible, and the conclusion follows. ∎

7 Let $a, b, c, x, y, z > 0$. Prove the inequality

$$\frac{ax}{a+x} + \frac{by}{b+y} + \frac{cz}{c+z} \leq \frac{(a+b+c)(x+y+z)}{a+b+c+x+y+z}.$$

Solution We'll use the following lemma.

Lemma 21.1 *For every $p, q, \alpha, \beta > 0$ we have*

$$\frac{pq}{p+q} \leq \frac{\alpha^2 p + \beta^2 q}{(\alpha+\beta)^2}.$$

Proof The given inequality is equivalent to $(\alpha p - \beta q)^2 \geq 0$. □

Now let $\alpha = x+y+z$, $\beta = a+b+c$, and applying Lemma 21.1, we obtain

$$\frac{ax}{a+x} \leq \frac{(x+y+z)^2 a + (a+b+c)^2 x}{(x+y+z+a+b+c)^2},$$

$$\frac{by}{b+y} \leq \frac{(x+y+z)^2 b + (a+b+c)^2 y}{(x+y+z+a+b+c)^2}$$

and

$$\frac{cz}{c+z} \leq \frac{(x+y+z)^2 c + (a+b+c)^2 z}{(x+y+z+a+b+c)^2}.$$

Adding these inequalities we get the required result. ■

8 Let $a, b, c \in \mathbb{R}^+$. Prove the inequality

$$\frac{2a}{a^2+bc} + \frac{2b}{b^2+ac} + \frac{2c}{c^2+ab} \leq \frac{a}{bc} + \frac{b}{ac} + \frac{c}{ab}.$$

Solution Notice that $\frac{2a}{a^2+bc} \leq \frac{1}{2}(\frac{1}{b} + \frac{1}{c})$, which is equivalent to

$$b(a-c)^2 + c(a-b)^2 \geq 0.$$

Also $\frac{1}{b} + \frac{1}{c} \leq \frac{1}{2}(\frac{2a}{bc} + \frac{b}{ac} + \frac{c}{ab})$, which is equivalent to

$$(a-b)^2 + (a-c)^2 \geq 0.$$

Hence

$$\frac{2a}{a^2+bc} \leq \frac{1}{4}\left(\frac{2a}{bc} + \frac{b}{ac} + \frac{c}{ab}\right). \tag{1}$$

Analogously, we obtain

$$\frac{2b}{b^2+ac} \le \frac{1}{4}\left(\frac{2b}{ac}+\frac{c}{ab}+\frac{a}{bc}\right), \tag{2}$$

$$\frac{2c}{c^2+ab} \le \frac{1}{4}\left(\frac{2c}{ab}+\frac{a}{bc}+\frac{b}{ac}\right). \tag{3}$$

Adding (1), (2) and (3) we obtain the required inequality.

Equality occurs if and only if $a=b=c$. ■

9 Let $a,b,c,x,y,z \in \mathbb{R}^+$ such that $a+x=b+y=c+z=1$. Prove the inequality

$$(abc+xyz)\left(\frac{1}{ay}+\frac{1}{bz}+\frac{1}{cx}\right) \ge 3.$$

Solution We have

$$abc+xyz = abc+(1-a)(1-b)(1-c) = (1-b)(1-c)+ac+ab-a.$$

So

$$\frac{abc+xyz}{a(1-b)} = \frac{1-c}{a}+\frac{c}{1-b}-1,$$

and analogously we obtain $\frac{abc+xyz}{b(1-c)}$ and $\frac{abc+xyz}{c(1-a)}$.

Hence

$$\begin{aligned}&(abc+xyz)\left(\frac{1}{ay}+\frac{1}{bz}+\frac{1}{cx}\right)\\ &= \frac{a}{1-c}+\frac{b}{1-a}+\frac{c}{1-b}+\frac{1-c}{a}+\frac{1-b}{c}+\frac{1-a}{b}-3 \ge 6-3=3.\end{aligned}$$ ■

10 Let $a_1,a_2,\ldots,a_n$ be positive real numbers and let $b_1,b_2,\ldots,b_n$ be their permutation. Prove the inequality

$$\frac{a_1^2}{b_1}+\frac{a_2^2}{b_2}+\cdots+\frac{a_n^2}{b_n} \ge a_1+a_2+\cdots+a_n.$$

Solution For each $x,y \in \mathbb{R}^+$ we have $\frac{x^2}{y} \ge 2x-y$.

Hence

$$\frac{a_i^2}{b_i} \ge 2a_i-b_i, \quad i=1,2,\ldots,n.$$

After summing for $i = 1, 2, \dots, n$ we obtain

$$\frac{a_1^2}{b_1} + \frac{a_2^2}{b_2} + \cdots + \frac{a_n^2}{b_n} \geq 2(a_1 + a_2 + \cdots + a_n) - (b_1 + b_2 + \cdots + b_n)$$
$$= a_1 + a_2 + \cdots + a_n,$$

and we are done.

Equality occurs if and only if $a_i = b_i, i = 1, 2, \dots, n$. ■

11 Let $x \in \mathbb{R}^+$. Find the minimum value of the expression $\frac{x^2+1}{x+1}$.

Solution Denote $A = \frac{x^2+1}{x+1}$.

We have

$$A = \frac{x^2 - 1 + 2}{x + 1} = (x - 1) + \frac{2}{x + 1} = \left((x + 1) + \frac{2}{x + 1}\right) - 2. \quad (1)$$

For any $a, b \geq 0$ we have $a + b \geq 2\sqrt{ab}$ (equality occurs iff $a = b$).

Now from (1) we get $A \geq 2\sqrt{2} - 2$.

Equality occurs if and only if $x = \sqrt{2} - 1$. ■

12 Let $a, b, c \in \mathbb{R}^+$ such that $abc = 1$. Prove the inequality

$$\frac{a}{(a+1)(b+1)} + \frac{b}{(b+1)(c+1)} + \frac{c}{(c+1)(a+1)} \geq \frac{3}{4}.$$

Solution After expanding we get

$$ab + ac + bc + a + b + c \geq 3(abc + 1)$$

i.e.

$$ab + ac + bc + a + b + c \geq 6.$$

Since

$$ab + ac + bc + a + b + c = \frac{1}{c} + \frac{1}{b} + \frac{1}{a} + a + b + c$$
$$= \left(\frac{1}{a} + a\right) + \left(\frac{1}{b} + b\right) + \left(\frac{1}{c} + c\right) \geq 2 + 2 + 2 = 6,$$

we are done.

Equality occurs if and only if $\frac{1}{a} + a = \frac{1}{b} + b = \frac{1}{c} + c = 1$, i.e. $a = b = c = 1$. ■

13 Let $x, y \geq 0$ be real numbers such that $y(y + 1) \leq (x + 1)^2$. Prove the inequality

$$y(y - 1) \leq x^2.$$

Solution If $0 \le y \le 1$, then $y(y-1) \le 0 \le x^2$.

Suppose that $y > 1$.

If $x + \frac{1}{2} \le y$, then

$$y(y-1) = y(y+1) - 2y \le (x+1)^2 - 2\left(x + \frac{1}{2}\right) = x^2.$$

If $x + \frac{1}{2} > y$ then we have $x > y - \frac{1}{2} > 0$, i.e.

$$x^2 > \left(y - \frac{1}{2}\right)^2 = y(y-1) + \frac{1}{4} > y(y-1).$$

■

14 Let $x, y \in \mathbb{R}^+$ such that $x^3 + y^3 \le x - y$. Prove that

$$x^2 + y^2 \le 1.$$

Solution From $x^3 + y^3 \le x - y$ we have

$$0 \le y \le x$$

and

$$0 \le x^3 \le x^3 + y^3 \le x - y \le x,$$

i.e.

$$x^3 \le x,$$

from where we deduce that $x \le 1$.

Thus $0 \le y \le x \le 1$.

Now we have $x(x+y) \le 1 \cdot 2 = 2$ and $xy(x+y) \le 2y$.

From $x^3 + y^3 \le x - y$ we obtain

$$(x+y)(x^2 - xy + y^2) \le x - y \quad \Leftrightarrow \quad x^2 - xy + y^2 \le \frac{x-y}{x+y}$$

$$\Leftrightarrow \quad x^2 + y^2 \le \frac{x-y}{x+y} + xy = \frac{x - y + xy(x+y)}{x+y} \le \frac{x - y + 2y}{x+y} = \frac{x+y}{x+y} = 1.$$

■

15 Let $a, b, x, y \in \mathbb{R}$ such that $ay - bx = 1$. Prove that

$$a^2 + b^2 + x^2 + y^2 + ax + by \ge \sqrt{3}.$$

Solution Let us denote $u = a^2 + b^2$, $v = x^2 + y^2$ and $w = ax + by$.

Then

$$\begin{aligned} uv &= (a^2 + b^2)(x^2 + y^2) = a^2x^2 + a^2y^2 + b^2x^2 + b^2y^2 \\ &= a^2x^2 + b^2y^2 + 2axby + a^2y^2 + b^2x^2 - 2axby \\ &= (ax + by)^2 + (ay - bx)^2 = w^2 + 1. \end{aligned}$$

From the obvious inequality $(t\sqrt{3}+1)^2 \geq 0$ we deduce

$$3t^2+1 \geq -2t\sqrt{3},$$

i.e.

$$4t^2+4 \geq 3-2t\sqrt{3}+t^2,$$

i.e.

$$4t^2+4 \geq (\sqrt{3}-t)^2. \tag{1}$$

Now we have

$$(u+v)^2 \geq 4uv = 4(w^2+1) \overset{(1)}{\geq} (\sqrt{3}-w)^2,$$

from which we get $u+v \geq \sqrt{3}-w$, which is equivalent to $u+v+w \geq \sqrt{3}$. ■

16 Let a,b,c,d be non-negative real numbers such that $a^2+b^2+c^2+d^2=1$. Prove the inequality

$$(1-a)(1-b)(1-c)(1-d) \geq abcd.$$

Solution We have $2cd \leq c^2+d^2 = 1-a^2-b^2$.

Hence

$$2(1-a)(1-b)-2cd \geq 2(1-a)(1-b)-1+a^2+b^2 = (1-a-b)^2 \geq 0,$$

i.e.

$$(1-a)(1-b) \geq cd. \tag{1}$$

Similarly we get

$$(1-c)(1-d) \geq ab. \tag{2}$$

After multiplying (1) and (2) we obtain $(1-a)(1-b)(1-c)(1-d) \geq abcd$, as required. Equality occurs iff $a=b=c=d=1/2$ or $a=1, b=c=d=0$ (up to permutation). ■

17 Let x, y be non-negative real numbers. Prove the inequality

$$4(x^9+y^9) \geq (x^2+y^2)(x^3+y^3)(x^4+y^4).$$

Solution Since the given inequality is symmetric we may assume that $x \geq y \geq 0$.

Let $a,b \in \mathbb{N}$. Then we have $x^a \geq y^a$ and $x^b \geq y^b$.

Hence

$$\begin{aligned}
&(x^a-y^a)(x^b-y^b) \geq 0 \\
\Leftrightarrow\quad & x^{a+b}+y^{a+b} \geq x^a y^b + x^b y^a \\
\Leftrightarrow\quad & 2(x^{a+b}+y^{a+b}) \geq (x^a+y^a)(x^b+y^b). \qquad (1)
\end{aligned}$$

For $a=2, b=3$ in (1) we get

$$2(x^5+y^5) \geq (x^2+y^2)(x^3+y^3). \tag{2}$$

For $a=5, b=4$ in (1) we get

$$2(x^9+y^9) \geq (x^5+y^5)(x^4+y^4). \tag{3}$$

From (2) and (3) we get

$$4(x^9+y^9) = 2 \cdot 2(x^9+y^9) \geq 2(x^5+y^5)(x^4+y^4) \geq (x^2+y^2)(x^3+y^3)(x^4+y^4),$$

and we are done. ■

18 Let $x, y, z \in \mathbb{R}^+$ such that $xyz=1$ and $\frac{1}{x}+\frac{1}{y}+\frac{1}{z} \geq x+y+z$. Prove that for any natural number n the inequality

$$\frac{1}{x^n}+\frac{1}{y^n}+\frac{1}{z^n} \geq x^n+y^n+z^n$$

is true.

Solution After setting $x=\frac{a}{b}$, $y=\frac{b}{c}$ and $z=\frac{c}{a}$, the initial condition

$$\frac{1}{x}+\frac{1}{y}+\frac{1}{z} \geq x+y+z$$

becomes

$$\begin{aligned} &\frac{b}{a}+\frac{c}{b}+\frac{a}{c} \geq \frac{a}{b}+\frac{b}{c}+\frac{c}{a} \\ \Leftrightarrow \quad & a^2b+b^2c+c^2a \geq ab^2+bc^2+ca^2 \\ \Leftrightarrow \quad & (a-b)(b-c)(c-a) \leq 0. \end{aligned}$$

Let $n \in \mathbb{N}$, and take $A=a^n, B=b^n, C=c^n$.

Then $a \geq b \Leftrightarrow A \geq B$ and $a \leq b \Leftrightarrow A \leq B$, etc.

So we have

$$\begin{aligned} &(A-B)(B-C)(C-A) \leq 0 \\ \Leftrightarrow \quad & \frac{B}{A}+\frac{C}{B}+\frac{A}{C} \geq \frac{A}{B}+\frac{B}{C}+\frac{C}{A} \\ \Leftrightarrow \quad & \frac{1}{x^n}+\frac{1}{y^n}+\frac{1}{z^n} \geq x^n+y^n+z^n. \end{aligned}$$

■

19 Let x, y, z be real numbers different from 1, such that $xyz=1$. Prove the inequality

$$\left(\frac{3-x}{1-x}\right)^2+\left(\frac{3-y}{1-y}\right)^2+\left(\frac{3-z}{1-z}\right)^2 > 7.$$

Solution Denote $A=(\frac{3-x}{1-x})^2+(\frac{3-y}{1-y})^2+(\frac{3-z}{1-z})^2-7$.

We have

$$A=\left(1+\frac{2}{1-x}\right)^2+\left(1+\frac{2}{1-y}\right)^2+\left(1+\frac{2}{1-z}\right)^2-7.$$

Let $\frac{1}{1-x}=a$, $\frac{1}{1-y}=b$, $\frac{1}{1-z}=c$.

Then $A=(1+2a)^2+(1+2b)^2+(1+2c)^2-7$, i.e.

$$A=4a^2+4b^2+4c^2+4a+4b+4c-4. \tag{1}$$

Furthermore, the condition $xyz=1$ is equivalent to $abc=(a-1)(b-1)(c-1)$, i.e.

$$a+b+c-1=ab+bc+ca. \tag{2}$$

Using (1) and (2) we get

$$A=4a^2+4b^2+4c^2+4(ab+bc+ca)=2((a+b)^2+(b+c)^2+(c+a)^2),$$

i.e. $A\geq 0$.

Equality occurs if and only if $a=b=c=0$, which is clearly impossible.

So we have strict inequality, i.e. $A>0$, i.e.

$$\left(\frac{3-x}{1-x}\right)^2+\left(\frac{3-y}{1-y}\right)^2+\left(\frac{3-z}{1-z}\right)^2-7>0$$

and we are done. ■

20 Let $x,y,z\leq 1$ be real numbers such that $x+y+z=1$. Prove the inequality

$$\frac{1}{1+x^2}+\frac{1}{1+y^2}+\frac{1}{1+z^2}\leq\frac{27}{10}.$$

Solution We'll prove that for every $t\leq 1$ we have $\frac{1}{1+t^2}\leq\frac{27}{50}(2-t)$.

The last inequality is equivalent to $(4-3t)(1-3t)^2\geq 0$, which is clearly true.

Hence

$$\begin{aligned}\frac{1}{1+x^2}+\frac{1}{1+y^2}+\frac{1}{1+z^2}&\leq\frac{27}{50}((2-x)+(2-y)+(2-z))\\&=\frac{27}{50}(6-(x+y+z))=\frac{27}{10}.\end{aligned}$$ ■

21 Let $a,b,c\in\mathbb{R}^+$. Prove the inequality

$$\frac{1}{a(1+b)}+\frac{1}{b(1+c)}+\frac{1}{c(1+a)}\geq\frac{3}{1+abc}.$$

Solution We can easily check the following identities

$$\frac{1+abc}{a(1+b)} = \frac{1+a}{a(1+b)} + \frac{b(1+c)}{1+b} - 1, \qquad \frac{1+abc}{b(1+c)} = \frac{1+b}{b(1+c)} + \frac{c(1+a)}{1+c} - 1$$

and

$$\frac{1+abc}{a(1+b)} = \frac{1+c}{c(1+a)} + \frac{a(1+b)}{1+a} - 1.$$

Adding these identities we obtain

$$\begin{aligned} &\frac{1+abc}{a(1+b)} + \frac{1+abc}{b(1+c)} + \frac{1+abc}{c(1+a)} \\ &\quad = \left(\frac{1+a}{a(1+b)} + \frac{a(1+b)}{1+a}\right) + \left(\frac{1+b}{b(1+c)} + \frac{b(1+c)}{1+b}\right) \\ &\qquad + \left(\frac{1+c}{c(1+a)} + \frac{c(1+a)}{1+c}\right) - 3 \geq 2+2+2-3 = 3, \end{aligned}$$

i.e.

$$\frac{1}{a(1+b)} + \frac{1}{b(1+c)} + \frac{1}{c(1+a)} \geq \frac{3}{1+abc}.$$

Equality occurs if and only if $a=b=c=1$. ■

22 Let x, y, z be positive real numbers. Prove the inequality

$$9(a+b)(b+c)(c+a) \geq 8(a+b+c)(ab+bc+ca).$$

Solution The given inequality is equivalent to $a(b-c)^2+b(c-a)^2+c(a-b)^2 \geq 0$, which is obviously true. Equality occurs iff $a=b=c$. ■

23 Let a, b, c be real numbers. Prove the inequality

$$(a^2+b^2+c^2)^2 \geq 3(a^3b+b^3c+c^3a).$$

Solution By the well-known inequality $(x+y+z)^2 \geq 3(xy+yz+zx)$ for

$$x = a^2+bc-ab, \qquad y = b^2+ca-bc, \qquad z = c^2+ab-ca,$$

we obtain the required inequality. ■

24 Let a, b, c be positive real numbers such that $a^2+b^2+c^2=3$. Prove the inequality

$$a^3(b+c)+b^3(c+a)+c^3(a+b) \leq 6.$$

Solution We'll show that

$$a^3(b+c)+b^3(c+a)+c^3(a+b)\le \frac{2}{3}(a^2+b^2+c^2)^2. \qquad (1)$$

Inequality (1) is equivalent to

$$\begin{aligned}&2(a^4+b^4+c^4)+4(a^2b^2+b^2c^2+c^2a^2)\\&\quad\ge 3ab(a^2+b^2)+3bc(b^2+c^2)+3ca(c^2+a^2).\end{aligned} \qquad (2)$$

We have

$$a^4+b^4+4a^2b^2\ge 3ab(a^2+b^2) \quad\Leftrightarrow\quad (a-b)^4+ab(a-b)^2\ge 0,$$

which is clearly true.

Analogously we get

$$b^4+c^4+4b^2c^2\ge 3bc(b^2+c^2) \quad\text{and}\quad c^4+a^4+4c^2a^2\ge 3ca(c^2+a^2).$$

Adding the last inequalities we get (2), i.e. (1).

Finally using $a^2+b^2+c^2=3$ we obtain the required result.

Equality holds if and only if $a=b=c$. ■

25 Let a, b, c be positive real numbers. Prove the inequality

$$\sqrt{\frac{a}{b+c}}+\sqrt{\frac{b}{c+a}}+\sqrt{\frac{c}{a+b}}>2.$$

Solution We'll show that $\sqrt{\frac{x}{y+z}}\ge\frac{2x}{x+y+z}$, for every $x, y, z\in\mathbb{R}^+$.

We have

$$\begin{aligned}&\sqrt{\frac{x}{y+z}}\ge\frac{2x}{x+y+z} \quad\Leftrightarrow\quad \frac{x}{y+z}\ge\left(\frac{2x}{x+y+z}\right)^2\\&\quad\Leftrightarrow\quad (x+y+z)^2\ge 4x(y+z) \quad\Leftrightarrow\quad (y+z-x)^2\ge 0,\end{aligned}$$

with equality iff $x=y+z$.

Now we easily obtain

$$\sqrt{\frac{a}{b+c}}+\sqrt{\frac{b}{c+a}}+\sqrt{\frac{c}{a+b}}\ge\frac{2(a+b+c)}{a+b+c}=2,$$

with equality if and only if $a=b+c, b=a+c, c=a+b$, i.e. $a=b=c=0$, which is impossible.

So we have strict inequality, i.e. $\sqrt{\frac{a}{b+c}}+\sqrt{\frac{b}{c+a}}+\sqrt{\frac{c}{a+b}}>2$, as required. ■

26 Let a, b, c be positive real numbers such that $a^2 + b^2 + c^2 = 3$. Prove the inequality

$$\frac{a}{b+2} + \frac{b}{c+2} + \frac{c}{a+2} \leq 1.$$

Solution The given inequality is equivalent to

$$ab^2 + bc^2 + ca^2 \leq 2 + abc.$$

We may assume that $a \geq b \geq c$ (since the inequality is cyclic we must also consider the case $c \geq b \geq a$, which is analogous).

Then we have $a(b-a)(b-c) \leq 0$ from which we have $a^2b + abc \geq ab^2 + ca^2$. Thus

$$ab^2 + bc^2 + ca^2 \leq a^2b + abc + bc^2.$$

We'll show that

$$a^2b + bc^2 \leq 2.$$

We have

$$a^2b + bc^2 \leq 2 \quad \Leftrightarrow \quad b(3-b^2) \leq 2 \quad \Leftrightarrow \quad (b-1)^2(b+2) \geq 0,$$

which is clearly true, and we are done.

Equality occurs iff $a = b = c = 1$ or $a = 0, b = 1, c = \sqrt{2}$ (over all permutations). ■

27 Let x, y, z be distinct non-negative real numbers. Prove the inequality

$$\frac{1}{(x-y)^2} + \frac{1}{(y-z)^2} + \frac{1}{(z-x)^2} \geq \frac{4}{xy+yz+zx}.$$

Solution If $a, b > 0$, then $\frac{1}{(a-b)^2} + \frac{1}{a^2} + \frac{1}{b^2} \geq \frac{4}{ab}$.

The last inequality is true since

$$\frac{1}{(a-b)^2} + \frac{1}{a^2} + \frac{1}{b^2} - \frac{4}{ab} = \frac{(a^2+b^2-3ab)^2}{a^2b^2(a-b)^2}.$$

Without loss of generality we may assume that $z = \min\{x, y, z\}$.

By the previous inequality for $a = x - z$ and $b = y - z$ we get

$$\frac{1}{(x-y)^2} + \frac{1}{(y-z)^2} + \frac{1}{(z-x)^2} \geq \frac{4}{(x-z)(y-z)}.$$

So it suffices to show that

$$\frac{4}{(x-z)(y-z)} \geq \frac{4}{xy+yz+zx},$$

i.e.

$$xy + yz + zx \geq (x - z)(y - z),$$

i.e.

$$2z(y + x) \geq z^2,$$

which is true since $z = \min\{x, y, z\}$. ■

28 Let a, b, c be non-negative real numbers. Prove the inequality

$$3(a^2 - a + 1)(b^2 - b + 1)(c^2 - c + 1) \geq 1 + abc + (abc)^2.$$

Solution Since

$$2(a^2 - a + 1)(b^2 - b + 1) = 1 + a^2b^2 + (a - b)^2 + (1 - a)^2(1 - b)^2$$

we deduce that

$$2(a^2 - a + 1)(b^2 - b + 1) \geq 1 + a^2b^2.$$

It follows that

$$3(a^2 - a + 1)(b^2 - b + 1)(c^2 - c + 1) \geq \frac{3}{2}(1 + a^2b^2)(c^2 - c + 1),$$

and it remains to prove that

$$3(1 + a^2b^2)(c^2 - c + 1) \geq 2(1 + abc + (abc)^2),$$

which is equivalent to the following quadratic in c

$$(3 + a^2b^2)c^2 - (3 + 2ab + 3a^2b^2)c + 1 + 3a^2b^2 \geq 0,$$

and clearly the last inequality is true, since $3 + a^2b^2 > 0$ and $D = -3(1 - ab)^4 \leq 0$.
Equality occurs iff $a = b = c = 1$. ■

29 Let $a, b \in \mathbb{R}, a \neq 0$. Prove the inequality

$$a^2 + b^2 + \frac{1}{a^2} + \frac{b}{a} \geq \sqrt{3}.$$

Solution We have

$$a^2 + b^2 + \frac{1}{a^2} + \frac{b}{a} = \left(b + \frac{1}{2a}\right)^2 + a^2 + \frac{3}{4a^2}. \quad (1)$$

Since $(b + \frac{1}{2a})^2 \geq 0$, using (1) we get

$$a^2 + b^2 + \frac{1}{a^2} + \frac{b}{a} \geq a^2 + \frac{3}{4a^2}. \quad (2)$$

Using $AM \geq GM$ we have

$$a^2 + \frac{3}{4a^2} \geq 2\sqrt{a^2 \frac{3}{4a^2}} = \sqrt{3}. \tag{3}$$

From (2) and (3) we get

$$a^2 + b^2 + \frac{1}{a^2} + \frac{b}{a} \geq \sqrt{3}.$$

Equality occurs iff $b + \frac{1}{2a} = 0$ and $a^2 = \frac{3}{4a^2}$, i.e. $a = \pm\sqrt[4]{\frac{3}{4}}$ and $b = \mp\frac{1}{2}\sqrt[4]{\frac{4}{3}}$. ■

30 Let $a, b, c \in \mathbb{R}^+$. Prove the inequality

$$\frac{a^2+1}{b+c} + \frac{b^2+1}{c+a} + \frac{c^2+1}{a+b} \geq 3.$$

Solution For each $x \in \mathbb{R}$ we have $x^2 + 1 \geq 2x$.

So we have

$$\frac{a^2+1}{b+c} + \frac{b^2+1}{c+a} + \frac{c^2+1}{a+b} \geq \frac{2a}{b+c} + \frac{2b}{c+a} + \frac{2c}{a+b}.$$

It's enough to prove that $\frac{a}{b+c} + \frac{b}{c+a} + \frac{c}{a+b} \geq \frac{3}{2}$, which is *Nesbitt's inequality*.

Equality occurs if and only if $a = b = c = 1$. ■

31 Let x, y, z be positive real numbers such that $xy + yz + zx = 5$. Prove the inequality

$$3x^2 + 3y^2 + z^2 \geq 10.$$

Solution Using the inequality $AM \geq GM$ we obtain

$$4x^2 + z^2 \geq 4xz, \qquad 4y^2 + z^2 \geq 4yz \quad \text{and} \quad 2x^2 + 2y^2 \geq 4xy.$$

Adding these inequalities and using $xy + yz + zx = 5$ we get the required inequality.

Equality occurs iff $x = y = 1, z = 2$. ■

32 Let a, b, c be positive real numbers such that $ab + bc + ca > a + b + c$. Prove the inequality

$$a + b + c > 3.$$

Solution We have

$$\begin{aligned}(a+b+c)^2 &= a^2 + b^2 + c^2 + 2(ab + ac + bc)\\ &\geq ab + ac + bc + 2(ab + ac + bc)\\ &= 3(ab + ac + bc) > 3(a + b + c),\end{aligned}$$

from which we get $a + b + c > 3$. ■

33 Let a, b be real numbers such that $9a^2 + 8ab + 7b^2 \le 6$. Prove that

$$7a + 5b + 12ab \le 9.$$

Solution By the inequality $AM \ge GM$ we have

$$\begin{aligned} 7a + 5b + 12ab &\le 7\left(a^2 + \frac{1}{4}\right) + 5\left(a^2 + \frac{1}{4}\right) + 12ab \\ &= 7a^2 + 5b^2 + 12ab + 3 \\ &= 9a^2 + 8ab + 7b^2 - 2a^2 + 4ab - 2b^2 + 3 \\ &= 9a^2 + 8ab + 7b^2 - 2(a-b)^2 + 3 \le 6 + 3 = 9, \end{aligned}$$

as required. Equality holds iff $a = b = 1/2$. ∎

34 Let $x, y, z \in \mathbb{R}^+$, such that $xyz \ge xy + yz + zx$. Prove the inequality

$$xyz \ge 3(x + y + z).$$

Solution Letting $\frac{1}{x} = a$, $\frac{1}{y} = b$, $\frac{1}{z} = c$, the initial condition $xyz \ge xy + yz + zx$ becomes

$$a + b + c \le 1. \tag{1}$$

We need to show that

$$xyz \ge 3(x + y + z) \quad \Leftrightarrow \quad 3(ab + bc + ca) \le 1. \tag{2}$$

Clearly

$$(a + b + c)^2 \ge 3(ab + bc + ca). \tag{3}$$

Now from (1) and (3) we obtain (2). ∎

35 Let $a, b, c \in \mathbb{R}^+$ with $a^2 + b^2 + c^2 = 3$. Prove the inequality

$$\frac{ab}{c} + \frac{bc}{a} + \frac{ca}{b} \ge 3.$$

Solution The given inequality is equivalent to

$$\begin{aligned} &\left(\frac{ab}{c} + \frac{bc}{a} + \frac{ca}{b}\right)^2 \ge 9 \\ \Leftrightarrow \quad &\frac{a^2b^2}{c^2} + \frac{b^2c^2}{a^2} + \frac{c^2a^2}{b^2} + 2(a^2 + b^2 + c^2) \ge 3(a^2 + b^2 + c^2), \end{aligned}$$

i.e.

$$\frac{a^2b^2}{c^2}+\frac{b^2c^2}{a^2}+\frac{c^2a^2}{b^2}\geq a^2+b^2+c^2.$$

Furthermore, applying $AM \geq GM$ we get

$$\frac{a^2b^2}{c^2}+\frac{b^2c^2}{a^2}\geq 2b^2,\qquad \frac{b^2c^2}{a^2}+\frac{c^2a^2}{b^2}\geq 2c^2,\qquad \frac{a^2b^2}{c^2}+\frac{c^2a^2}{b^2}\geq 2a^2.$$

After adding these inequalities we obtain

$$\frac{a^2b^2}{c^2}+\frac{b^2c^2}{a^2}+\frac{c^2a^2}{b^2}\geq a^2+b^2+c^2$$

and we are done. ■

36 Let a, b, c be positive real numbers such that $a+b+c=\sqrt{abc}$. Prove the inequality

$$ab+bc+ca\geq 9(a+b+c).$$

Solution By the inequality $AM \geq GM$ we have

$$\sqrt{abc}=a+b+c\geq 3\sqrt[3]{abc},$$

which implies

$$abc\geq 3^6 \quad \text{and} \quad a+b+c=\sqrt{abc}\geq\sqrt{3^6}=3^3. \tag{1}$$

Once more, the inequality $AM \geq GM$ gives us

$$ab+bc+ca\geq 3\sqrt[3]{(abc)^2},$$

i.e.

$$(ab+bc+ca)^3\geq 3^3(abc)^2=3^3(a+b+c)^4\overset{(1)}{\geq} 3^6(a+b+c)^3.$$

Hence

$$ab+bc+ca\geq 9(a+b+c),$$

as required.

Equality occurs if and only if $a=b=c=9$. ■

37 Let a, b, c be positive real numbers such that $abc\geq 1$. Prove the inequality

$$\left(a+\frac{1}{a+1}\right)\left(b+\frac{1}{b+1}\right)\left(c+\frac{1}{c+1}\right)\geq\frac{27}{8}.$$

Solution By the inequality $AM \geq GM$ we have

$$\frac{a+1}{4}+\frac{1}{a+1}\geq 2\sqrt{\frac{a+1}{4}\cdot\frac{1}{a+1}}=1 \quad \text{and} \quad \frac{3a}{4}+\frac{3}{4}\geq 2\sqrt{\frac{3a}{4}\cdot\frac{3}{4}}=\frac{3}{2}\sqrt{a}.$$

Adding these two inequalities we get

$$a+\frac{1}{a+1}\geq\frac{3}{2}\sqrt{a}.$$

Analogously we obtain

$$b+\frac{1}{b+1}\geq\frac{3}{2}\sqrt{b} \quad \text{and} \quad c+\frac{1}{c+1}\geq\frac{3}{2}\sqrt{c}.$$

Multiplying the last three inequalities gives us

$$\left(a+\frac{1}{a+1}\right)\left(b+\frac{1}{b+1}\right)\left(c+\frac{1}{c+1}\right)\geq\frac{27}{8}\sqrt{abc}\geq\frac{27}{8},$$

as required.

Equality occurs iff $a=b=c=1$. ■

38 Let $a,b,c,d\in\mathbb{R}^+$ such that $a^2+b^2+c^2+d^2=4$. Prove the inequality

$$a+b+c+d\geq ab+bc+cd+da.$$

Solution We have

$$a+b+c+d\geq ab+bc+cd+da \quad \Leftrightarrow \quad a+b+c+d\geq(a+c)(b+d),$$

i.e.

$$\frac{1}{a+c}+\frac{1}{b+d}\geq 1.$$

Since $AM\geq HM$ we have

$$\frac{1}{a+c}+\frac{1}{b+d}\geq\frac{4}{a+b+c+d}. \tag{1}$$

Applying $QM\geq AM$ we have

$$\frac{a+b+c+d}{4}\leq\sqrt{\frac{a^2+b^2+c^2+d^2}{4}}=1,$$

i.e.

$$a+b+c+d\leq 4.$$

Now by (1) we get

$$\frac{1}{a+c}+\frac{1}{b+d}\geq\frac{4}{a+b+c+d}\geq\frac{4}{4}=1.$$

Equality holds if and only if $a=b=c=d=1$. ∎

39 Let $a,b,c\in(-3,3)$ such that $\frac{1}{3+a}+\frac{1}{3+b}+\frac{1}{3+c}=\frac{1}{3-a}+\frac{1}{3-b}+\frac{1}{3-c}$. Prove the inequality

$$\frac{1}{3+a}+\frac{1}{3+b}+\frac{1}{3+c}\geq 1.$$

Solution By the inequality $AM\geq HM$ we have

$$((3+a)+(3+b)+(3+c))\left(\frac{1}{3+a}+\frac{1}{3+b}+\frac{1}{3+c}\right)\geq 9 \tag{1}$$

and

$$((3-a)+(3-b)+(3-c))\left(\frac{1}{3-a}+\frac{1}{3-b}+\frac{1}{3-c}\right)\geq 9$$

$$\Leftrightarrow \quad ((3-a)+(3-b)+(3-c))\left(\frac{1}{3+a}+\frac{1}{3+b}+\frac{1}{3+c}\right)\geq 9. \tag{2}$$

After adding (1) and (2) we obtain

$$18\left(\frac{1}{3+a}+\frac{1}{3+b}+\frac{1}{3+c}\right)\geq 18,\quad \text{i.e.}\quad \frac{1}{3+a}+\frac{1}{3+b}+\frac{1}{3+c}\geq 1.$$ ∎

40 Let $a,b,c\in\mathbb{R}^+$ such that $a^2+b^2+c^2=3$. Prove the inequality

$$\frac{1}{a+bc+abc}+\frac{1}{b+ca+bca}+\frac{1}{c+ab+cab}\geq 1.$$

Solution By $AM\geq HM$ we have:

$$\frac{1}{a+bc+abc}+\frac{1}{b+ca+bca}+\frac{1}{c+ab+cab}$$
$$\geq\frac{9}{a+b+c+ab+bc+ca+3abc}. \tag{1}$$

Using the well known inequalities:

$$a^2+b^2+c^2\geq ab+bc+ac\quad\text{and}\quad (a+b+c)^2\geq 3(a^2+b^2+c^2)$$

and according to $a^2+b^2+c^2=3$, we deduce

$$ab+bc+ca\leq 3\quad\text{and}\quad a+b+c\leq 3. \tag{2}$$

By $AM \geq GM$ we have $a^2+b^2+c^2 \geq 3\sqrt[3]{(abc)^2}$ and since $a^2+b^2+c^2=3$ we easily deduce that

$$abc \leq 1 \tag{3}$$

Now according to (1), (2) and (3) we obtain

$$\begin{aligned}&\frac{1}{a+bc+abc}+\frac{1}{b+ca+bca}+\frac{1}{c+ab+cab}\\&\quad\geq \frac{9}{a+b+c+ab+bc+ca+3abc} \geq \frac{9}{3+3+3}=1\end{aligned}$$

Equality occurs if and only if $a=b=c=1$. ■

41 Let $a,b,c \in \mathbb{R}^+$ such that $a+b+c=3$. Prove the inequality.

$$\frac{a^2b^2+a^2+b^2}{ab+1}+\frac{b^2c^2+b^2+c^2}{bc+1}+\frac{c^2a^2+c^2+a^2}{ca+1} \geq \frac{9}{2}.$$

Solution Let $a,b \in \mathbb{R}^+$ then we have

$$\begin{aligned}&(a-1)^2(b-1)^2 \geq 0\\&\Leftrightarrow\quad a^2b^2-2a^2b+a^2-2ab^2+4ab-2a+b^2-2b+1 \geq 0\\&\Leftrightarrow\quad a^2b^2+a^2+b^2 \geq 2a^2b+2ab^2+2a+2b-4ab-1\\&\Leftrightarrow\quad a^2b^2+a^2+b^2 \geq 2a(ab+1)+2b(ab+1)-4(ab+1)+3\\&\qquad\qquad = (ab+1)(2a+2b-4)+3.\end{aligned}$$

Hence

$$\frac{a^2b^2+a^2+b^2}{ab+1} \geq 2a+2b-4+\frac{3}{ab+1}. \tag{1}$$

Similarly we obtain

$$\frac{b^2c^2+b^2+c^2}{bc+1} \geq 2b+2c-4+\frac{3}{bc+1} \tag{2}$$

and

$$\frac{c^2a^2+c^2+a^2}{ca+1} \geq 2c+2a-4+\frac{3}{ca+1}. \tag{3}$$

Adding (1), (2) and (3) gives us

$$\begin{aligned}&\frac{a^2b^2+a^2+b^2}{ab+1}+\frac{b^2c^2+b^2+c^2}{bc+1}+\frac{c^2a^2+c^2+a^2}{ca+1}\\&\geq 4(a+b+c)-12+\frac{3}{ab+1}+\frac{3}{bc+1}+\frac{3}{ca+1}\\&=\frac{3}{ab+1}+\frac{3}{bc+1}+\frac{3}{ca+1}. \qquad (4)\end{aligned}$$

Applying $AM \geq HM$ we obtain

$$\frac{1}{1+ab}+\frac{1}{1+bc}+\frac{1}{1+ca}\geq\frac{9}{3+ab+bc+ca}. \qquad (5)$$

Using the well known inequality $(a+b+c)^2 \geq 3(ab+bc+ca)$ and $a+b+c=3$ we deduce

$$ab+bc+ca\leq 3. \qquad (6)$$

Finally by (4), (5) and (6) we obtain

$$\begin{aligned}&\frac{a^2b^2+a^2+b^2}{ab+1}+\frac{b^2c^2+b^2+c^2}{bc+1}+\frac{c^2a^2+c^2+a^2}{ca+1}\\&\geq\frac{3}{ab+1}+\frac{3}{bc+1}+\frac{3}{ca+1}\geq\frac{27}{3+ab+bc+ca}\geq\frac{27}{3+3}=\frac{9}{2}.\end{aligned}$$

Equality occurs iff $a=b=c=1$. ■

42 Let a, b, c, d be positive real numbers such that $a^2+b^2+c^2+d^2=4$. Prove the inequality

$$\frac{a^2+b^2+3}{a+b}+\frac{b^2+c^2+3}{b+c}+\frac{c^2+d^2+3}{c+d}+\frac{d^2+a^2+3}{d+a}\geq 10.$$

Solution Observe that for any real numbers x, y we have

$$x^2+xy+y^2=\left(x+\frac{y}{2}\right)^2+\frac{3y^2}{4}\geq 0,$$

equality achieves if and only if $x=y=0$.

Hence $(a-1)^2+(a-1)(b-1)+(b-1)^2\geq 0$, which is equivalent to

$$a^2+b^2+ab-3a-3b+3\geq 0,$$

from which we obtain

$$a^2+b^2+3\geq 3a+3b-ab,$$

i.e.

$$\frac{a^2+b^2+3}{a+b} \geq 3 - \frac{ab}{a+b}.$$

By $AM \geq GM$ we easily deduce that

$$\frac{a+b}{4} \geq \frac{ab}{a+b}.$$

Therefore by previous inequality we get

$$\frac{a^2+b^2+3}{a+b} \geq 3 - \frac{a+b}{4}.$$

Similarly we obtain

$$\frac{b^2+c^2+3}{b+c} \geq 3 - \frac{b+c}{4}, \qquad \frac{c^2+d^2+3}{c+d} \geq 3 - \frac{c+d}{4} \quad \text{and}$$

$$\frac{d^2+a^2+3}{d+a} \geq 3 - \frac{d+a}{4}.$$

Adding the last four inequality yields

$$\frac{a^2+b^2+3}{a+b} + \frac{b^2+c^2+3}{b+c} + \frac{c^2+d^2+3}{c+d} + \frac{d^2+a^2+3}{d+a} \geq 12 - \frac{a+b+c+d}{2}. \tag{1}$$

According to inequality $QM \geq AM$ we deduce that

$$\sqrt{\frac{a^2+b^2+c^2+d^2}{4}} \geq \frac{a+b+c+d}{4}$$

and since $a^2+b^2+c^2+d^2=4$ we obtain

$$a+b+c+d \leq 4. \tag{2}$$

By (1) and (2) we get

$$\begin{aligned}\frac{a^2+b^2+3}{a+b} + \frac{b^2+c^2+3}{b+c} + \frac{c^2+d^2+3}{c+d} + \frac{d^2+a^2+3}{d+a} &\geq 12 - \frac{a+b+c+d}{2} \\ &\geq 12 - \frac{4}{2} = 10,\end{aligned}$$

as required.

Equality occurs if and only if $a=b=c=d=1$. ■

43 Let a, b, c be positive real numbers. Prove the inequality

$$\frac{1}{ab(a+b)} + \frac{1}{bc(b+c)} + \frac{1}{ca(c+a)} \geq \frac{9}{2(a^3+b^3+c^3)}.$$

Solution According to the obvious inequality $(a+b)(a-b)^2 \geq 0$ we get the inequality

$$a^3+b^3 \geq ab(a+b).$$

Thus

$$\frac{1}{ab(a+b)} \geq \frac{1}{a^3+b^3}.$$

Similarly we get

$$\frac{1}{bc(b+c)} \geq \frac{1}{b^3+c^3} \quad \text{and} \quad \frac{1}{ca(c+a)} \geq \frac{1}{c^3+a^3}.$$

After adding the last three inequalities we obtain

$$\frac{1}{ab(a+b)}+\frac{1}{bc(b+c)}+\frac{1}{ca(c+a)} \geq \frac{1}{a^3+b^3}+\frac{1}{b^3+c^3}+\frac{1}{c^3+a^3}. \qquad (1)$$

Now since $AM \geq HM$ we have

$$\begin{aligned}\frac{1}{a^3+b^3}+\frac{1}{b^3+c^3}+\frac{1}{c^3+a^3} &\geq \frac{9}{(a^3+b^3)+(b^3+c^3)+(c^3+a^3)}\\ &= \frac{9}{2(a^3+b^3+c^3)}. \qquad (2)\end{aligned}$$

From (1) and (2) we get the required inequality.

Equality holds if and only if $a=b=c$. ∎

44 Let $a,b,c \in \mathbb{R}^+$ such that $a\sqrt{bc}+b\sqrt{ca}+c\sqrt{ab} \geq 1$. Prove the inequality

$$a+b+c \geq \sqrt{3}.$$

Solution We have

$$\begin{aligned}1 &\leq a\sqrt{bc}+b\sqrt{ca}+c\sqrt{ab} \leq a\frac{b+c}{2}+b\frac{c+a}{2}+c\frac{a+b}{2}\\ &= ab+ac+bc \leq \frac{(a+b+c)^2}{3},\end{aligned}$$

i.e.

$$(a+b+c)^2 \geq 3 \quad \Leftrightarrow \quad a+b+c \geq \sqrt{3}.$$ ∎

45 Let a,b,c be positive real numbers such that $abc=1$. Prove the inequality

$$\frac{b+c}{\sqrt{a}}+\frac{c+a}{\sqrt{b}}+\frac{a+b}{\sqrt{c}} \geq \sqrt{a}+\sqrt{b}+\sqrt{c}+3.$$

Solution By $AM \geq GM$ we get

$$\begin{aligned}
&\frac{b+c}{\sqrt{a}}+\frac{c+a}{\sqrt{b}}+\frac{a+b}{\sqrt{c}}\\
&\geq 2\sqrt{\frac{bc}{a}}+2\sqrt{\frac{ca}{b}}+2\sqrt{\frac{ab}{c}}\\
&=\left(\sqrt{\frac{bc}{a}}+\sqrt{\frac{ca}{b}}\right)+\left(\sqrt{\frac{ca}{b}}+\sqrt{\frac{ab}{c}}\right)+\left(\sqrt{\frac{ab}{c}}+\sqrt{\frac{bc}{a}}\right)\\
&\geq 2(\sqrt{a}+\sqrt{b}+\sqrt{c})\geq \sqrt{a}+\sqrt{b}+\sqrt{c}+3\sqrt[6]{abc}\\
&=\sqrt{a}+\sqrt{b}+\sqrt{c}+3.
\end{aligned}$$

■

46 Let x, y, z be positive real numbers such that $x+y+z=4$. Prove the inequality

$$\frac{1}{2xy+xz+yz}+\frac{1}{xy+2xz+yz}+\frac{1}{xy+xz+2yz}\leq\frac{1}{xyz}.$$

Solution By $AM \geq HM$ we have that $\frac{1}{a}+\frac{1}{b}\geq\frac{4}{a+b}$, for any $a, b \in \mathbb{R}^+$.
Therefore

$$\begin{aligned}
\frac{1}{2xy+xz+yz}&=\frac{1}{(xy+xz)+(xy+yz)}\leq\frac{1}{4}\left(\frac{1}{xy+xz}+\frac{1}{xy+yz}\right)\\
&\leq\frac{1}{4}\left(\frac{1}{4}\left(\frac{1}{xy}+\frac{1}{xz}\right)+\frac{1}{4}\left(\frac{1}{xy}+\frac{1}{yz}\right)\right)\\
&=\frac{1}{16}\left(\frac{2}{xy}+\frac{1}{xz}+\frac{1}{yz}\right)=\frac{2z+y+x}{16xyz}.
\end{aligned}$$

Similarly,

$$\frac{1}{xy+2xz+yz}\leq\frac{z+2y+x}{16xyz}\quad\text{and}\quad\frac{1}{xy+xz+2yz}\leq\frac{z+y+2x}{16xyz}.$$

Adding the three inequalities yields that

$$\frac{1}{2xy+xz+yz}+\frac{1}{xy+2xz+yz}+\frac{1}{xy+xz+2yz}\leq\frac{1}{16}\left(\frac{4(x+y+z)}{xyz}\right)=\frac{1}{xyz}.$$

Equality occurs iff $x=y=z=4/3$. ■

47 Let $a, b, c \in \mathbb{R}^+$. Prove the inequality

$$abc\geq(a+b-c)(b+c-a)(c+a-b).$$

Solution Setting $a+b-c=x, b+c-a=y, c+a-b=z$ the inequality becomes

$$(x+y)(y+z)(z+x) \geq 8xyz.$$

Let us assume that $x \leq 0$. Then $c \geq a+b$, and clearly y and z are positive and the right-hand side of the given inequality is negative or zero, but the left-hand side is positive, i.e. the inequality holds.

So we may assume that $x, y, z > 0$. Then using $AM \geq GM$ we get

$$(x+y)(y+z)(z+x) \geq 2\sqrt{xy} \cdot 2\sqrt{yz} \cdot 2\sqrt{xz} = 8xyz$$

and we are done. ■

48 Let a, b, c be positive real numbers such that $a+b+c=3$. Prove the inequality

$$abc + \frac{12}{ab+bc+ac} \geq 5.$$

Solution Recalling the well-known inequality $abc \geq (b+c-a)(c+a-b)(a+b-c)$ (Problem 47) we obtain

$$\begin{aligned} &abc \geq (3-2a)(3-2b)(3-2c) \\ &\Leftrightarrow \quad abc \geq 27 - 18(a+b+c) + 12(ab+bc+ca) - 8abc \\ &\Leftrightarrow \quad 3abc \geq 4(ab+bc+ca) - 9 \\ &\Leftrightarrow \quad abc \geq \frac{4(ab+bc+ca)}{3} - 3. \end{aligned}$$

Therefore we have

$$abc + \frac{12}{ab+bc+ac} \geq \frac{4(ab+bc+ca)}{3} + \frac{12}{ab+bc+ac} - 3 \geq 8 - 3 = 5,$$

where the last inequality follows since $AM \geq GM$. ■

49 Let a, b, c be positive real numbers such that $abc = 1$. Prove that

$$\left(a - 1 + \frac{1}{b}\right)\left(b - 1 + \frac{1}{c}\right)\left(c - 1 + \frac{1}{a}\right) \leq 1.$$

Solution Since $abc = 1$, it is natural to take $a = \frac{x}{y}, b = \frac{y}{z}, c = \frac{z}{x}$ where $x, y, z > 0$. Now the given inequality becomes

$$\left(\frac{x}{y} - 1 + \frac{z}{y}\right)\left(\frac{y}{z} - 1 + \frac{x}{z}\right)\left(\frac{z}{x} - 1 + \frac{y}{x}\right) \leq 1 \quad \text{i.e.}$$

$$(x+y-z)(z+x-y)(y+z-x) \leq xyz,$$

which is true (Problem 47). Equality occurs iff $a = b = c = 1$. ■

50 Let a, b, c be positive real numbers such that $abc = 1$. Prove the inequality

$$\frac{1}{1+a+b} + \frac{1}{1+b+c} + \frac{1}{1+c+a} \leq \frac{1}{2+a} + \frac{1}{2+b} + \frac{1}{2+c}.$$

Solution Let $x = a+b+c$ and $y = ab+ac+bc$.

Clearly $x, y \geq 3$ (these are immediate consequences of $AM \geq GM$).

Now the given inequality is equivalent to

$$\frac{3+4x+y+x^2}{2x+y+x^2+xy} \leq \frac{12+4x+y}{9+4x+2y},$$

i.e.

$$3x^2y + xy^2 + 6xy - 5x^2 - y^2 - 24x - 3y - 27 \geq 0,$$

i.e.

$$(3x^2y - 5x^2 - 12x) + (xy^2 - y^2 - 3x - 3y) + (6xy - 9x - 27) \geq 0,$$

which is true since $x, y \geq 3$. ■

51 Let $a, b, c > 0$. Prove the inequality

$$(a+b)^2 + (a+b+4c)^2 \geq \frac{100abc}{a+b+c}.$$

Solution Since $AM \geq GM$ we have

$$\begin{aligned}(a+b)^2 + (a+b+4c)^2 &= (a+b)^2 + (a+2c+b+2c)^2 \\ &\geq 4ab + (2\sqrt{2ac} + 2\sqrt{2bc})^2, \quad \text{i.e.}\end{aligned}$$

$$(a+b)^2 + (a+b+4c)^2 \geq 4ab + 8ac + 8bc + 16c\sqrt{ab}.$$

Now

$$\begin{aligned}&\frac{(a+b)^2 + (a+b+4c)^2}{abc}(a+b+c) \\ &\quad \geq \frac{4ab + 8ac + 8bc + 16c\sqrt{ab}}{abc}(a+b+c) \\ &\quad = \left(\frac{4}{c} + \frac{8}{b} + \frac{8}{a} + \frac{16}{\sqrt{ab}}\right)(a+b+c) \\ &\quad = 8\left(\frac{1}{2c} + \frac{1}{b} + \frac{1}{a} + \frac{1}{\sqrt{ab}} + \frac{1}{\sqrt{ab}}\right)\left(\frac{a}{2} + \frac{a}{2} + \frac{b}{2} + \frac{b}{2} + c\right).\end{aligned}$$

Using the last inequality and $AM \geq GM$ once more we obtain

$$\frac{(a+b)^2 + (a+b+4c)^2}{abc}(a+b+c) \geq 8 \cdot 5\sqrt[5]{\frac{1}{2a^2b^2c}} \cdot 5\sqrt[5]{\frac{a^2b^2c}{16}} = 100,$$

i.e.

$$(a+b)^2+(a+b+4c)^2 \geq \frac{100abc}{a+b+c}.$$

Equality occurs if and only if $a=b=2c$. ■

52 Let $a,b,c>0$ such that $abc=1$. Prove the inequality

$$\frac{1+ab}{1+a}+\frac{1+bc}{1+b}+\frac{1+ac}{1+a} \geq 3.$$

Solution Since $abc=1$ we have

$$\frac{1+ab}{1+a}=\frac{abc+ab}{1+a}=\frac{ab(c+1)}{a+1}$$

and similarly

$$\frac{1+bc}{1+b}=\frac{bc(a+1)}{b+1} \quad \text{and} \quad \frac{1+ca}{1+c}=\frac{ca(b+1)}{c+1}.$$

Now by $AM \geq GM$ we obtain

$$\begin{aligned}\frac{1+ab}{1+a}+\frac{1+bc}{1+b}+\frac{1+ca}{1+c} &= \frac{ab(c+1)}{a+1}+\frac{bc(a+1)}{b+1}+\frac{ca(b+1)}{c+1}\\ &\geq 3\sqrt[3]{\frac{ab(c+1)}{a+1}\cdot\frac{bc(a+1)}{b+1}\cdot\frac{ca(b+1)}{c+1}}\\ &= 3\sqrt[3]{(abc)^2}=3.\end{aligned}$$

Equality occurs iff $a=b=c=1$. ■

53 Let a,b,c be real numbers such that $ab+bc+ca=1$. Prove the inequality

$$\left(a+\frac{1}{b}\right)^2+\left(b+\frac{1}{c}\right)^2+\left(c+\frac{1}{a}\right)^2 \geq 16.$$

Solution 1 We have

$$\begin{aligned}&\left(a+\frac{1}{b}\right)^2+\left(b+\frac{1}{c}\right)^2+\left(c+\frac{1}{a}\right)^2\\ &\quad = a^2+b^2+c^2+\frac{1}{a^2}+\frac{1}{b^2}+\frac{1}{c^2}+2\left(\frac{a}{b}+\frac{b}{c}+\frac{c}{a}\right)\\ &\quad = a^2+\frac{1}{a^2}+b^2+\frac{1}{b^2}+c^2+\frac{1}{c^2}+2\left(\frac{a}{b}+\frac{b}{c}+\frac{c}{a}\right)\end{aligned}$$

$$= a^2 + \frac{ab+bc+ca}{a^2} + b^2 + \frac{ab+bc+ca}{b^2} + c^2 + \frac{ab+bc+ca}{c^2}$$

$$+ 2\left(\frac{a}{b} + \frac{b}{c} + \frac{c}{a}\right)$$

$$= (a^2+b^2+c^2) + \left(\frac{b}{a} + \frac{c}{b} + \frac{a}{c}\right) + 3\left(\frac{c}{a} + \frac{a}{b} + \frac{b}{c}\right) + \left(\frac{bc}{a^2} + \frac{ca}{b^2} + \frac{ab}{c^2}\right)$$

$$\geq ab + bc + ca + 3 + 9 + 3 = 1 + 3 + 9 + 3 = 16.$$

Clearly, equality occurs iff $a = b = c = 1/\sqrt{3}$. ∎

Solution 2 By well-known inequality $x^2 + y^2 + z^2 \geq xy + yz + zx$ we have

$$\left(a+\frac{1}{b}\right)^2 + \left(b+\frac{1}{c}\right)^2 + \left(c+\frac{1}{a}\right)^2 \geq \left(a+\frac{1}{b}\right)\left(b+\frac{1}{c}\right) + \left(b+\frac{1}{c}\right)\left(c+\frac{1}{a}\right)$$
$$+ \left(c+\frac{1}{a}\right)\left(a+\frac{1}{b}\right),$$

i.e.

$$\left(a+\frac{1}{b}\right)^2 + \left(b+\frac{1}{c}\right)^2 + \left(c+\frac{1}{a}\right)^2 \geq ab + bc + ca + \frac{a}{c} + \frac{b}{a} + \frac{c}{b} + 3$$
$$+ \frac{1}{ab} + \frac{1}{bc} + \frac{1}{ca}. \tag{1}$$

Using $AM \geq GM$ and $AM \geq HM$ we get $\frac{a}{c} + \frac{b}{a} + \frac{c}{b} \geq 3\sqrt[3]{\frac{a}{c} \cdot \frac{b}{a} \cdot \frac{c}{b}} = 3$ and $\frac{1}{ab} + \frac{1}{bc} + \frac{1}{ca} \geq \frac{9}{ab+bc+ca} = \frac{9}{1} = 9$, respectively.

By last two inequalities and (1) we obtain

$$\left(a+\frac{1}{b}\right)^2 + \left(b+\frac{1}{c}\right)^2 + \left(c+\frac{1}{a}\right)^2 \geq 1 + 3 + 3 + 9 = 16,$$

as required. ∎

Solution 3 By $QM \geq AM$ we have

$$\left(a+\frac{1}{b}\right)^2 + \left(b+\frac{1}{c}\right)^2 + \left(c+\frac{1}{a}\right)^2 \geq \frac{(a+b+c+\frac{1}{a}+\frac{1}{b}+\frac{1}{c})^2}{3}. \tag{1}$$

By well-known $(a+b+c)^2 \geq 3(ab+bc+ca)$ and $ab+bc+ca = 1$ we obtain

$$a + b + c \geq \sqrt{3}. \tag{2}$$

According to $AM \geq GM$ we have

$$1 = ab + bc + ca \geq 3\sqrt[3]{(abc)^2},$$

i.e.

$$\frac{1}{\sqrt[3]{abc}} \geq \sqrt{3}.$$

By $AM \geq GM$ and previous inequality we have

$$\frac{1}{a} + \frac{1}{b} + \frac{1}{c} \geq \frac{3}{\sqrt[3]{abc}} \geq 3\sqrt{3}. \tag{3}$$

Finally by (1), (2) and (3) we get

$$\left(a + \frac{1}{b}\right)^2 + \left(b + \frac{1}{c}\right)^2 + \left(c + \frac{1}{a}\right)^2 \geq \frac{(\sqrt{3} + 3\sqrt{3})^2}{3} = 16,$$

as required. Equality occurs iff $a = b = c = 1/\sqrt{3}$. ■

54 Let a, b, c be positive real numbers such that $abc \geq 1$. Prove the inequality

$$a + b + c \geq \frac{1+a}{1+b} + \frac{1+b}{1+c} + \frac{1+c}{1+a}.$$

Solution We have

$$\begin{aligned}
&a + b + c - \frac{1+a}{1+b} - \frac{1+b}{1+c} - \frac{1+c}{1+a} \\
&= \left(1 + a - \frac{1+a}{1+b}\right) + \left(1 + b - \frac{1+b}{1+c}\right) + \left(1 + c - \frac{1+c}{1+a}\right) - 3 \\
&= (1+a)\left(1 - \frac{1}{1+b}\right) + (1+b)\left(1 - \frac{1}{1+c}\right) + (1+c)\left(1 - \frac{1}{1+a}\right) - 3 \\
&= \frac{(1+a)b}{1+b} + \frac{(1+b)c}{1+c} + \frac{(1+c)a}{1+a} - 3 \\
&\geq 3\sqrt[3]{\frac{(1+a)b}{1+b} \cdot \frac{(1+b)c}{1+c} \cdot \frac{(1+c)a}{1+a}} - 3 \\
&= 3\sqrt[3]{abc} - 3 \geq 0 \quad (abc \geq 1).
\end{aligned}$$

Equality occurs iff $a = b = c = 1$. ■

55 Let $a, b \in \mathbb{R}^+$. Prove the inequality

$$\left(a^2 + b + \frac{3}{4}\right)\left(b^2 + a + \frac{3}{4}\right) \geq \left(2a + \frac{1}{2}\right)\left(2b + \frac{1}{2}\right).$$

Solution For any $x \in \mathbb{R}$ we have $x^2 + \frac{1}{4} \geq x$.

So it follows that

$$\left(a^2+b+\frac{3}{4}\right)\left(b^2+a+\frac{3}{4}\right) \geq \left(a+b+\frac{1}{2}\right)\left(a+b+\frac{1}{2}\right) = \left(a+b+\frac{1}{2}\right)^2$$
$$= \left(\frac{2a+2b+1}{2}\right)^2 = \left(\frac{2a+\frac{1}{2}+2b+\frac{1}{2}}{2}\right)^2$$
$$\overset{A\geq G}{\geq} \left(2a+\frac{1}{2}\right)\left(2b+\frac{1}{2}\right).$$

■

56 Let $a, b, c \in \mathbb{R}^+$ such that $abc = 1$. Prove the inequality

$$\frac{a}{a^2+2} + \frac{b}{b^2+2} + \frac{c}{c^2+2} \leq 1.$$

Solution By the well-known inequality $x^2 + 1 \geq 2x, \forall x \in \mathbb{R}$, we have

$$\frac{a}{a^2+2} + \frac{b}{b^2+2} + \frac{c}{c^2+2} = \frac{a}{a^2+1+1} + \frac{b}{b^2+1+1} + \frac{c}{c^2+1+1}$$
$$\leq \frac{a}{2a+1} + \frac{b}{2b+1} + \frac{c}{2c+1}$$
$$= \frac{1}{2+\frac{1}{a}} + \frac{1}{2+\frac{1}{b}} + \frac{1}{2+\frac{1}{c}} = A.$$

The inequality $A \leq 1$ is equivalent to

$$\left(2+\frac{1}{b}\right)\left(2+\frac{1}{c}\right) + \left(2+\frac{1}{a}\right)\left(2+\frac{1}{c}\right) + \left(2+\frac{1}{a}\right)\left(2+\frac{1}{b}\right)$$
$$\leq \left(2+\frac{1}{a}\right)\left(2+\frac{1}{b}\right)\left(2+\frac{1}{c}\right),$$

i.e.

$$4 \leq \frac{1}{ab} + \frac{1}{ac} + \frac{1}{bc} + \frac{1}{abc}, \quad \text{i.e.} \quad 3 \leq \frac{1}{ab} + \frac{1}{ac} + \frac{1}{bc},$$

which is true since $\frac{1}{ab} + \frac{1}{ac} + \frac{1}{bc} \geq 3\sqrt[3]{(\frac{1}{abc})^3} = 3$.

Equality occurs iff $a = b = c = 1$. ■

57 Let $x, y, z > 0$ be real numbers such that $x + y + z = xyz$. Prove the inequality

$$(x-1)(y-1)(z-1) \leq 6\sqrt{3} - 10.$$

Solution Since $x < xyz$ we have $yz > 1$ and analogously $xz > 1$ and $xy > 1$. At most one of x, y, z can be less than 1.

Let $x \leq 1, y \geq 1, z \geq 1$. Then we have $(x-1)(y-1)(z-1) \leq 0$, so the given inequality holds.

So it's enough to consider the case when $x \geq 1, y \geq 1, z \geq 1$.

Let $x-1=a, y-1=b, z-1=c$.

Then a, b, c are non-negative and since $x=a+1, y=b+1, z=c+1$ we obtain

$$a+1+b+1+c+1=(a+1)(b+1)(c+1), \quad \text{i.e.}$$

$$abc+ab+bc+ca=2. \tag{1}$$

Let $x=\sqrt[3]{abc}$, so we have

$$ab+bc+ca \geq 3\sqrt[3]{(abc)^2}=3x^2. \tag{2}$$

Combine (1) and (2) we have

$$x^3+3x^2 \leq abc+ab+bc+ca=2 \quad \Leftrightarrow \quad (x+1)(x^2+2x-2) \leq 0,$$

so we must have $x^2+2x-2 \leq 0$ and we easily deduce that $x \leq \sqrt{3}-1$, i.e. we get

$$x^3 \leq (\sqrt{3}-1)^3=6\sqrt{3}-10$$

and we are done. ■

58 Let $a, b, c \in (1,2)$ be real numbers. Prove the inequality

$$\frac{b\sqrt{a}}{4b\sqrt{c}-c\sqrt{a}}+\frac{c\sqrt{b}}{4c\sqrt{a}-a\sqrt{b}}+\frac{a\sqrt{c}}{4a\sqrt{b}-b\sqrt{c}} \geq 1.$$

Solution Since $a, b, c \in (1,2)$ we have

$$4b\sqrt{c}-c\sqrt{a}>4\sqrt{c}-2\sqrt{c}=2\sqrt{c}>0.$$

Analogously we get $4c\sqrt{a}-a\sqrt{b}>0$ and $4a\sqrt{b}-b\sqrt{c}>0$.

We'll prove that

$$\frac{b\sqrt{a}}{4b\sqrt{c}-c\sqrt{a}} \geq \frac{a}{a+b+c}. \tag{1}$$

Since $4b\sqrt{c}-c\sqrt{a}>0$ inequality (1) is

$$b(a+b+c) \geq \sqrt{a}(4b\sqrt{c}-c\sqrt{a})$$

$$\Leftrightarrow \quad (a+b)(b+c) \geq 4b\sqrt{ac},$$

which is clearly true ($AM \geq GM$).

Similarly we deduce that

$$\frac{c\sqrt{b}}{4c\sqrt{a}-a\sqrt{b}} \geq \frac{b}{a+b+c} \tag{2}$$

and

$$\frac{a\sqrt{c}}{4a\sqrt{b}-b\sqrt{c}} \geq \frac{c}{a+b+c}. \tag{3}$$

Adding (1), (2) and (3) we get the required result. ∎

59 Let $a, b, c \in \mathbb{R}^+$ such that $a+b+c=3$. Prove the inequality

$$\sqrt{a(b+c)}+\sqrt{b(c+a)}+\sqrt{c(a+b)} \geq 3\sqrt{2abc}.$$

Solution We have

$$\sqrt{ab+ac} \geq \frac{\sqrt{2}}{2}(\sqrt{ab}+\sqrt{ac}).$$

Analogously

$$\sqrt{bc+ba} \geq \frac{\sqrt{2}}{2}(\sqrt{bc}+\sqrt{ba}) \quad \text{and} \quad \sqrt{ca+cb} \geq \frac{\sqrt{2}}{2}(\sqrt{ca}+\sqrt{cb}).$$

So it suffices to show that

$$\sqrt{2}(\sqrt{ab}+\sqrt{ac}+\sqrt{bc}) \geq 3\sqrt{2abc},$$

i.e.

$$\sqrt{ab}+\sqrt{ac}+\sqrt{bc} \geq 3\sqrt{abc}. \tag{1}$$

By $AM \geq GM$ we have

$$\sqrt{ab}+\sqrt{ac}+\sqrt{bc} \geq 3\sqrt[3]{abc} \geq 3\sqrt{abc}$$

where the last inequality is true since

$$\sqrt[3]{abc} \leq \frac{a+b+c}{3} = 1, \quad \text{i.e.} \quad abc \leq 1.$$ ∎

60 Let a, b, c be positive real numbers such that $a+b+c=1$. Prove the inequality

$$\sqrt{a+bc}+\sqrt{b+ca}+\sqrt{c+ab} \leq 2.$$

Solution Since $a+b+c=1$ we have $a+bc=a(a+b+c)+bc=(a+b)(a+c)$ i.e.

$$\sqrt{a+bc}=\sqrt{(a+b)(a+c)} \leq \frac{(a+b)+(a+c)}{2} = \frac{2a+b+c}{2}.$$

Similarly we obtain

$$\sqrt{b+ca} \leq \frac{2b+c+a}{2} \quad \text{and} \quad \sqrt{c+ab} \leq \frac{2c+a+b}{2}.$$

After adding the last three inequalities, we obtain

$$\begin{aligned}\sqrt{a+bc}+\sqrt{b+ca}+\sqrt{c+ab} &\le \frac{2a+b+c}{2}+\frac{2b+c+a}{2}+\frac{2c+a+b}{2}\\ &= 2(a+b+c)=2.\end{aligned}$$

Equality occurs iff $a=b=c=1/3$. ■

61 Let a, b, c be positive real numbers such that $a+b+c+1=4abc$. Prove that

$$\frac{b^2+c^2}{a}+\frac{c^2+a^2}{b}+\frac{a^2+b^2}{c}\ge 2(ab+bc+ca).$$

Solution By the well-known inequalities:

$$x^2+y^2\ge 2xy \quad \text{and} \quad 3(x^2+y^2+z^2)\ge (x+y+z)^2,$$

we obtain

$$\begin{aligned}&\frac{b^2+c^2}{a}+\frac{c^2+a^2}{b}+\frac{a^2+b^2}{c}\\ &\quad\ge \frac{2bc}{a}+\frac{2ca}{b}+\frac{2ab}{c}=\frac{2((bc)^2+(ca)^2+(ab)^2)}{abc}\ge \frac{2(bc+ca+ab)^2}{3abc}. \quad (1)\end{aligned}$$

We have

$$(ab+bc+ca)^2\ge 3((ab)(bc)+(bc)(ca)+(ca)(ab))=3abc(a+b+c),$$

i.e.

$$ab+bc+ca\ge\sqrt{3abc(a+b+c)}. \quad (2)$$

Also

$$4abc=a+b+c+1\ge 4\sqrt[4]{abc}$$

i.e.

$$abc\ge 1. \quad (3)$$

Therefore

$$a+b+c=4abc-1=3abc+abc-1\overset{(3)}{\ge} 3abc. \quad (4)$$

By (1), (2) and (4) we obtain

$$\begin{aligned}&\frac{b^2+c^2}{a}+\frac{c^2+a^2}{b}+\frac{a^2+b^2}{c}\\ &\quad\ge \frac{2(bc+ca+ab)^2}{3abc}\ge\frac{2(ab+bc+ca)\sqrt{3abc(a+b+c)}}{3abc}\\ &\quad\ge \frac{2(ab+bc+ca)\sqrt{(3abc)^2}}{3abc}=2(ab+bc+ca).\end{aligned}$$

■

62 Let $a, b, c \in (-1, 1)$ be real numbers such that $ab + bc + ac = 1$. Prove the inequality

$$6\sqrt[3]{(1-a^2)(1-b^2)(1-c^2)} \le 1 + (a+b+c)^2.$$

Solution Since $a, b, c \in (-1, 1)$ we have $1 - a^2, 1 - b^2, 1 - c^2 > 0$.

By $AM \ge GM$ we get

$$\begin{aligned} 6\sqrt[3]{(1-a^2)(1-b^2)(1-c^2)} &= 2 \cdot 3\sqrt[3]{(1-a^2)(1-b^2)(1-c^2)} \\ &\le 2(1-a^2+1-b^2+1-c^2) \\ &= 2(3-(a^2+b^2+c^2)) \\ &= 6-2(a^2+b^2+c^2). \end{aligned}$$

We'll show that

$$6 - 2(a^2+b^2+c^2) \le 1 + (a+b+c)^2.$$

This inequality is equivalent to

$$6 - 2(a^2+b^2+c^2) \le 1 + a^2 + b^2 + c^2 + 2$$

i.e.

$$3 \le 3(a^2+b^2+c^2)$$

i.e.

$$a^2 + b^2 + c^2 \ge 1,$$

which is true since $a^2 + b^2 + c^2 \ge ab + bc + ac = 1$.

Equality holds iff $a = b = c = \pm\frac{1}{\sqrt{3}}$. ■

63 Let a, b, c, d be positive real numbers such that $a^2 + b^2 + c^2 + d^2 = 1$. Prove the inequality

$$\sqrt{1-a} + \sqrt{1-b} + \sqrt{1-c} + \sqrt{1-d} \ge \sqrt{a} + \sqrt{b} + \sqrt{c} + \sqrt{d}.$$

Solution First we'll show that

$$a + b + c + d \le 2. \tag{1}$$

We have

$$\frac{a+b+c+d}{4} \le \sqrt{\frac{a^2+b^2+c^2+d^2}{4}} = \frac{1}{2}$$

i.e.

$$a + b + c + d \le 2.$$

Furthermore

$$\sqrt{1-a}-\sqrt{a}=\frac{(\sqrt{1-a}-\sqrt{a})(\sqrt{1-a}+\sqrt{a})}{\sqrt{1-a}+\sqrt{a}}=\frac{1-2a}{\sqrt{1-a}+\sqrt{a}}. \tag{2}$$

By $AM \leq QM$ we have

$$\frac{\sqrt{1-a}+\sqrt{a}}{2}\leq\sqrt{\frac{1-a+a}{2}}=\frac{1}{\sqrt{2}}, \quad \text{i.e.} \quad \frac{1}{\sqrt{1-a}+\sqrt{a}}\geq\frac{1}{\sqrt{2}}. \tag{3}$$

Using (2) and (3) we deduce

$$\sqrt{1-a}-\sqrt{a}\geq\frac{1-2a}{\sqrt{2}}.$$

Similarly

$$\sqrt{1-b}-\sqrt{b}\geq\frac{1-2b}{\sqrt{2}}, \qquad \sqrt{1-c}-\sqrt{c}\geq\frac{1-2c}{\sqrt{2}} \quad \text{and}$$
$$\sqrt{1-d}-\sqrt{d}\geq\frac{1-2d}{\sqrt{2}}.$$

So it follows that

$$\sqrt{1-a}-\sqrt{a}+\sqrt{1-b}-\sqrt{b}+\sqrt{1-c}-\sqrt{c}+\sqrt{1-d}-\sqrt{d}$$
$$\geq\frac{4-2(a+b+c+d)}{\sqrt{2}}\overset{(1)}{\geq}0,$$

as required. ■

64 Let x, y, z be positive real numbers such that $xyz=1$. Prove the inequality

$$\frac{1}{(x+1)^2+y^2+1}+\frac{1}{(y+1)^2+z^2+1}+\frac{1}{(z+1)^2+x^2+1}\leq\frac{1}{2}.$$

Solution We have

$$\frac{1}{(x+1)^2+y^2+1}=\frac{1}{2+x^2+y^2+2x}\leq\frac{1}{2(1+x+xy)}.$$

Similarly

$$\frac{1}{(y+1)^2+z^2+1}\leq\frac{1}{2(1+y+yz)} \quad \text{and} \quad \frac{1}{(z+1)^2+x^2+1}\leq\frac{1}{2(1+z+zx)}.$$

So we have

$$\frac{1}{(x+1)^2+y^2+1}+\frac{1}{(y+1)^2+z^2+1}+\frac{1}{(z+1)^2+x^2+1}$$
$$\leq \frac{1}{2}\left(\frac{1}{1+x+xy}+\frac{1}{1+y+yz}+\frac{1}{1+z+zx}\right).$$

We'll show that

$$\frac{1}{1+x+xy}+\frac{1}{1+y+yz}+\frac{1}{1+z+zx}=1,$$

from which we'll deduce the required result.

We have

$$\begin{aligned}
&\frac{1}{1+x+xy}+\frac{1}{1+y+yz}+\frac{1}{1+z+zx}\\
&\quad=\frac{xyz}{xyz+x+xy}+\frac{1}{1+y+yz}+\frac{1}{1+z+zx}\\
&\quad=\frac{yz}{yz+1+y}+\frac{1}{1+y+yz}+\frac{y}{y+yz+1}\\
&\quad=\frac{1+y+yz}{1+y+yz}=1,
\end{aligned}$$

as required. ∎

65 Let $a, b, c \in \mathbb{R}^+$. Prove the inequality

$$\sqrt{\frac{a^3}{a^3+(b+c)^3}}+\sqrt{\frac{a^3}{a^3+(b+c)^3}}+\sqrt{\frac{a^3}{a^3+(b+c)^3}}\geq 1.$$

Solution We'll prove that for any $x, y, z \in \mathbb{R}^+$ we have

$$\sqrt{\frac{x^3}{x^3+(y+z)^3}}\geq \frac{x^2}{x^2+y^2+z^2}. \tag{1}$$

We have

$$\begin{aligned}
&\sqrt{\frac{x^3}{x^3+(y+z)^3}}\geq \frac{x^2}{x^2+y^2+z^2}\\
&\quad\Leftrightarrow\quad \frac{x^3}{x^3+(y+z)^3}\geq \frac{x^4}{(x^2+y^2+z^2)^2}\\
&\quad\Leftrightarrow\quad 2x^2(y^2+z^2)+(y^2+z^2)^2\geq x(y+z)^3.
\end{aligned} \tag{2}$$

By $AM \le QM$ we have

$$2(y^2+z^2) \ge (y+z)^2,$$

i.e.

$$8(y^2+z^2)^3 \ge (y+z)^6.$$

Using $AM \ge GM$ and the previous result we get

$$2x^2(y^2+z^2)+(y^2+z^2)^2 \ge 2\sqrt{2x^2(y^2+z^2)^3} \ge 2\sqrt{\frac{2x^2(y+z)^6}{8}} = x(y+z)^3,$$

so we prove (2), i.e. (1).

By (1) we have

$$\sqrt{\frac{a^3}{a^3+(b+c)^3}}+\sqrt{\frac{a^3}{a^3+(b+c)^3}}+\sqrt{\frac{a^3}{a^3+(b+c)^3}}$$
$$\ge \frac{a^2}{a^2+b^2+c^2}+\frac{b^2}{a^2+b^2+c^2}+\frac{c^2}{a^2+b^2+c^2}=1. \quad \blacksquare$$

66 Let $x, y, z \in \mathbb{R}^+$. Prove the inequality

$$(x+y+z)^2(xy+yz+zx)^2 \le 3(x^2+xy+y^2)(y^2+yz+z^2)(z^2+zx+x^2).$$

Solution We have

$$x^2+xy+y^2 = \frac{3}{4}(x+y)^2+\frac{1}{4}(x-y)^2 \ge \frac{3}{4}(x+y)^2,$$

similarly

$$y^2+yz+z^2 \ge \frac{3}{4}(y+z)^2 \quad \text{and} \quad z^2+zx+x^2 \ge \frac{3}{4}(z+x)^2.$$

Hence

$$3(x^2+xy+y^2)(y^2+yz+z^2)(z^2+zx+x^2) \ge 3\left(\frac{3}{4}\right)^3(x+y)^2(y+z)^2(z+x)^2$$
$$= \frac{81}{64}((x+y)(y+z)(z+x))^2.$$

We'll show that

$$\frac{81}{64}((x+y)(y+z)(z+x))^2 \ge (x+y+z)^2(xy+yz+zx)^2,$$

i.e.

$$\frac{9}{8}(x+y)(y+z)(z+x) \ge (x+y+z)(xy+yz+zx),$$

i.e.

$$9(x+y)(y+z)(z+x) \geq 8(x+y+z)(xy+yz+zx), \qquad (1)$$

from which we'll obtain the desired inequality.

Let's note that

$$(x+y)(y+z)(z+x) = (x+y+z)(xy+yz+zx) - xyz.$$

Now by (1) we get

$$9(x+y)(y+z)(z+x) \geq 8((x+y)(y+z)(z+x) + xyz),$$

i.e.

$$(x+y)(y+z)(z+x) \geq 8xyz,$$

which is clearly true since

$$x+y \geq 2\sqrt{xy}, \qquad y+z \geq 2\sqrt{yz}, \qquad z+x \geq 2\sqrt{zx}.$$

Equality occurs if and only if $x = y = z$. ∎

67 Let a, b, c be real numbers such that $a+b+c=3$. Prove the inequality

$$2(a^2b^2 + b^2c^2 + c^2a^2) + 3 \leq 3(a^2+b^2+c^2).$$

Solution Without loss of generality we may assume $a \geq b \geq c$.

Let's denote $u = \frac{a+b}{2}$ and $v = \frac{a-b}{2}$.

We easily obtain $a = u+v$ and $b = u-v$.

We have $ab = u^2 - v^2 \geq c^2$ which implies $2u^2 - 2c^2 - v^2 \geq 0$.

Now we have

$$\begin{aligned} a^2b^2 + b^2c^2 + c^2a^2 &= c^2(a^2+b^2) + a^2b^2 = c^2(2u^2+2v^2) + (u^2-v^2)^2 \\ &= -v^2(2u^2 - 2c^2 - v^2) + u^4 + 2c^2u^2 \leq u^4 + 2c^2u^2. \qquad (1) \end{aligned}$$

Also

$$a^2+b^2+c^2 = 2u^2 + 2v^2 + c^2 \geq 2u^2 + c^2. \qquad (2)$$

We'll show that

$$2(u^4 + 2c^2u^2) + 3 \leq 3(2u^2 + c^2). \qquad (3)$$

From $a+b+c=3$ we have $c = 3-2u$.

Now inequality (3) is equivalent to

$$\begin{aligned} &2u^4 + 4(3-2u)^2u^2 + 3 \leq 6u^2 + 3(3-2u)^2 \\ &\quad\Leftrightarrow\quad 3u^4 - 8u^3 + 3u^2 + 6u - 4 \leq 0 \quad\Leftrightarrow\quad (u-1)^2(3u^2 - 2u - 4) \leq 0. \end{aligned}$$

Since $2u \leq 3$ we easily deduce that $3u^2 - 2u - 4 \leq 0$. So inequality (3) holds.

Combining (1), (2) and (3) we obtain the required result.

Equality holds if and only if $a = b = c = 1$. ∎

68 Let a, b, c, d be positive real numbers. Prove the inequality

$$\frac{a-b}{b+c} + \frac{b-c}{c+d} + \frac{c-d}{d+a} + \frac{d-a}{a+b} \geq 0.$$

Solution Applying $AM \geq HM$ we have

$$\begin{aligned}
&\frac{a-b}{b+c} + \frac{b-c}{c+d} + \frac{c-d}{d+a} + \frac{d-a}{a+b} \\
&\quad = \frac{a+c}{b+c} + \frac{b+d}{c+d} + \frac{c+a}{d+a} + \frac{d+b}{a+b} - 4 \\
&\quad = (a+c)\left(\frac{1}{b+c} + \frac{1}{d+a}\right) + (b+d)\left(\frac{1}{c+d} + \frac{1}{a+b}\right) - 4 \\
&\quad \geq \frac{4(a+c)}{a+b+c+d} + \frac{4(b+d)}{a+b+c+d} - 4 = 0.
\end{aligned}$$

∎

69 Let $a, b, c \in \mathbb{R}^+$ such that $a+b+c = 1$. Prove the inequality

$$\frac{a}{(b+c)^2} + \frac{b}{(c+a)^2} + \frac{c}{(a+b)^2} \geq \frac{9}{4}.$$

Solution We'll use the following well known inequalities:

For any $a, b, c > 0$ we have $\frac{a}{b+c} + \frac{b}{c+a} + \frac{c}{a+b} \geq \frac{3}{2}$ (*Nesbit's*) and for any $x, y, z \geq 0$ we have

$$x^2 + y^2 + z^2 \geq \frac{(x+y+z)^2}{3}.$$

Now we obtain

$$\begin{aligned}
\frac{a}{(b+c)^2} + \frac{b}{(c+a)^2} + \frac{c}{(a+b)^2} &= \frac{a(a+b+c)}{(b+c)^2} + \frac{b(a+b+c)}{(c+a)^2} + \frac{c(a+b+c)}{(a+b)^2} \\
&= \left(\frac{a}{b+c}\right)^2 + \left(\frac{b}{c+a}\right)^2 + \left(\frac{c}{a+b}\right)^2 + \frac{a}{b+c} \\
&\quad + \frac{b}{c+a} + \frac{c}{a+b}.
\end{aligned}$$

Using previous well-known inequalities we have

$$\begin{aligned}
&\frac{a}{(b+c)^2}+\frac{b}{(c+a)^2}+\frac{c}{(a+b)^2}\\
&\quad\geq\frac{1}{3}\left(\frac{a}{b+c}+\frac{b}{c+a}+\frac{c}{a+b}\right)^2+\frac{a}{b+c}+\frac{b}{c+a}+\frac{c}{a+b}\\
&\quad\geq\frac{1}{3}\left(\frac{3}{2}\right)^2+\frac{3}{2}=\frac{9}{4}.
\end{aligned}$$

■

70 Let $a,b,c\in\mathbb{R}^+$ such that $abc=1$. Prove the inequality

$$\frac{a^3c}{(b+c)(c+a)}+\frac{b^3a}{(c+a)(a+b)}+\frac{c^3b}{(a+b)(b+c)}\geq\frac{3}{4}.$$

Solution Clearing denominators gives us

$$\begin{aligned}
&4(a^4c+b^4a+c^4a+a^3cb+b^3ac+c^3ba)\\
&\quad\geq 3(a^2b+a^2c+b^2a+b^2c+c^2a+c^2b+2abc),
\end{aligned}$$

i.e.

$$4(a^4c+b^4a+c^4a+a^2+b^2+c^2)\geq 3(a^2b+a^2c+b^2a+b^2c+c^2a+c^2b+2).$$

By $AM\geq GM$ and $abc=1$ we have

$$\begin{aligned}
&4(a^4c+b^4a+c^4a+a^2+b^2+c^2)\\
&\quad=(a^4c+a^2+b^4a)+(b^4a+b^2+c^4b)+(c^4b+c^2+a^4c)+(a^4c+a^2+c^2)\\
&\qquad+(b^4a+b^2+a^2)+(c^4b+c^2+b^2)+(a^4c+b^4a+c^4b)+(a^2+b^2+c^2)\\
&\quad\geq 3\sqrt[3]{a^6b^3}+3\sqrt[3]{b^6c^3}+3\sqrt[3]{c^6a^3}+3\sqrt[3]{a^6c^3}+3\sqrt[3]{c^6b^3}+3\sqrt[3]{a^5b^5c^5}\\
&\qquad+3\sqrt[3]{a^2b^2c^2}\\
&\quad=3(a^2b+a^2c+b^2a+b^2c+c^2a+c^2b+2),
\end{aligned}$$

and we are done. ■

71 Let $a,b,c>0$ be real numbers such that $abc=1$. Prove that

$$(a+b)(b+c)(c+a)\geq 4(a+b+c-1).$$

Solution Using the identity

$$(a+b)(b+c)(c+a)=(a+b+c)(ab+bc+ca)-1,$$

the given inequality becomes

$$ab+bc+ca+\frac{3}{a+b+c}\geq 4.$$

By $AM \geq GM$ we have

$$ab+bc+ca+\frac{3}{a+b+c}=\frac{3(ab+bc+ca)}{3}+\frac{3}{a+b+c}\geq 4\sqrt[4]{\frac{(ab+bc+ca)^3}{9(a+b+c)}}.$$

So it's enough to show that

$$(ab+bc+ca)^3\geq 9(a+b+c). \tag{1}$$

By $AM \geq GM$ and $abc=1$ we get

$$ab+bc+ca\geq 3\sqrt[3]{(abc)^2}=3. \tag{2}$$

Furthermore, since $(x+y+z)^2\geq 3(xy+yz+zx)$, we deduce

$$(ab+bc+ca)^2\geq 3((ab)(bc)+(bc)(ca)+(ca)(ab))=3(a+b+c). \tag{3}$$

By (2) and (3) we obtain $(ab+bc+ca)^3\geq 9(a+b+c)$, i.e. (1) is true. ■

72 Let a, b, c be positive real numbers such that $abc=1$. Prove the inequality

$$1+\frac{3}{a+b+c}\geq\frac{6}{ab+bc+ca}.$$

Solution Let $x=\frac{1}{a}, y=\frac{1}{b}, z=\frac{1}{c}$. Then clearly $xyz=1$.

The given inequality becomes

$$1+\frac{3}{xy+yz+zx}\geq\frac{6}{x+y+z}.$$

Using the well-known inequality $(x+y+z)^2\geq 3(xy+yz+zx)$ we deduce

$$1+\frac{3}{xy+yz+zx}\geq 1+\frac{9}{(x+y+z)^2}.$$

So it's enough to prove that

$$1+\frac{9}{(x+y+z)^2}\geq\frac{6}{x+y+z}.$$

The last inequality is equivalent to $(1-\frac{3}{x+y+z})^2\geq 0$, and clearly holds. ■

73 Let x, y, z be positive real numbers such that $x^2+y^2+z^2=xyz$. Prove the following inequalities:

1° $xyz \geq 27$
2° $xy + yz + zx \geq 27$
3° $x + y + z \geq 9$
4° $xy + yz + zx \geq 2(x + y + z) + 9$.

Solution

1° Using $AM \geq GM$ we get

$$xyz = x^2 + y^2 + z^2 \geq 3\sqrt[3]{(xyz)^2}, \quad \text{i.e.} \quad (xyz)^3 \geq 27(xyz)^2,$$

which implies

$$xyz \geq 27.$$

2° By $AM \geq GM$ we get $xy + yz + zx \geq 3\sqrt[3]{(xyz)^2} \geq 3\sqrt[3]{27^2} = 27$.
3° By $AM \geq GM$ and 1° we get $x + y + z \geq 3\sqrt[3]{xyz} \geq 3\sqrt[3]{27} = 9$.
4° Note that $x^2 + y^2 + z^2 = xyz$ implies $x^2 < xyz$, i.e. $x < yz$; analogously $y < zx$ and $z < xy$.

So $xy < yz \cdot zx$, i.e. $z^2 > 1$, from which we deduce that $z > 1$; analogously $x > 1$ and $y > 1$. So all three numbers are greater than 1.

Let's denote $a = x - 1, b = y - 1, c = z - 1$. Then $a, b, c > 0$ and clearly $x = a + 1, y = b + 1, z = c + 1$.

Now the initial condition $x^2 + y^2 + z^2 = xyz$ becomes

$$a^2 + b^2 + c^2 + a + b + c + 2 = abc + ab + bc + ca. \tag{1}$$

If we set $q = ab + bc + ca$ we have

$$a^2 + b^2 + c^2 \geq q, \qquad a + b + c \geq \sqrt{3q} \quad \text{and} \quad abc \leq \left(\frac{q}{3}\right)^{3/2} = \frac{(3q)^{3/2}}{27}.$$

Finally by (1) and the last three inequalities we obtain

$$q + \sqrt{3q} + 2 \leq a^2 + b^2 + c^2 + a + b + c + 2 = abc + ab + bc + ca \leq \frac{(3q)^{3/2}}{27} + q,$$

i.e.

$$\sqrt{3q} + 2 \leq \frac{(3q)^{3/2}}{27}. \tag{2}$$

Denote $\sqrt{3q} = A$. Then inequality (2) is equivalent to

$$A + 2 \leq \frac{A^3}{27} \quad \Leftrightarrow \quad (A - 6)(A + 3)^2 \geq 0,$$

from which we deduce that we must have $\sqrt{3q} = A \geq 6$, i.e. $q \geq 12$.

Hence

$$ab+bc+ca \ge 12 \quad \Leftrightarrow \quad (x-1)(y-1)+(y-1)(z-1)+(z-1)(x-1) \ge 12,$$

from which we obtain $xy+yz+zx \ge 2(x+y+z)+9$, and we are done. ■

74 Let a, b, c be real numbers such that $a^3+b^3+c^3-3abc=1$. Prove the inequality

$$a^2+b^2+c^2 \ge 1.$$

Solution Observe that

$$\begin{aligned}1 = a^3+b^3+c^3-3abc &= (a+b+c)(a^2+b^2+c^2-ab-bc-ca)\\ &= \frac{(a+b+c)}{2}((a-b)^2+(b-c)^2+(c-a)^2).\end{aligned}$$

Since $(a-b)^2+(b-c)^2+(c-a)^2 \ge 0$ we must have $a+b+c>0$.

According to

$$(a+b+c)(a^2+b^2+c^2-ab-bc-ca)=1$$

we deduce

$$(a+b+c)\left(a^2+b^2+c^2-\frac{(a+b+c)^2-a^2-b^2-c^2}{2}\right)=1$$

and easily find

$$a^2+b^2+c^2=\frac{1}{3}\left((a+b+c)^2+\frac{2}{a+b+c}\right).$$

Since $a+b+c>0$ we may use $AM \ge GM$ as follows

$$a^2+b^2+c^2=\frac{1}{3}\left((a+b+c)^2+\frac{1}{a+b+c}+\frac{1}{a+b+c}\right) \ge 1,$$

as required.

Equality occurs iff $a+b+c=1$. ■

75 Let $a, b, c, d \in \mathbb{R}^+$ such that $\frac{1}{1+a^4}+\frac{1}{1+b^4}+\frac{1}{1+c^4}+\frac{1}{1+d^4}=1$. Prove that

$$abcd \ge 3.$$

Solution We'll use the following substitutions

$$\frac{1}{1+a^4}=x, \qquad \frac{1}{1+b^4}=y, \qquad \frac{1}{1+c^4}=z, \qquad \frac{1}{1+d^4}=t.$$

Then we obtain $x+y+z+t=1$ and $a^4=\frac{1-x}{x}, b^4=\frac{1-y}{y}, c^4=\frac{1-z}{z}, d^4=\frac{1-t}{t}$.

We need to show that

$$a^4b^4c^4d^4 \geq 81,$$

i.e.

$$\frac{1-x}{x}\cdot\frac{1-y}{y}\cdot\frac{1-z}{z}\cdot\frac{1-t}{t}\geq 81.$$

Applying $AM \geq GM$ we have

$$\begin{aligned}\frac{1-x}{x}\cdot\frac{1-y}{y}\cdot\frac{1-z}{z}\cdot\frac{1-t}{t} &= \frac{y+z+t}{x}\cdot\frac{x+z+t}{y}\cdot\frac{x+y+t}{z}\cdot\frac{x+y+z}{t}\\ &\geq \frac{3\sqrt[3]{yzt}}{x}\cdot\frac{3\sqrt[3]{xzt}}{y}\cdot\frac{3\sqrt[3]{xyt}}{z}\cdot\frac{3\sqrt[3]{xyz}}{t}=81,\end{aligned}$$

as desired. ■

76 Let a, b, c be non-negative real numbers. Prove the inequality

$$\sqrt{\frac{ab+bc+ca}{3}}\leq\sqrt[3]{\frac{(a+b)(b+c)(c+a)}{8}}.$$

Solution The given inequality is homogenous, so we may assume that $ab+bc+ca=3$.

Then clearly

$$(a+b+c)^2\geq 3(ab+bc+ca)=9,\quad \text{i.e.}\quad a+b+c\geq 3$$

and

$$1=\frac{ab+bc+ca}{3}\geq\sqrt[3]{(abc)^2},\quad \text{i.e.}\quad abc\leq 1.$$

So we need to prove that

$$\sqrt[3]{\frac{(a+b)(b+c)(c+a)}{8}}\geq 1,\quad \text{i.e.}\quad (a+b)(b+c)(c+a)\geq 8.$$

We have

$$\begin{aligned}(a+b)(b+c)(c+a)&=(a+b+c)(ab+bc+ca)-abc\\ &=3(a+b+c)-abc\geq 9-1=8,\end{aligned}$$

and we are done.

Equality holds iff $a=b=c$. ■

77 Let a, b, c, d be positive real numbers such that $a+b+c+d=1$. Prove that

$$16(abc+bcd+cda+dab)\leq 1.$$

Solution We'll show that

$$16(abc+bcd+cda+dab)\le(a+b+c+d)^3.$$

Applying $AM \ge GM$ gives us

$$\begin{aligned}16(abc+bcd+cda+dab) &= 16ab(c+d)+16cd(a+b)\\ &\le 4(a+b)^2(c+d)+4(c+d)^2(a+b)\\ &= 4(c+d)(a+b)(a+b+c+d)\\ &\le (a+b+c+d)^3.\end{aligned}$$

It is obvious that equality holds if and only if $a=b=c=d=1/4$. ■

78 Let a,b,c,d,e be positive real numbers such that $a+b+c+d+e=5$. Prove the inequality

$$abc+bcd+cde+dea+eab\le 5.$$

Solution Without loss of generality, we may assume that $e=\min\{a,b,c,d,e\}$.
By $AM \ge GM$, we have

$$\begin{aligned}abc+bcd+cde+dea+eab &= e(a+c)(b+d)+bc(a+d-e)\\ &\le e\left(\frac{a+c+b+d}{2}\right)^2+\left(\frac{b+c+a+d-e}{3}\right)^3\\ &= \frac{e(5-e)^2}{4}+\frac{(5-2e)^3}{27}.\end{aligned}$$

So it suffices to prove that

$$\frac{e(5-e)^2}{4}+\frac{(5-2e)^3}{27}\le 5,$$

which can be rewrite as $(e-1)^2(e+8)\ge 0$, which is obviously true.
Equality holds if and only if $a=b=c=d=e=1$. ■

79 Let $a,b,c>0$ be real numbers. Prove the inequality

$$\frac{a}{b}+\frac{b}{c}+\frac{c}{a}\ge\frac{a+b}{b+c}+\frac{b+c}{c+a}+1.$$

Solution Let $x=\frac{a}{b}$, $y=\frac{c}{b}$.
Then we get

$$\frac{c}{a}=\frac{y}{x},\qquad \frac{a+b}{b+c}=\frac{x+1}{y+1},\qquad \frac{b+c}{c+a}=\frac{y+1}{x+y},$$

and the given inequality becomes

$$x^3y^2 + x^2 + x + y^3 + y^2 \geq x^2y + 2xy + 2xy^2. \qquad (1)$$

Using $AM \geq GM$ we obtain

$$\frac{x^3y^2 + x}{2} \geq x^2y, \qquad \frac{x^3y^2 + x + y^3 + y^2}{2} \geq 2xy^2 \quad \text{and} \quad x^2 + y^2 \geq 2xy.$$

After adding the last three inequalities we obtain inequality (1).

Equality occurs iff $x = y = 1$, i.e. iff $a = b = c$. ■

80 Let $a, b, c > 0$ be real numbers such that $abc = 1$. Prove the inequality

$$\left(1 + \frac{a}{b}\right)\left(1 + \frac{b}{c}\right)\left(1 + \frac{c}{a}\right) \geq 2(1 + a + b + c).$$

Solution The given inequality is equivalent to

$$\frac{a}{b} + \frac{a}{c} + \frac{b}{a} + \frac{b}{c} + \frac{c}{a} + \frac{c}{b} \geq 2(a + b + c).$$

Furthermore

$$\frac{a}{b} + \frac{a}{c} + 1 = \frac{a}{b} + \frac{a}{c} + abc \geq 3\sqrt[3]{a^3} = 3a.$$

Analogously

$$\frac{b}{a} + \frac{b}{c} + 1 \geq 3b \quad \text{and} \quad \frac{c}{a} + \frac{c}{b} + 1 \geq 3c.$$

So

$$\frac{a}{b} + \frac{a}{c} + \frac{b}{a} + \frac{b}{c} + \frac{c}{a} + \frac{c}{b} + 3 \geq 3(a + b + c). \qquad (1)$$

It is enough to show that $a + b + c \geq 3$.

We have $a + b + c \geq 3\sqrt[3]{abc} = 3$, and finally from (1) we obtain

$$\frac{a}{b} + \frac{a}{c} + \frac{b}{a} + \frac{b}{c} + \frac{c}{a} + \frac{c}{b} + 3 \geq 2(a + b + c) + (a + b + c) \geq 2(a + b + c) + 3,$$

i.e.

$$\frac{a}{b} + \frac{a}{c} + \frac{b}{a} + \frac{b}{c} + \frac{c}{a} + \frac{c}{b} \geq 2(a + b + c).$$

Equality holds iff $a = b = c = 1$. ■

81 Let a, b, c be positive real numbers such that $a + b + c \geq \frac{1}{a} + \frac{1}{b} + \frac{1}{c}$. Prove the inequality

$$a + b + c \geq \frac{3}{a + b + c} + \frac{2}{abc}.$$

Solution By $AM \geq HM$ we get

$$a+b+c \geq \frac{1}{a}+\frac{1}{b}+\frac{1}{c} \geq \frac{9}{a+b+c},$$

i.e.

$$\frac{a+b+c}{3} \geq \frac{3}{a+b+c}. \tag{1}$$

We will prove that

$$\frac{2(a+b+c)}{3} \geq \frac{2}{abc}, \tag{2}$$

i.e.

$$a+b+c \geq \frac{3}{abc}.$$

Using the well-known inequality $(xy+yz+zx)^2 \geq 3(xy+yz+zx)$ we obtain

$$(a+b+c)^2 \geq \left(\frac{1}{a}+\frac{1}{b}+\frac{1}{c}\right)^2 \geq 3\left(\frac{1}{ab}+\frac{1}{bc}+\frac{1}{ca}\right) = 3\frac{a+b+c}{abc},$$

i.e.

$$a+b+c \geq \frac{3}{abc}.$$

After adding (1) and (2) we get the required inequality. ■

82 Let a, b, c, d be positive real numbers such that $abcd = 1$. Prove the inequality

$$\frac{1+ab}{1+a}+\frac{1+bc}{1+b}+\frac{1+cd}{1+c}+\frac{1+da}{1+d} \geq 4.$$

Solution Clearly $cd = \frac{1}{ab}$ and $ad = \frac{1}{bc}$.

Now we have

$$\begin{aligned} A &= \frac{1+ab}{1+a}+\frac{1+bc}{1+b}+\frac{1+cd}{1+c}+\frac{1+da}{1+d} \\ &= \frac{1+ab}{1+a}+\frac{1+bc}{1+b}+\frac{1+1/ab}{1+c}+\frac{1+1/bc}{1+d} \\ &= (1+ab)\left(\frac{1}{1+a}+\frac{1}{ab+abc}\right)+(1+bc)\left(\frac{1}{1+b}+\frac{1}{bc+bcd}\right). \end{aligned} \tag{1}$$

By $AM \geq HM$ and (1) we deduce

$$\begin{aligned}
A &= (1+ab)\left(\frac{1}{1+a}+\frac{1}{ab+abc}\right)+(1+bc)\left(\frac{1}{1+b}+\frac{1}{bc+bcd}\right)\\
&\geq \frac{4(1+ab)}{1+a+ab+abc}+\frac{4(1+bc)}{1+b+bc+bcd}\\
&=4\left(\frac{1+ab}{1+a+ab+abc}+\frac{1+bc}{1+b+bc+bcd}\right)\\
&=4\left(\frac{1+ab}{1+a+ab+abc}+\frac{a+abc}{a+ab+abc+abcd}\right). \qquad (2)
\end{aligned}$$

Since $abcd=1$ from (2) we obtain $A \geq 4$, as required. ■

83 Let $a,b,c \in \mathbb{R}^+$. Prove the inequality

$$\frac{1}{b(a+b)}+\frac{1}{c(b+c)}+\frac{1}{a(c+a)} \geq \frac{27}{2(a+b+c)^2}.$$

Solution Applying $AM \geq GM$ we have

$$\frac{1}{b(a+b)}+\frac{1}{c(b+c)}+\frac{1}{a(c+a)} \geq 3\sqrt[3]{\frac{1}{abc(a+b)(b+c)(c+a)}} \qquad (1)$$

and

$$a+b+c \geq 3\sqrt[3]{abc}, \quad \text{i.e.} \quad \frac{1}{\sqrt[3]{abc}} \geq \frac{3}{a+b+c}. \qquad (2)$$

Furthermore

$$a+b+c=\frac{1}{2}((a+b)+(b+c)+(c+a)) \geq \frac{3}{2}\sqrt[3]{(a+b)(b+c)(c+a)},$$

i.e.

$$\frac{1}{\sqrt[3]{(a+b)(b+c)(c+a)}} \geq \frac{3}{2(a+b+c)}. \qquad (3)$$

Combining (2), (3) and (1) we get

$$\begin{aligned}
\frac{1}{b(a+b)}+\frac{1}{c(b+c)}+\frac{1}{a(c+a)} &\geq 3\cdot\sqrt[3]{\frac{1}{abc(a+b)(b+c)(c+a)}}\\
&\geq 3\cdot\frac{3}{a+b+c}\cdot\frac{3}{2(a+b+c)}=\frac{27}{2(a+b+c)^2}.
\end{aligned}$$

■

84 Let a, b, c be positive real numbers such that $a+b+c=3$. Prove the inequality

$$\frac{a^2}{b^2-2b+3}+\frac{b^2}{c^2-2c+3}+\frac{c^2}{a^2-2a+3}\geq\frac{3}{2}.$$

Solution Since $a+b+c=3$ by $QM \geq AM$ we have

$$(b-1)^2=((1-a)+(1-c))^2\leq 2((a-1)^2+(c-1)^2).$$

Hence

$$(b-1)^2\leq\frac{2}{3}((a-1)^2+(b-1)^2+(c-1)^2)=\frac{2}{3}(a^2+b^2+c^2-3).$$

So we have

$$b^2-2b+3=(b-1)^2+2\leq\frac{2}{3}(a^2+b^2+c^2-3)+2=\frac{2}{3}(a^2+b^2+c^2),$$

which implies

$$\frac{a^2}{b^2-2b+3}\geq\frac{a^2}{\frac{2}{3}(a^2+b^2+c^2)}=\frac{3a^2}{2(a^2+b^2+c^2)}.$$

Similarly we get

$$\frac{b^2}{c^2-2c+3}\geq\frac{3b^2}{2(a^2+b^2+c^2)}\quad\text{and}\quad\frac{c^2}{a^2-2a+3}\geq\frac{3c^2}{2(a^2+b^2+c^2)}.$$

By adding the last three inequalities we obtain the required inequality. ■

85 Let a, b, c be positive real numbers such that $ab+bc+ca=3$. Prove the inequality

$$\frac{1}{1+a^2(b+c)}+\frac{1}{1+b^2(c+a)}+\frac{1}{1+c^2(a+b)}\leq\frac{1}{abc}.$$

Solution Observe that

$$\frac{1}{1+a^2(b+c)}=\frac{1}{1+a(ab+ac)}=\frac{1}{1+a(3-bc)}=\frac{1}{3a+1-abc}.$$

By $AM \geq GM$ we get

$$1=\frac{ab+bc+ca}{3}\geq\sqrt[3]{(abc)^2}.$$

Thus

$$abc\leq 1.$$

Therefore

$$\frac{1}{1+a^2(b+c)} = \frac{1}{3a+1-abc} \leq \frac{1}{3a}.$$

Similarly,

$$\frac{1}{1+b^2(c+a)} \leq \frac{1}{3b} \quad \text{and} \quad \frac{1}{1+c^2(a+b)} \leq \frac{1}{3c}.$$

Now we have

$$\frac{1}{1+a^2(b+c)} + \frac{1}{1+b^2(c+a)} + \frac{1}{1+c^2(a+b)}$$
$$\leq \frac{1}{3}\left(\frac{1}{a}+\frac{1}{b}+\frac{1}{c}\right) = \frac{1}{3}\left(\frac{ab+bc+ca}{abc}\right) = \frac{1}{abc}.$$

Equality holds iff $a=b=c=1$. ■

86 Let a, b, c be positive real numbers such that $a+b+c=1$. Prove the inequality

$$a\sqrt[3]{1+b-c} + b\sqrt[3]{1+c-a} + a\sqrt[3]{1+a-b} \leq 1.$$

Solution Note that $1+b-c = a+b+c+b-c = a+2b \geq 0$.

Now by $GM \leq AM$ we have

$$a\sqrt[3]{1+b-c} \leq a\frac{1+b-c+1+1}{3} = a + \frac{a(b-c)}{3}.$$

Similarly

$$b\sqrt[3]{1+c-a} \leq b + \frac{b(c-a)}{3} \quad \text{and} \quad c\sqrt[3]{1+a-b} \leq c + \frac{c(a-b)}{3}.$$

Adding these three inequalities we get

$$a\sqrt[3]{1+b-c} + b\sqrt[3]{1+c-a} + c\sqrt[3]{1+a-b} \leq a+b+c = 1.$$

Equality occurs iff $a=b=c=1/3$. ■

87 Let $a, b, c \in \mathbb{R}^+$ such that $a+b+c=1$. Prove the inequality

$$\frac{1-2ab}{c} + \frac{1-2bc}{a} + \frac{1-2ca}{b} \geq 7.$$

Solution We have

$$\frac{1-2ab}{c} + \frac{1-2bc}{a} + \frac{1-2ca}{b}$$
$$= \frac{(a+b+c)^2-2ab}{c} + \frac{(a+b+c)^2-2bc}{a} + \frac{(a+b+c)^2-2ca}{b}$$

$$= \frac{a^2+b^2+c^2+2bc+2ac}{c} + \frac{a^2+b^2+c^2+2ac+2ab}{a}$$
$$+ \frac{a^2+b^2+c^2+2ab+2bc}{b}$$
$$= (a^2+b^2+c^2)\left(\frac{1}{a}+\frac{1}{b}+\frac{1}{c}\right) + 4(a+b+c)$$
$$= (a^2+b^2+c^2)\left(\frac{1}{a}+\frac{1}{b}+\frac{1}{c}\right) + 4. \qquad (1)$$

By $QM \geq AM$ we get

$$a^2+b^2+c^2 \geq \frac{(a+b+c)^2}{3} = \frac{1}{3} \quad \text{and} \quad \frac{1}{a}+\frac{1}{b}+\frac{1}{c} \geq \frac{9}{a+b+c} = 9.$$

Finally, from previous inequalities and (1) we obtain

$$\frac{1-2ab}{c} + \frac{1-2bc}{a} + \frac{1-2ca}{b} = (a^2+b^2+c^2)\left(\frac{1}{a}+\frac{1}{b}+\frac{1}{c}\right) + 4 \geq \frac{9}{3} + 4 = 7.$$

■

88 Let a, b, c be non-negative real numbers such that $a^2+b^2+c^2=1$. Prove the inequality

$$\frac{1-ab}{7-3ac} + \frac{1-ab}{7-3ac} + \frac{1-ab}{7-3ac} \geq \frac{1}{3}.$$

Solution First we'll show that

$$\frac{1}{7-3ab} + \frac{1}{7-3bc} + \frac{1}{7-3ca} \leq \frac{1}{2}. \qquad (1)$$

By $AM \geq HM$ we have

$$\frac{1}{7-3ab} = \frac{1}{3(1-ab)+2+2} \leq \frac{1}{9}\left(\frac{1}{3(1-ab)} + 1\right).$$

Similarly we get

$$\frac{1}{7-3bc} \leq \frac{1}{9}\left(\frac{1}{3(1-bc)} + 1\right) \quad \text{and} \quad \frac{1}{7-3ca} \leq \frac{1}{9}\left(\frac{1}{3(1-ca)} + 1\right).$$

So it follows that

$$\frac{1}{7-3ab} + \frac{1}{7-3bc} + \frac{1}{7-3ca} \leq \frac{1}{27}\left(\frac{1}{1-ab} + \frac{1}{1-bc} + \frac{1}{1-ca}\right) + \frac{1}{3}. \qquad (2)$$

Recalling the well-known *Vasile Cirtoaje's* inequality

$$\frac{1}{1-ab} + \frac{1}{1-bc} + \frac{1}{1-ca} \leq \frac{9}{2},$$

by (2) we obtain

$$\frac{1}{7-3ab}+\frac{1}{7-3bc}+\frac{1}{7-3ca}\leq\frac{1}{2}.$$

Since $a^2+b^2+c^2=1$ we have $a, b, c \leq 1$ and then clearly

$$7-3ab, 7-3ab, 7-3ab>0,$$

so by $AM \geq GM$ we have

$$\frac{7-3ab}{7-3ac}+\frac{7-3ab}{7-3ac}+\frac{7-3ab}{7-3ac}\geq 3. \tag{3}$$

Finally by (2) and (3) we have

$$\begin{aligned}&\frac{3-3ab}{7-3ac}+\frac{3-3ab}{7-3ac}+\frac{3-3ab}{7-3ac}\\ &\quad=\left(\frac{7-3ab}{7-3ac}+\frac{7-3ab}{7-3ac}+\frac{7-3ab}{7-3ac}\right)-4\left(\frac{1}{7-3ab}+\frac{1}{7-3bc}+\frac{1}{7-3ca}\right)\\ &\quad\geq 3-2=1,\end{aligned}$$

i.e.

$$\frac{1-ab}{7-3ac}+\frac{1-ab}{7-3ac}+\frac{1-ab}{7-3ac}\geq\frac{1}{3},$$

as required. ■

89 Let $x, y, z \in \mathbb{R}^+$ such that $x+y+z=1$. Prove the inequality

$$\frac{xy}{\sqrt{\frac{1}{3}+z^2}}+\frac{zx}{\sqrt{\frac{1}{3}+y^2}}+\frac{yz}{\sqrt{\frac{1}{3}+x^2}}\leq\frac{1}{2}.$$

Solution We have

$$\begin{aligned}\frac{1}{3}+x^2&=\frac{1}{3}(x+y+z)^2+x^2=\frac{x^2+y^2+z^2+2(xy+yz+zx)}{3}+x^2\\ &\geq\frac{xy+yz+zx+2(xy+yz+zx)}{3}+x^2=xy+yz+zx+x^2\\ &=(x+y)(x+z).\end{aligned}$$

Now we get

$$\frac{yz}{\sqrt{\frac{1}{3}+x^2}}\leq\frac{yz}{\sqrt{(x+y)(x+z)}}\overset{HM\leq GM}{\leq}\frac{yz}{2}\left(\frac{1}{x+y}+\frac{1}{x+z}\right). \tag{1}$$

Analogously

$$\frac{xy}{\sqrt{\frac{1}{3}+z^2}} \leq \frac{xy}{2}\left(\frac{1}{z+x}+\frac{1}{z+y}\right) \tag{2}$$

and

$$\frac{zx}{\sqrt{\frac{1}{3}+y^2}} \leq \frac{zx}{2}\left(\frac{1}{y+z}+\frac{1}{y+x}\right). \tag{3}$$

Adding (1), (2) and (3) we obtain

$$\begin{aligned} L &\leq \frac{xy}{2}\left(\frac{1}{z+x}+\frac{1}{z+y}\right)+\frac{zx}{2}\left(\frac{1}{y+z}+\frac{1}{y+x}\right)+\frac{yz}{2}\left(\frac{1}{x+y}+\frac{1}{x+z}\right) \\ &= \frac{1}{2}\left(\frac{xy+yz}{x+z}+\frac{xy+zx}{y+z}+\frac{yz+zx}{y+x}\right)=\frac{x+y+z}{2}=\frac{1}{2}. \end{aligned}$$

■

90 Let a, b, c be positive real numbers such that $a+b+c=1$. Prove the inequality

$$\frac{a-bc}{a+bc}+\frac{b-ca}{b+ca}+\frac{c-ab}{c+ab} \leq \frac{3}{2}.$$

Solution Note that

$$1-\frac{a-bc}{a+bc}=\frac{2bc}{1-b-c+bc}=\frac{2bc}{(1-b)(1-c)}=\frac{2bc}{(c+a)(a+b)},$$

i.e.

$$\frac{a-bc}{a+bc}=1-\frac{2bc}{(c+a)(a+b)}.$$

Similarly we get

$$\frac{b-ca}{b+ca}=1-\frac{2ca}{(c+b)(b+a)} \quad \text{and} \quad \frac{c-ab}{c+ab}=1-\frac{2ab}{(b+c)(c+a)}.$$

Now the given inequality becomes

$$1-\frac{2bc}{(c+a)(a+b)}+1-\frac{2ca}{(c+b)(b+a)}1-\frac{2ab}{(b+c)(c+a)} \leq \frac{3}{2}$$

or

$$\frac{2bc}{(c+a)(a+b)}+\frac{2ca}{(c+b)(b+a)}+\frac{2ab}{(b+c)(c+a)} \geq \frac{3}{2}.$$

After expanding we get the equivalent form as follows

$$4(bc(b+c)+ca(c+a)+ab(a+b)) \geq 3(a+b)(b+c)(c+a),$$

i.e.

$$ab+bc+ac \geq 9abc, \quad \text{i.e.} \quad \frac{1}{a}+\frac{1}{b}+\frac{1}{c} \geq 9,$$

which is true since

$$\frac{1}{a}+\frac{1}{b}+\frac{1}{c} \geq \frac{9}{a+b+c} = 9 \quad (AM \geq HM).$$

Equality occurs iff $a=b=c=1/3$. ■

91 Let a, b, c be positive real numbers such that $abc=1$. Prove the inequality

$$\sqrt{\frac{a+b}{a+1}}+\sqrt{\frac{b+c}{c+1}}+\sqrt{\frac{c+a}{a+1}} \geq 3.$$

Solution By $AM \geq GM$ we get

$$\sqrt{\frac{a+b}{a+1}}+\sqrt{\frac{b+c}{c+1}}+\sqrt{\frac{c+a}{a+1}} \geq 3\sqrt[6]{\frac{(a+b)(b+c)(c+a)}{(a+1)(b+1)(c+1)}}.$$

So it suffices to prove that

$$\frac{(a+b)(b+c)(c+a)}{(a+1)(b+1)(c+1)} \geq 1,$$

i.e.

$$(a+b)(b+c)(c+a) \geq (a+1)(b+1)(c+1).$$

Since $abc=1$ we need to prove that

$$ab(a+b)+bc(b+c)+ca(c+a) \geq a+b+c+ab+bc+ca. \tag{1}$$

According to $AM \geq GM$ we have

$$\begin{aligned}&2(ab(a+b)+bc(b+c)+ca(c+a))+(ab+bc+ca)\\&\quad=\sum_{cyc}(a^2b+a^2b+a^2c+a^2c+bc) \geq 5\sum_{cyc} a = 5(a+b+c)\end{aligned} \tag{2}$$

and

$$\begin{aligned}&2(ab(a+b)+bc(b+c)+ca(c+a))+(a+b+c)\\&\quad=\sum_{cyc}(a^2b+a^2b+b^2a+b^2a+c) \geq 5\sum_{cyc} ab = 5(ab+bc+ca).\end{aligned} \tag{3}$$

After adding (2) and (3) we obtain

$$\begin{aligned}&4(ab(a+b)+bc(b+c)+ca(c+a))+(ab+bc+ca)+(a+b+c)\\&\quad\geq 5(ab+bc+ca)+5(a+b+c).\end{aligned}$$

Hence we have proved (1), as required. Equality holds iff $a=b=c=1$. ■

92 Let $x, y, z \geq 0$ be real numbers such that $xy+yz+zx=1$. Prove the inequality

$$\frac{x}{1+x^2}+\frac{y}{1+y^2}+\frac{z}{1+z^2}\leq\frac{3\sqrt{3}}{4}.$$

Solution We have

$$1+x^2=xy+yz+zx+x^2=(x+y)(x+z).$$

Analogously we obtain

$$1+y^2=(y+x)(y+z)\quad\text{and}\quad 1+z^2=(z+x)(z+y).$$

Therefore

$$\begin{aligned}\frac{x}{1+x^2}+\frac{y}{1+y^2}+\frac{z}{1+z^2}&=\frac{x}{(x+y)(x+z)}+\frac{y}{(y+x)(y+z)}+\frac{z}{(z+x)(z+y)}\\&=\frac{x(y+z)+y(x+z)+z(x+y)}{(x+y)(y+z)(z+x)}\\&=\frac{2}{(x+y)(y+z)(z+x)}.\end{aligned}\tag{1}$$

It is easy to show that

$$(x+y)(y+z)(z+x)=x+y+z-xyz.\tag{2}$$

Due to the well-known inequality $(x+y+z)^2\geq 3(xy+yz+zx)$ we obtain

$$(x+y+z)^2\geq 3(xy+yz+zx)=3,\quad\text{i.e.}\quad x+y+z\geq\sqrt{3}.\tag{3}$$

Applying $AM \geq GM$ it follows that

$$xy+yz+zx\geq 3\sqrt[3]{(xyz)^2},$$

i.e.

$$\frac{1}{27}\geq(xyz)^2\quad\Leftrightarrow\quad\frac{1}{3\sqrt{3}}\geq xyz.\tag{4}$$

Using (3) and (4) we obtain

$$x + y + z - xyz \geq \sqrt{3} - \frac{1}{3\sqrt{3}} = \frac{8}{3\sqrt{3}}. \tag{5}$$

Finally using (1), (2) and (5) we get

$$\frac{x}{1+x^2} + \frac{y}{1+y^2} + \frac{z}{1+z^2} = \frac{2}{(x+y)(y+z)(z+x)} = \frac{2}{x+y+z-xyz} \leq \frac{3\sqrt{3}}{4}.$$

Equality occurs iff $x = y = z = \frac{1}{\sqrt{3}}$. ∎

93 Let a, b, c be non-negative real numbers such that $ab + bc + ca = 1$. Prove the inequality

$$\frac{1}{1+a} + \frac{1}{1+b} + \frac{1}{1+c} \geq \frac{3\sqrt{3}}{\sqrt{3}+1}.$$

Solution After some algebraic calculations we get

$$\begin{aligned} &\frac{4+2(a+b+c)}{2+a+b+c+abc} \geq \frac{3\sqrt{3}}{\sqrt{3}+1} \\ &\Leftrightarrow \quad 2(2+a+b+c)(\sqrt{3}+1) \geq 3\sqrt{3}(2+a+b+c+abc) \\ &\Leftrightarrow \quad 2+(a+b+c) \geq \frac{3\sqrt{3}}{2-\sqrt{3}}abc, \end{aligned}$$

i.e.

$$2 + (a+b+c) \geq 3\sqrt{3}(2+\sqrt{3})abc. \tag{1}$$

Applying $AM \geq GM$ we obtain

$$1 = ab + bc + ca \geq 3\sqrt[3]{(abc)^2},$$

i.e.

$$\frac{1}{3\sqrt{3}} \geq abc. \tag{2}$$

Also we have

$$a + b + c \geq \sqrt{3}. \tag{3}$$

Using (2) and (3) we get

$$2 + (a+b+c) \geq 2 + \sqrt{3} \geq 3\sqrt{3}(2+\sqrt{3})abc,$$

i.e. we have shown inequality (1), as desired.

Equality holds if and only if $a = b = c = 1/\sqrt{3}$. ∎

94 Let a, b, c be non-negative real numbers such that $ab + bc + ca = 1$. Prove the inequality

$$\frac{a^2}{1+a} + \frac{b^2}{1+b} + \frac{c^2}{1+c} \geq \frac{\sqrt{3}}{\sqrt{3}+1}.$$

Solution Using $\frac{x^2}{1+x} = x - 1 + \frac{1}{1+x}$ we have

$$\frac{a^2}{1+a} + \frac{b^2}{1+b} + \frac{c^2}{1+c} = a + b + c - 3 + \frac{1}{1+a} + \frac{1}{1+b} + \frac{1}{1+c}.$$

Now using the result from Problem 89 and the inequality $a + b + c \geq \sqrt{3}$ we obtain

$$\begin{aligned}\frac{a^2}{1+a} + \frac{b^2}{1+b} + \frac{c^2}{1+c} &= a + b + c - 3 + \frac{1}{1+a} + \frac{1}{1+b} + \frac{1}{1+c}\\ &\geq \sqrt{3} - 3 + \frac{3\sqrt{3}}{\sqrt{3}+1} = \frac{\sqrt{3}}{\sqrt{3}+1}.\end{aligned}$$

Equality occurs if and only if $a = b = c = 1/\sqrt{3}$. ■

95 Let $a, b, c \in \mathbb{R}^+$ such that $(a + b)(b + c)(c + a) = 8$. Prove the inequality

$$\frac{a+b+c}{3} \geq \sqrt[27]{\frac{a^3+b^3+c^3}{3}}.$$

Solution We have

$$\begin{aligned}(a+b+c)^3 &= a^3 + b^3 + c^3 + 3(a+b)(b+c)(c+a)\\ &= a^3 + b^3 + c^3 + 24 = a^3 + b^3 + c^3 + \underbrace{3 + \cdots + 3}_{8}\\ &\geq 9\sqrt[9]{(a^3+b^3+c^3)3^8}\end{aligned}$$

$$\Leftrightarrow \left(\frac{a+b+c}{3}\right)^3 \geq \sqrt[9]{\frac{a^3+b^3+c^3}{3}}, \quad \text{i.e.} \quad \frac{a+b+c}{3} \geq \sqrt[27]{\frac{a^3+b^3+c^3}{3}}.$$

■

96 Find the maximum value of $\frac{x^4-x^2}{x^6+2x^3-1}$, where $x \in \mathbb{R}, x > 1$.

Solution We have

$$\frac{x^4 - x^2}{x^6 + 2x^3 - 1} = \frac{x - \frac{1}{x}}{x^3 + 2 - \frac{1}{x^3}} = \frac{x - \frac{1}{x}}{(x - \frac{1}{x})^3 + 2 + 3(x - \frac{1}{x})}. \quad (1)$$

We'll show that

$$\left(x-\frac{1}{x}\right)^3+2\geq 3\left(x-\frac{1}{x}\right).$$

Since $x>1$ we have $1>\frac{1}{x}$, i.e. $x-\frac{1}{x}>0$.

From $AM\geq GM$ we get

$$\left(x-\frac{1}{x}\right)^3+2=\left(x-\frac{1}{x}\right)^3+1+1\geq 3\sqrt[3]{\left(x-\frac{1}{x}\right)^3\cdot 1\cdot 1}=3\left(x-\frac{1}{x}\right).$$

Now in (1) we obtain

$$\frac{x^4-x^2}{x^6+2x^3-1}=\frac{x-\frac{1}{x}}{x^3+2-\frac{1}{x^3}}=\frac{x-\frac{1}{x}}{(x-\frac{1}{x})^3+2+3(x-\frac{1}{x})}\leq\frac{x-\frac{1}{x}}{3(x-\frac{1}{x})+3(x-\frac{1}{x})}$$
$$=\frac{1}{6}.$$

■

97 Let a,b,c be positive real numbers. Prove the inequality

$$\frac{a+\sqrt{ab}+\sqrt[3]{abc}}{3}\leq\sqrt[3]{a\cdot\frac{a+b}{2}\cdot\frac{a+b+c}{3}}.$$

Solution Applying $AM\geq GM$ we get

$$\sqrt[3]{ab\cdot\frac{a+b}{2}}\geq\sqrt[3]{ab\cdot\sqrt{ab}}=\sqrt{ab}.$$

So

$$a+\sqrt{ab}+\sqrt[3]{abc}\leq a+\sqrt[3]{ab\cdot\frac{a+b}{2}}+\sqrt[3]{abc}.$$

Now, it is enough to show that

$$a+\sqrt[3]{ab\cdot\frac{a+b}{2}}+\sqrt[3]{abc}\leq 3\sqrt[3]{a\cdot\frac{a+b}{2}\cdot\frac{a+b+c}{3}}.$$

Another application of $AM\geq GM$ gives us

$$\sqrt[3]{1\cdot\frac{2a}{a+b}\cdot\frac{3a}{a+b+c}}\leq\frac{1+\frac{2a}{a+b}+\frac{3a}{a+b+c}}{3},\qquad\sqrt[3]{1\cdot 1\cdot\frac{3b}{a+b+c}}\leq\frac{2+\frac{3b}{a+b+c}}{3}$$

and

$$\sqrt[3]{1\cdot\frac{2b}{a+b}\cdot\frac{3c}{a+b+c}}\leq\frac{1+\frac{2b}{a+b}+\frac{3c}{a+b+c}}{3}.$$

Adding, we obtain

$$\sqrt[3]{\frac{2a}{a+b}\cdot\frac{3a}{a+b+c}}+\sqrt[3]{\frac{3b}{a+b+c}}+\sqrt[3]{\frac{2b}{a+b}\cdot\frac{3c}{a+b+c}}\leq 3,$$

i.e.

$$\sqrt[3]{\frac{1}{a}\cdot\frac{2}{a+b}\cdot\frac{3}{a+b+c}}\left(a+\sqrt[3]{ab\cdot\frac{a+b}{2}}+\sqrt[3]{abc}\right)\leq 3,$$

i.e.

$$a+\sqrt[3]{ab\cdot\frac{a+b}{2}}+\sqrt[3]{abc}\leq 3\sqrt[3]{a\cdot\frac{a+b}{2}\cdot\frac{a+b+c}{3}}.$$

■

98 Let a, b, c be positive real numbers such that $abc(a+b+c)=3$. Prove the inequality

$$(a+b)(b+c)(c+a)\geq 8.$$

Solution We have

$$\begin{aligned}A&=(a+b)(b+c)(c+a)=(ab+ac+b^2+bc)(c+a)\\&=(b(a+b+c)+ac)(c+a)=\left(\frac{3}{ac}+ac\right)(c+a).\end{aligned}$$

By $AM\geq GM$ we obtain

$$\begin{aligned}A&=\left(\frac{3}{ac}+ac\right)(c+a)=\left(\frac{1}{ac}+\frac{1}{ac}+\frac{1}{ac}+ac\right)(c+a)\\&\geq 4\sqrt[4]{\frac{ac}{(ac)^3}}\cdot 2\sqrt{ac}=4\frac{1}{\sqrt{ac}}\cdot 2\sqrt{ac}=8.\end{aligned}$$

Equality occurs iff $a=c$ and $\frac{1}{ac}=ac$, i.e. $a=c=1$, and then we easily get $b=1$.

■

99 Let a, b, c be positive real numbers. Prove the inequality

$$\sqrt{\frac{2a}{b+c}}+\sqrt{\frac{2b}{c+a}}+\sqrt{\frac{2c}{a+b}}\leq 3.$$

Solution Applying $AM\geq HM$ we get

$$\sqrt{1\cdot\frac{2a}{b+c}}\leq\frac{2}{1+\frac{b+c}{2a}}=\frac{4a}{2a+b+c}.$$

Analogously we obtain

$$\sqrt{\frac{2b}{c+a}} \leq \frac{4b}{a+2b+c} \quad \text{and} \quad \sqrt{\frac{2c}{a+b}} \leq \frac{4c}{a+b+2c}.$$

So it is enough to prove that

$$4\left(\frac{a}{2a+b+c}+\frac{b}{a+2b+c}+\frac{c}{a+b+2c}\right) \leq 3,$$

i.e.

$$\frac{a}{2a+b+c}+\frac{b}{a+2b+c}+\frac{c}{a+b+2c} \leq \frac{3}{4}. \tag{1}$$

Since the last inequality is homogeneous we can assume that $a+b+c=1$.
Now inequality (1) becomes

$$\frac{a}{1+a}+\frac{b}{1+b}+\frac{c}{1+c} \leq \frac{3}{4}, \quad \text{i.e.} \quad 5(ab+bc+ca)+9abc \leq 2. \tag{2}$$

By the well-known inequality $3(ab+bc+ac) \leq (a+b+c)^2$ and $AM \geq GM$ we obtain $ab+bc+ac \leq \frac{1}{3}$ and $abc \leq \frac{1}{27}$. Now it is quite easy to prove inequality (2), as desired. ∎

100 Let $a, b, c \in \mathbb{R}^+$ such that $ab+bc+ca=1$. Prove the inequality

$$\frac{1}{a(a+b)}+\frac{1}{b(b+c)}+\frac{1}{c(c+a)} \geq \frac{9}{2}.$$

Solution The given inequality is equivalent to

$$\frac{c(a+b)+ab}{a(a+b)}+\frac{a(b+c)+bc}{b(b+c)}+\frac{b(c+a)+ac}{c(c+a)} \geq \frac{9}{2},$$

i.e.

$$\frac{a}{b}+\frac{b}{c}+\frac{c}{a}+\frac{b}{a+b}+\frac{c}{b+c}+\frac{a}{c+a} \geq \frac{9}{2}$$

$$\Leftrightarrow \quad \frac{a+b}{b}+\frac{b+c}{c}+\frac{c+a}{a}+\frac{b}{a+b}+\frac{c}{b+c}+\frac{a}{c+a} \geq \frac{15}{2}. \tag{1}$$

We have

$$\frac{a+b}{b}+\frac{b+c}{c}+\frac{c+a}{a}+\frac{b}{a+b}+\frac{c}{b+c}+\frac{a}{c+a}$$
$$= \frac{a+b}{4b}+\frac{b+c}{4c}+\frac{c+a}{4a}+\frac{b}{a+b}+\frac{c}{b+c}+\frac{a}{c+a}$$

$$+\frac{3}{4}\left(\frac{a+b}{b}+\frac{b+c}{c}+\frac{c+a}{a}\right)$$

$$\geq 6\sqrt[6]{\frac{a+b}{4b}\cdot\frac{b+c}{4c}\cdot\frac{c+a}{4a}\cdot\frac{b}{a+b}\cdot\frac{c}{b+c}\cdot\frac{a}{c+a}}+\frac{3}{4}\left(\frac{a}{b}+\frac{b}{c}+\frac{c}{a}+3\right)$$

$$\geq 3+\frac{3}{4}\left(3\sqrt[3]{\frac{a}{b}\cdot\frac{b}{c}\cdot\frac{c}{a}}+3\right)=3+\frac{18}{4}=\frac{15}{2},$$

as required. ■

101 Let $0 \leq a \leq b \leq c \leq 1$ be real numbers. Prove that

$$a^2(b-c)+b^2(c-b)+c^2(1-c)\leq \frac{108}{529}.$$

Solution Using $AM \geq GM$ we have

$$\begin{aligned}
a^2(b-c)+b^2(c-b)+c^2(1-c) &\leq 0+\frac{1}{2}(b\cdot b\cdot(2c-2b))+c^2(1-c)\\
&\leq \frac{1}{2}\left(\frac{b+b+2c-2b}{3}\right)^3+c^2(1-c)\\
&= c^2\left(\frac{4c}{27}+1-c\right)=c^2\left(1-\frac{23c}{27}\right)\\
&= \left(\frac{54}{23}\right)^2\left(\frac{23c}{54}\right)\left(\frac{23c}{54}\right)\left(1-\frac{23c}{27}\right)\\
&\leq \left(\frac{54}{23}\right)^2\left(\frac{1}{3}\right)^3=\frac{108}{529}.
\end{aligned}$$

■

102 Let $a, b, c \in \mathbb{R}^+$ such that $a+b+c=1$. Prove the inequality

$$S=a^4b+b^4c+c^4a\leq \frac{256}{3125}.$$

Solution Without loss of generality we can assume that $a=\max\{a,b,c\}$.

So it follows that

$$b^4c\leq a^3bc \quad \text{and} \quad c^4a\leq c^2a^3\leq ca^4.$$

Since $\frac{3c}{4}\geq\frac{c}{2}$ we obtain

$$\begin{aligned}
S&=a^4b+b^4c+\frac{c^4a}{2}+\frac{c^4a}{2}\leq a^4b+a^3bc+\frac{ca^4}{2}+\frac{c^2a^3}{2}\\
&=a^3b(a+c)+\frac{a^3c}{2}(a+c)=a^3(a+c)\left(b+\frac{c}{2}\right)\leq a^3(a+c)\left(b+\frac{3c}{4}\right). \quad (1)
\end{aligned}$$

Now using (1) and $AM \geq GM$ we get

$$S \leq a^3(a+c)\left(b+\frac{3c}{4}\right) = 4^4 \cdot \frac{a}{4} \cdot \frac{a}{4} \cdot \frac{a}{4} \cdot \frac{a+c}{4} \cdot \left(b+\frac{3c}{4}\right)$$

$$\leq 4^4\left(\frac{\frac{a}{4}+\frac{a}{4}+\frac{a}{4}+\frac{a+c}{4}+(b+\frac{3c}{4})}{5}\right)^5 = 4^4\left(\frac{a+b+c}{5}\right)^5 = \frac{256}{3125}. \quad \blacksquare$$

103 Let $a, b, c > 0$ be real numbers. Prove the inequality

$$\frac{a^2}{b^2}+\frac{b^2}{c^2}+\frac{c^2}{a^2} \geq \frac{a}{b}+\frac{b}{c}+\frac{c}{a}.$$

Solution Let $\frac{a}{b}=x$, $\frac{b}{c}=y$, $\frac{c}{a}=z$. Then it is clear that $xyz=1$, and the given inequality becomes

$$x^2+y^2+z^2 \geq x+y+z.$$

From $QM \geq AM$ we have

$$\sqrt{\frac{x^2+y^2+z^2}{3}} \geq \frac{x+y+z}{3},$$

i.e.

$$x^2+y^2+z^2 \geq \frac{(x+y+z)^2}{3} \geq \frac{3\sqrt[3]{xyz}(x+y+x)}{3} = x+y+z. \quad \blacksquare$$

104 Prove that for all positive real numbers a, b, c we have

$$\frac{a^3}{b^2}+\frac{b^3}{c^2}+\frac{c^3}{a^2} \geq a+b+c.$$

Solution Using $AM \geq GM$ we get

$$\frac{a^3}{b^2}+2b = \frac{a^3}{b^2}+b+b \geq 3\sqrt[3]{\frac{a^3}{b^2}\cdot b \cdot b} = 3a.$$

Analogously we have

$$\frac{b^3}{c^2}+2c \geq 3b \quad \text{and} \quad \frac{c^3}{a^2}+2a \geq 3c.$$

Adding these three inequalities we obtain

$$\frac{a^3}{b^2}+\frac{b^3}{c^2}+\frac{c^3}{a^2}+2(a+b+c) \geq 3(a+b+c),$$

as required. Equality holds iff $a=b=c$. $\blacksquare$

105 Prove that for all positive real numbers a, b, c we have

$$\frac{a^3}{b^2} + \frac{b^3}{c^2} + \frac{c^3}{a^2} \geq \frac{a^2}{b} + \frac{b^2}{c} + \frac{c^2}{a}.$$

Solution Using $AM \geq GM$ we get

$$\frac{a^3}{b^2} + a \geq 2\sqrt{\frac{a^3}{b^2} \cdot a} = 2\frac{a^2}{b}.$$

Analogously we have

$$\frac{b^3}{c^2} + b \geq 2\frac{b^2}{c} \quad \text{and} \quad \frac{c^3}{a^2} + c \geq 2\frac{c^2}{a}.$$

Adding these three inequalities we obtain

$$\frac{a^3}{b^2} + \frac{b^3}{c^2} + \frac{c^3}{a^2} + (a+b+c) \geq 2\left(\frac{a^2}{b} + \frac{b^2}{c} + \frac{c^2}{a}\right). \tag{1}$$

According to Exercise 2.12 (Chap. 2) we have that

$$\frac{a^2}{b} + \frac{b^2}{c} + \frac{c^2}{a} \geq a+b+c. \tag{2}$$

Now using (1) and (2) we obtain

$$\frac{a^3}{b^2} + \frac{b^3}{c^2} + \frac{c^3}{a^2} + (a+b+c) \geq 2\left(\frac{a^2}{b} + \frac{b^2}{c} + \frac{c^2}{a}\right) \geq \frac{a^2}{b} + \frac{b^2}{c} + \frac{c^2}{a} + (a+b+c),$$

and equality holds iff $a = b = c$. ■

106 Prove that for all positive real numbers a, b, c we have

$$\frac{a^3}{b} + \frac{b^3}{c} + \frac{c^3}{a} \geq ab + bc + ca.$$

Solution Using $AM \geq GM$ we get

$$\frac{a^3}{b} + \frac{b^3}{c} + bc \geq 3\sqrt[3]{\frac{a^3}{b} \cdot \frac{b^3}{c} \cdot bc} = 3ab.$$

Analogously we have

$$\frac{b^3}{c} + \frac{c^3}{a} + ca \geq 3bc \quad \text{and} \quad \frac{c^3}{a} + \frac{a^3}{b} + ab \geq 3ca.$$

Adding these three inequalities we obtain

$$2\left(\frac{a^3}{b}+\frac{b^3}{c}+\frac{c^3}{a}\right)+ab+bc+ca\geq 3(ab+bc+ca),$$

from which follows the desired inequality. Equality holds iff $a=b=c$. ■

107 Prove that for all positive real numbers a, b, c we have

$$\frac{a^5}{b^3}+\frac{b^5}{c^3}+\frac{c^5}{a^3}\geq a^2+b^2+c^2.$$

Solution Using $AM \geq GM$ we get

$$2\frac{a^5}{b^3}+3b^2=\frac{a^5}{b^3}+\frac{a^5}{b^3}+b^2+b^2+b^2\geq 5\sqrt[5]{\frac{a^5}{b^3}\cdot\frac{a^5}{b^3}\cdot b^2\cdot b^2\cdot b^2}=5a^2.$$

Analogously we have

$$2\frac{b^5}{c^3}+3c^2\geq 5b^2 \quad\text{and}\quad 2\frac{c^5}{a^3}+3a^2\geq 5c^2.$$

Adding these three inequalities we obtain

$$2\left(\frac{a^5}{b^3}+\frac{b^5}{c^3}+\frac{c^5}{a^3}\right)+3(a^2+b^2+c^2)\geq 5(a^2+b^2+c^2),$$

i.e.

$$\frac{a^5}{b^3}+\frac{b^5}{c^3}+\frac{c^5}{a^3}\geq a^2+b^2+c^2.$$

Equality holds iff $a=b=c$. ■

108 Let $a, b, c \in \mathbb{R}^+$ such that $a+b+c=3$. Prove the inequality

$$\frac{a^3}{b(2c+a)}+\frac{b^3}{c(2a+b)}+\frac{c^3}{a(2b+c)}\geq 1.$$

Solution We'll show that

$$\frac{a^3}{b(2c+a)}+\frac{b^3}{c(2a+b)}+\frac{c^3}{a(2b+c)}\geq \frac{a+b+c}{3},$$

from which, with the initial condition, will follow the desired inequality.

Using $AM \geq GM$ we get

$$\frac{9a^3}{b(2c+a)}+3b+(2c+a)\geq 3\sqrt[3]{\frac{9a^3}{b(2c+a)}\cdot 3b\cdot(2c+a)}=9a.$$

Analogously we have

$$\frac{9b^3}{c(2a+b)}+3c+(2a+b)\geq 3b \quad \text{and} \quad \frac{9c^3}{a(2b+c)}+3a+(2b+c)\geq 3c.$$

Adding the last three inequalities we obtain

$$9\left(\frac{a^3}{b(2c+a)}+\frac{b^3}{c(2a+b)}+\frac{c^3}{a(2b+c)}\right)+6(a+b+c)\geq 9(a+b+c),$$

i.e.

$$\frac{a^3}{b(2c+a)}+\frac{b^3}{c(2a+b)}+\frac{c^3}{a(2b+c)}\geq \frac{a+b+c}{3}=\frac{3}{3}=1.$$

■

109 Let $a,b,c\in\mathbb{R}^+$ and $a^2+b^2+c^2=3$. Prove the inequality

$$\frac{a^3}{b+2c}+\frac{b^3}{c+2a}+\frac{c^3}{a+2b}\geq 1.$$

Solution We'll prove that

$$\frac{a^3}{b+2c}+\frac{b^3}{c+2a}+\frac{c^3}{a+2b}\geq \frac{a^2+b^2+c^2}{3},$$

from which since $a^2+b^2+c^2=3$, we'll obtain the required result.

Applying $AM\geq GM$ we get

$$\frac{9a^3}{b+2c}+a(b+2c)\geq 2\sqrt{\frac{9a^3}{b+2c}\cdot a\cdot(b+2c)}=6a^2.$$

Analogously we deduce

$$\frac{9b^3}{c+2a}+b(c+2a)\geq 6b^2 \quad \text{and} \quad \frac{9c^3}{a+2b}+c(a+2b)\geq 6c^2.$$

Adding the last three inequalities we obtain

$$9\left(\frac{a^3}{b+2c}+\frac{b^3}{c+2a}+\frac{c^3}{a+2b}\right)+3(ab+bc+ca)\geq 6(a^2+b^2+c^2),$$

i.e.

$$\frac{a^3}{b+2c}+\frac{b^3}{c+2a}+\frac{c^3}{a+2b}\geq \frac{6(a^2+b^2+c^2)-3(ab+bc+ca)}{9}. \tag{1}$$

Using the well-known inequality

$$a^2+b^2+c^2\geq ab+bc+ca,$$

according to (1) we obtain

$$\frac{a^3}{b+2c}+\frac{b^3}{c+2a}+\frac{c^3}{a+2b}\geq\frac{3(a^2+b^2+c^2)}{9}=\frac{a^2+b^2+c^2}{3}=\frac{3}{3}=1. \qquad\blacksquare$$

110 Let a, b, c be positive real numbers such that $a^2+b^2+c^2=3$. Prove the inequality

$$\frac{1}{a^3+2}+\frac{1}{b^3+2}+\frac{1}{c^3+2}\geq 1.$$

Solution We have

$$\frac{1}{a^3+2}=\frac{1}{2}\left(1-\frac{a^3}{a^3+2}\right)=\frac{1}{2}\left(1-\frac{a^3}{a^3+1+1}\right)\geq\frac{1}{2}\left(1-\frac{a^3}{3a}\right)=\frac{1}{2}\left(1-\frac{a^2}{3}\right).$$

Therefore

$$\begin{aligned}\frac{1}{a^3+2}+\frac{1}{b^3+2}+\frac{1}{c^3+2}&\geq\frac{1}{2}\left(1-\frac{a^2}{3}\right)+\frac{1}{2}\left(1-\frac{b^2}{3}\right)+\frac{1}{2}\left(1-\frac{c^2}{3}\right)\\&=\frac{3}{2}-\frac{a^2+b^2+c^2}{6}=1.\end{aligned}$$

Equality holds iff $a=b=c=1$. ■

111 Let $a, b, c\in\mathbb{R}^+$ such that $a+b+c=1$. Prove the inequality

$$\frac{a^3}{a^2+b^2}+\frac{b^3}{b^2+c^2}+\frac{c^3}{c^2+a^2}\geq\frac{1}{2}.$$

Solution Clearly we have

$$\frac{a^2+b^2}{2}\geq ab \quad\text{i.e.}\quad \frac{ab}{a^2+b^2}\leq\frac{1}{2}.$$

Therefore

$$\frac{a^3}{a^2+b^2}=a-b\frac{ab}{a^2+b^2}\geq a-\frac{b}{2}.$$

Analogously

$$\frac{b^3}{b^2+c^2}\geq b-\frac{c}{2}\quad\text{and}\quad\frac{c^3}{c^2+a^2}\geq c-\frac{a}{2}.$$

After adding these and using $a+b+c=1$ we obtain

$$\frac{a^3}{a^2+b^2}+\frac{b^3}{b^2+c^2}+\frac{c^3}{c^2+a^2}\geq a+b+c-\frac{a+b+c}{2}=\frac{a+b+c}{2}=\frac{1}{2}. \qquad\blacksquare$$

112 Let a, b, c be positive real numbers such that $a+b+c=3$. Prove the inequality

$$\frac{1}{1+2a^2b}+\frac{1}{1+2b^2c}+\frac{1}{1+2c^2a}\geq 1.$$

Solution Note that

$$\frac{1}{1+2a^2b}=1-\frac{2a^2b}{1+2a^2b}=1-\frac{2a^2b}{1+a^2b+a^2b}\geq 1-\frac{2a^2b}{3\sqrt[3]{a^4b^2}}$$
$$=1-\frac{2\sqrt[3]{a^2b}}{3}\geq 1-\frac{2(2a+b)}{9}.$$

After adding these inequalities for all variables we get

$$\frac{1}{1+2a^2b}+\frac{1}{1+2b^2c}+\frac{1}{1+2c^2a}\geq 3-\frac{6(a+b+c)}{9}=3-2=1,$$

as required.

Equality holds iff $a=b=c=1$. ∎

113 Let a, b, c, d be positive real numbers such that $a+b+c+d=4$. Prove the inequality

$$\frac{a}{1+b^2c}+\frac{b}{1+c^2d}+\frac{c}{1+d^2a}+\frac{d}{1+a^2b}\geq 2.$$

Solution Applying $AM \geq GM$ we have

$$\frac{a}{1+b^2c}=a-\frac{ab^2c}{1+b^2c}\geq a-\frac{ab^2c}{2b\sqrt{c}}=a-\frac{ab\sqrt{c}}{2}\geq a-\frac{b\sqrt{a\cdot ac}}{2}$$
$$\geq a-\frac{b(a+ac)}{4},$$

i.e.

$$\frac{a}{1+b^2c}\geq a-\frac{1}{4}(ab+abc).$$

Analogously we obtain

$$\frac{b}{1+c^2d}\geq b-\frac{1}{4}(bc+bcd),\qquad \frac{c}{1+d^2a}\geq c-\frac{1}{4}(cd+cda),$$
$$\frac{d}{1+a^2b}\geq d-\frac{1}{4}(da+dab).$$

Adding these three inequalities we obtain

$$\frac{a}{1+b^2c}+\frac{b}{1+c^2d}+\frac{c}{1+d^2a}+\frac{d}{1+a^2b}$$
$$\geq (a+b+c+d)-\frac{1}{4}(ab+bc+cd+da+abc+bcd+cda+dab). \tag{1}$$

One more use of $AM \geq GM$ give us

$$ab+bc+cd+da \leq \frac{1}{4}(a+b+c+d)^2=4 \tag{2}$$

and

$$abc+bcd+cda+dab \leq \frac{1}{16}(a+b+c+d)^3=4. \tag{3}$$

From (1), (2) and (3) it follows that

$$\frac{a}{1+b^2c}+\frac{b}{1+c^2d}+\frac{c}{1+d^2a}+\frac{d}{1+a^2b} \geq 4-2=2,$$

as desired.

Equality holds if and only if $a=b=c=d=1$. ■

114 Let a, b, c, d be positive real numbers. Prove the inequality

$$\frac{a^3}{a^2+b^2}+\frac{b^3}{b^2+c^2}+\frac{c^3}{c^2+d^2}+\frac{d^3}{d^2+a^2} \geq \frac{a+b+c+d}{2}.$$

Solution Using $AM \geq GM$ we get

$$\frac{a^3}{a^2+b^2}=a-\frac{ab^2}{a^2+b^2} \geq a-\frac{ab^2}{2ab}=a-\frac{b}{2}.$$

Analogously

$$\frac{b^3}{b^2+c^2} \geq b-\frac{c}{2}, \qquad \frac{c^3}{c^2+d^2} \geq c-\frac{d}{2}, \qquad \frac{d^3}{d^2+a^2} \geq d-\frac{a}{2}.$$

Adding these inequalities give us the required inequality. ■

115 Let a, b, c be positive real numbers such that $a+b+c=3$. Prove the inequality

$$\frac{a^2}{a+2b^2}+\frac{b^2}{b+2c^2}+\frac{c^2}{c+2a^2} \geq 1.$$

Solution Applying $AM \geq GM$ we get

$$\frac{a^2}{a+2b^2}=a-\frac{2ab^2}{a+2b^2} \geq a-\frac{2ab^2}{3\sqrt[3]{ab^4}}=a-\frac{2(ab)^{2/3}}{3}.$$

Analogously we obtain

$$\frac{b^2}{b+2c^2} \geq b - \frac{2(bc)^{2/3}}{3} \quad \text{and} \quad \frac{c^2}{c+2a^2} \geq c - \frac{2(ca)^{2/3}}{3}.$$

Adding these three inequalities gives us

$$\frac{a^2}{a+2b^2} + \frac{b^2}{b+2c^2} + \frac{c^2}{c+2a^2} \geq (a+b+c) - \frac{2}{3}((ab)^{2/3} + (bc)^{2/3} + (ca)^{2/3}).$$

So it is enough to show that

$$(a+b+c) - \frac{2}{3}((ab)^{2/3} + (bc)^{2/3} + (ca)^{2/3}) \geq 1,$$

i.e.

$$(ab)^{2/3} + (bc)^{2/3} + (ca)^{2/3} \leq 3. \tag{1}$$

Applying $AM \geq GM$ we get

$$\begin{aligned}(ab)^{2/3} + (bc)^{2/3} + (ca)^{2/3} &\leq \frac{(a+ab+b) + (b+bc+c) + (c+ca+a)}{3} \\ &= \frac{2(a+b+c) + (ab+bc+ca)}{3} \\ &\leq \frac{2(a+b+c) + (a+b+c)^2/3}{3} = \frac{2 \cdot 3 + 3^2/3}{3} = 3,\end{aligned}$$

i.e. we have proved (1), and we are done.

Equality holds iff $a = b = c = 1$. ∎

116 Let a, b, c be positive real numbers such that $a+b+c=3$. Prove the inequality

$$\frac{a^2}{a+2b^3} + \frac{b^2}{b+2c^3} + \frac{c^2}{c+2a^3} \geq 1.$$

Solution Applying $AM \geq GM$ gives us

$$\frac{a^2}{a+2b^3} = a - \frac{2ab^3}{a+2b^3} \geq a - \frac{2ab^3}{3\sqrt[3]{ab^4}} = a - \frac{2ba^{2/3}}{3}.$$

Analogously

$$\frac{b^2}{b+2c^3} \geq b - \frac{2cb^{2/3}}{3} \quad \text{and} \quad \frac{c^2}{c+2a^3} \geq c - \frac{2ac^{2/3}}{3}.$$

Adding these three inequalities implies

$$\frac{a^2}{a+2b^2} + \frac{b^2}{b+2c^2} + \frac{c^2}{c+2a^2} \geq (a+b+c) - \frac{2}{3}(ba^{2/3} + cb^{2/3} + ac^{2/3}).$$

So it is enough to prove that

$$(a+b+c) - \frac{2}{3}(ba^{2/3} + cb^{2/3} + ac^{2/3}) \geq 1,$$

i.e.

$$ba^{2/3} + cb^{2/3} + ac^{2/3} \leq 3. \tag{1}$$

After another application of $AM \geq GM$ we get

$$\begin{aligned} ba^{2/3} + cb^{2/3} + ac^{2/3} &\leq \frac{b(2a+1) + c(2b+1) + a(2c+1)}{3} \\ &= \frac{a+b+c+2(ab+bc+ca)}{3} \\ &\leq \frac{(a+b+c) + (a+b+c)^2/3}{3} = \frac{3+2\cdot 3^2/3}{3} = 3, \end{aligned}$$

i.e. we have proved (1), and we are done.

Equality holds iff $a=b=c=1$. ■

117 Let a, b, c be positive real numbers such that $a^2+b^2+c^2=3$. Find the minimum value of the expression

$$a+b+c+\frac{16}{a+b+c}.$$

Solution By the inequality $AM \geq GM$ we get

$$a+b+c+\frac{16}{a+b+c} \geq 2\sqrt{16} = 8,$$

with equality if and only if $a+b+c = \frac{16}{a+b+c}$ from which we deduce that $a+b+c = 4$ and then

$$16 = (a+b+c)^2 \leq 3(a^2+b^2+c^2) = 9,$$

a contradiction.

We estimate that the minimal value occurs when $a=b=c$, i.e. $a=b=c=1$.

Let $a+b+c = \frac{\alpha}{a+b+c}$. Thus $\alpha = 9$ at the point of incidence $a=b=c=1$.

Therefore let us rewrite the given expression as follows

$$a+b+c+\frac{9}{a+b+c}+\frac{7}{a+b+c}. \tag{1}$$

Applying $AM \geq GM$ and $3(a^2+b^2+c^2) \geq (a+b+c)^2$ we have

$$a+b+c+\frac{9}{a+b+c} \geq 2\sqrt{9} = 6 \tag{2}$$

and

$$\frac{1}{a+b+c} \geq \frac{1}{\sqrt{3(a^2+b^2+c^2)}} = \frac{1}{3}. \tag{3}$$

By (1), (2) and (3) we obtain

$$a+b+c+\frac{16}{a+b+c} = a+b+c+\frac{9}{a+b+c}+\frac{7}{a+b+c} \geq 6+\frac{7}{3} = \frac{25}{3},$$

with equality if and only if $a=b=c=1$. ■

118 Let $a,b,c \geq 0$ be real numbers such that $a^2+b^2+c^2=1$. Find the minimal value of the expression

$$A = a+b+c+\frac{1}{abc}.$$

Solution By $AM \geq GM$ we obtain

$$A = a+b+c+\frac{1}{abc} \geq 4\sqrt[4]{abc\cdot\frac{1}{abc}} = 4,$$

with equality iff $a=b=c=\frac{1}{abc}$, i.e. $a=b=c=1$.

Thus $a^2+b^2+c^2 = 3 \neq 1$, a contradiction.

Since A is a symmetrical expression in a,b and c, we estimate that min A occurs at the incidence point $a=b=c$, i.e. $a=b=c=1/\sqrt{3}$.

Hence at the incidence point we have $a=b=c=\frac{1}{\alpha abc}=\frac{1}{\sqrt{3}}$, and it follows that $\alpha = \frac{1}{a^2bc} = \frac{1}{(1/\sqrt{3})^4} = 9$.

Therefore

$$\begin{aligned} A = a+b+c+\frac{1}{abc} &= a+b+c+\frac{1}{9abc}+\frac{8}{9abc} \\ &\geq 4\sqrt[4]{abc\cdot\frac{1}{9abc}}+\frac{8}{9abc} = 4\sqrt[4]{\frac{1}{9}}+\frac{8}{9abc}. \end{aligned} \tag{1}$$

By $QM \geq GM$ we obtain

$$\sqrt{\frac{a^2+b^2+c^2}{3}} \geq 3\sqrt[3]{abc}, \quad \text{i.e.} \quad \sqrt{\frac{1}{3}} \geq 3\sqrt[3]{abc}.$$

Hence

$$\frac{1}{abc} \geq 3\sqrt{3}. \tag{2}$$

By (1) and (2) we get

$$A \geq \frac{4}{\sqrt{3}} + 3\sqrt{3}\cdot\frac{8}{9} = 4\sqrt{3}.$$

So min $A = 4\sqrt{3}$, and it occurs iff $a=b=c=1/\sqrt{3}$. ■

119 Let a, b, c be positive real numbers such that $a+b+c=6$. Prove the inequality

$$\sqrt[3]{ab+bc}+\sqrt[3]{bc+ca}+\sqrt[3]{ca+ab}+\sqrt[3]{\frac{9}{4}(a^2+b^2+c^2)}\leq 9.$$

Solution Analogously as in the first solution of Exercise 5.13 we obtain that

$$\sqrt[3]{ab+bc}+\sqrt[3]{bc+ca}+\sqrt[3]{ca+ab}\leq \frac{1}{4}\left(\frac{2(ab+bc+ca)+48}{3}\right). \tag{1}$$

At the point of incidence $a=b=c=2$ we have $a^2+b^2+c^2=12$.

Therefore by $AM \geq GM$ we have

$$\begin{aligned}\sqrt[3]{\frac{9}{4}(a^2+b^2+c^2)}&=\sqrt[3]{\frac{9(a^2+b^2+c^2)\cdot 12\cdot 12}{4\cdot 12\cdot 12}}=\frac{1}{4}\sqrt[3]{(a^2+b^2+c^2)\cdot 12\cdot 12}\\&\leq \frac{1}{4}\left(\frac{a^2+b^2+c^2+24}{3}\right).\end{aligned} \tag{2}$$

By (1) and (2) we obtain

$$\begin{aligned}&\sqrt[3]{ab+bc}+\sqrt[3]{bc+ca}+\sqrt[3]{ca+ab}+\sqrt[3]{\frac{9}{2}(a^2+b^2+c^2)}\\&\quad\leq \frac{1}{4}\left(\frac{2(ab+bc+ca)+48}{3}\right)+\frac{1}{4}\left(\frac{a^2+b^2+c^2+24}{3}\right)\\&\quad=\frac{1}{12}((a+b+c)^2+72)=\frac{6^2+72}{12}=9,\end{aligned}$$

as required.

Equality occurs if and only if $a=b=c=2$. ■

120 Let $a, b, c \in \mathbb{R}^+$ such that $a+2b+3c\geq 20$. Prove the inequality

$$S=a+b+c+\frac{3}{a}+\frac{9}{2b}+\frac{4}{c}\geq 13.$$

Solution $S=13$ at the point $a=2, b=3, c=4$.

Using $AM\geq GM$ we get

$$a+\frac{4}{a}\geq 2\sqrt{a\cdot\frac{4}{a}}=4,\qquad b+\frac{9}{b}\geq 2\sqrt{b\cdot\frac{9}{b}}=6,\qquad c+\frac{16}{c}\geq 2\sqrt{c\cdot\frac{16}{c}}=8,$$

i.e.

$$\frac{3}{4}\left(a+\frac{4}{a}\right)\geq 3,\qquad \frac{1}{2}\left(b+\frac{9}{b}\right)\geq 3\quad\text{and}\quad \frac{1}{4}\left(c+\frac{16}{c}\right)\geq 2.$$

Adding the last three inequalities we have

$$\frac{3}{4}a + \frac{1}{2}b + \frac{1}{4}c + \frac{3}{a} + \frac{9}{2b} + \frac{4}{c} \geq 8. \tag{1}$$

Using $a + 2b + 3c \geq 20$ we obtain

$$\frac{1}{4}a + \frac{1}{2}b + \frac{3}{4}c \geq 5. \tag{2}$$

Finally, after adding (1) and (2) we get

$$a + b + c + \frac{3}{a} + \frac{9}{2b} + \frac{4}{c} \geq 13,$$

as desired. ■

121 Let $a, b, c \in \mathbb{R}^+$. Prove the inequality

$$S = 30a + 3b^2 + \frac{2c^3}{9} + 36\left(\frac{1}{ab} + \frac{1}{bc} + \frac{1}{ca}\right) \geq 84.$$

Solution $S = 84$ at the point $a = 1, b = 2, c = 3$.

From $AM \geq GM$ we obtain

$$2 \cdot a + \frac{b^2}{4} + 2 \cdot \frac{2}{ab} = a + a + \frac{b^2}{4} + \frac{2}{ab} + \frac{2}{ab} \geq 5\sqrt[5]{a^2 \cdot \frac{b^2}{4} \cdot \left(\frac{2}{ab}\right)^2} = 5,$$

$$3 \cdot \frac{b^2}{4} + 2 \cdot \frac{c^3}{27} + 6 \cdot \frac{6}{bc} \geq 11\sqrt[11]{\left(\frac{b^2}{4}\right)^3 \cdot \left(\frac{c^3}{27}\right)^2 \cdot \left(\frac{6}{bc}\right)^6} = 11,$$

$$\frac{c^3}{27} + 3 \cdot a + 3 \cdot \frac{3}{ca} \geq 7\sqrt[7]{\frac{c^3}{27} \cdot a^3 \cdot \left(\frac{3}{ca}\right)^3} = 7,$$

i.e.

$$9\left(2a + \frac{b^2}{4} + \frac{4}{ab}\right) \geq 45, \qquad \frac{3b^2}{4} + \frac{2c^3}{27} + \frac{36}{bc} \geq 11, \qquad 4\left(\frac{c^3}{27} + 3a + \frac{9}{ca}\right) \geq 28.$$

After adding these three inequalities we get

$$30a + 3b^2 + \frac{2c^3}{9} + 36\left(\frac{1}{ab} + \frac{1}{bc} + \frac{1}{ca}\right) \geq 84.$$

■

122 Let $a, b, c \in \mathbb{R}^+$ such that $ac \geq 12$ and $bc \geq 8$. Prove the inequality

$$S = a + b + c + 2\left(\frac{1}{ab} + \frac{1}{bc} + \frac{1}{ca}\right) + \frac{8}{abc} \geq \frac{121}{12}.$$

Solution $S=\frac{121}{12}$ at the point $a=3, b=2, c=4$.

Use of $AM \geq GM$ gives us

$$\frac{a}{3}+\frac{b}{2}+\frac{6}{ab}\geq 3,\qquad \frac{b}{2}+\frac{c}{4}+\frac{8}{bc}\geq 3,\qquad \frac{c}{4}+\frac{a}{3}+\frac{12}{ca}\geq 3\quad\text{and}$$
$$\frac{a}{3}+\frac{b}{2}+\frac{c}{4}+\frac{24}{abc}\geq 4,$$

i.e.

$$\frac{a}{3}+\frac{b}{2}+\frac{6}{ab}\geq 3,\qquad 4\left(\frac{b}{2}+\frac{c}{4}+\frac{8}{bc}\right)\geq 12,\qquad 7\left(\frac{c}{4}+\frac{a}{3}+\frac{12}{ca}\right)\geq 21,$$
$$\frac{a}{3}+\frac{b}{2}+\frac{c}{4}+\frac{24}{abc}\geq 4.$$

After adding these three inequalities we get

$$3(a+b+c)+\frac{6}{ab}+\frac{32}{bc}+\frac{84}{ca}+\frac{24}{abc}\geq 40. \tag{1}$$

Also, since $ac\geq 12$ and $bc\geq 8$ we obtain

$$\frac{1}{ac}\leq\frac{1}{12}\quad\text{and}\quad\frac{1}{bc}\leq\frac{1}{8},$$

so from (1) it follows that

$$40\leq 3S+\frac{26}{bc}+\frac{78}{ca}\leq 3S+\frac{26}{12}+\frac{78}{8},\quad\text{i.e.}\quad S\geq\frac{121}{12}.$$ ■

123 Let $a,b,c,d>0$ be real numbers. Determine the minimal value of the expression

$$A=\left(1+\frac{2a}{3b}\right)\left(1+\frac{2b}{3c}\right)\left(1+\frac{2c}{3d}\right)\left(1+\frac{2d}{3a}\right).$$

Solution By $AM\geq GM$ we get

$$A\geq 2\sqrt{\frac{2a}{3b}}\cdot 2\sqrt{\frac{2b}{3c}}\cdot 2\sqrt{\frac{2c}{3d}}\cdot 2\sqrt{\frac{2d}{3a}}=8,$$

with equality if and only if $\frac{2a}{3b}=\frac{2b}{3c}=\frac{2c}{3d}=\frac{2d}{3a}=1$.

Hence $2(a+b+c+d)=3(a+b+c+d)$, i.e. $2=3$, which is impossible.

Since A is a symmetrical expression in a,b,c and d, the minimum (maximum) occurs at the incidence point $a=b=c=d>0$, and then

$$A=\left(1+\frac{2}{3}\right)^4=\frac{625}{81}.$$

We have

$$1+\frac{2a}{3b}=\frac{1}{3}+\frac{1}{3}+\frac{1}{3}+\frac{a}{3b}+\frac{a}{3b}\geq 5\sqrt[5]{\left(\frac{1}{3}\right)^3\left(\frac{a}{3b}\right)^2}=\frac{5}{3}\left(\frac{a}{b}\right)^{2/5}.$$

Similarly we get

$$1+\frac{2b}{3c}\geq\frac{5}{3}\left(\frac{b}{c}\right)^{2/5},\qquad 1+\frac{2c}{3d}\geq\frac{5}{3}\left(\frac{c}{d}\right)^{2/5}\quad\text{and}\quad 1+\frac{2d}{3a}\geq\frac{5}{3}\left(\frac{d}{a}\right)^{2/5}.$$

If we multiply the above inequalities we obtain $A\geq\frac{625}{81}$.

Equality holds if and only if $a=b=c=d>0$. ■

124 Let $a,b,c>0$ be real numbers such that $a^2+b^2+c^2=12$. Determine the maximal value of the expression

$$A=a\sqrt[3]{b^2+c^2}+b\sqrt[3]{c^2+a^2}+c\sqrt[3]{a^2+b^2}.$$

Solution Since A is a symmetrical expression with respect to a,b and c, max A occurs when $a=b=c>0$, i.e. $a=b=c=2$.

Hence

$$2a^2=2b^2=2c^2=8$$

and

$$b^2+c^2=c^2+a^2=a^2+b^2=8.$$

By $AM\geq GM$ we have

$$\begin{aligned}a\sqrt[3]{b^2+c^2}&=\sqrt[3]{a^3(b^2+c^2)}=\sqrt[6]{a^6(b^2+c^2)^2}=\frac{1}{2}\sqrt[6]{(2a^2)^3\cdot(b^2+c^2)^2\cdot 8}\\&=\frac{1}{2}\sqrt[6]{8(2a^2)(2a^2)(2a^2)(b^2+c^2)(b^2+c^2)}\\&\leq\frac{1}{2}\cdot\frac{8+6a^2+2(b^2+c^2)}{6}=\frac{4+3a^2+b^2+c^2}{6}.\end{aligned}$$

Similarly

$$b\sqrt[3]{c^2+a^2}\leq\frac{4+a^2+3b^2+c^2}{6}\quad\text{and}\quad c\sqrt[3]{a^2+b^2}\leq\frac{4+a^2+b^2+3c^2}{6}.$$

After adding the last three inequalities we get

$$A\leq\frac{12+5(a^2+b^2+c^2)}{6}=\frac{12+5\cdot 12}{6}=12,$$

with equality if and only if $a=b=c=2$. ■

125 Let $a, b, c \geq 0$ such that $a + b + c = 3$. Prove the inequality

$$(a^2 - ab + b^2)(b^2 - bc + c^2)(c^2 - ca + a^2) \leq 12.$$

Solution Without loss of generality we may assume that $a \geq b \geq c \geq 0$.
So it follows that

$$0 \leq b^2 - bc + c^2 \leq b^2 \quad \text{and} \quad 0 \leq c^2 - ca + a^2 \leq a^2,$$

i.e. we obtain

$$(b^2 - bc + c^2)(c^2 - ca + a^2) \leq a^2 b^2.$$

Now we have

$$\begin{aligned}
&(a^2 - ab + b^2)(b^2 - bc + c^2)(c^2 - ca + a^2) \\
&\quad \leq a^2 b^2 (a^2 - ab + b^2) \\
&\quad = \frac{4}{9} \cdot \frac{3ab}{2} \cdot \frac{3ab}{2} \cdot (a^2 - ab + b^2) \leq \frac{4}{9} \cdot \left(\frac{1}{3} \left(\frac{3ab}{2} + \frac{3ab}{2} + (a^2 - ab + b^2) \right) \right)^3 \\
&\quad = \frac{4}{9} \left(\frac{(a+b)^2}{3} \right)^3 \leq \frac{4}{9} \left(\frac{(a+b+c)^2}{3} \right)^3 = \frac{4}{9} \left(\frac{3^2}{3} \right)^3 = 12.
\end{aligned}$$

■

126 Let a, b, c be positive real numbers. Prove the inequality

$$(a^5 - a^2 + 3)(b^5 - b^2 + 3)(c^5 - c^2 + 3) \geq (a + b + c)^3.$$

Solution For every positive real number x, we have that $x^2 - 1$ and $x^3 - 1$ have the same signs, and because of this $x^5 - x^3 - x^2 + 1 = (x^2 - 1)(x^3 - 1) \geq 0$, i.e. we obtain

$$x^5 - x^2 + 3 \geq x^3 + 2.$$

Now we get

$$(a^5 - a^2 + 3)(b^5 - b^2 + 3)(c^5 - c^2 + 3) \geq (a^3 + 2)(b^3 + 2)(c^3 + 2).$$

So it is enough to show that

$$(a^3 + 2)(b^3 + 2)(c^3 + 2) \geq (a + b + c)^3. \tag{1}$$

After a little algebra we obtain that (1) is equivalent to

$$\begin{aligned}
&a^3 b^3 c^3 + 3(a^3 + b^3 + c^3) + 2(a^3 b^3 + b^3 c^3 + c^3 a^3) + 8 \\
&\quad \geq 3(a^2 b + b^2 a + b^2 c + c^2 b + c^2 a + a^2 c) + 6abc.
\end{aligned} \tag{2}$$

Using $AM \geq GM$ we can easily obtain the following inequalities

$$a^3 + a^3b^3 + 1 \geq 3a^2b, \qquad a^3 + a^3c^3 + 1 \geq 3a^2c, \qquad b^3 + a^3b^3 + 1 \geq 3b^2a,$$
$$b^3 + b^3c^3 + 1 \geq 3b^2c, \qquad c^3 + c^3a^3 + 1 \geq 3c^2a, \qquad c^3 + c^3b^3 + 1 \geq 3c^2b,$$
$$a^3b^3c^3 + a^3 + b^3 + c^3 + 1 + 1 \geq 6abc.$$

After adding the previous inequalities we obtain inequality (2), as desired. ∎

127 Let $x, y, z \in \mathbb{R}^+$ such that $x + y + z = 1$. Prove the inequality

$$\frac{xy}{\sqrt{1+z^2}} + \frac{zx}{\sqrt{1+y^2}} + \frac{yz}{\sqrt{1+x^2}} \leq \frac{1}{\sqrt{10}}.$$

Solution We have

$$\sqrt{1+z^2} = \sqrt{9 \cdot \left(\frac{1}{3}\right)^2 + z^2} = \sqrt{\underbrace{\frac{1}{3^2} + \cdots + \frac{1}{3^2}}_{9} + z^2} \overset{K \geq A}{\geq} \frac{1}{\sqrt{10}} \Bigg(\underbrace{\frac{1}{3} + \cdots + \frac{1}{3}}_{9} + z\Bigg)$$
$$= \frac{3+z}{\sqrt{10}},$$

i.e. we obtain that

$$\frac{xy}{\sqrt{1+z^2}} \leq \sqrt{10}\frac{xy}{3+z}.$$

Analogously we obtain

$$\frac{yz}{\sqrt{1+x^2}} \leq \sqrt{10}\frac{yz}{3+x} \quad \text{and} \quad \frac{zx}{\sqrt{1+y^2}} \leq \sqrt{10}\frac{zx}{3+y}.$$

So it is enough to prove that

$$\sqrt{10}\left(\frac{xy}{3+z} + \frac{zx}{3+y} + \frac{yz}{3+x}\right) \leq \frac{1}{\sqrt{10}},$$

i.e.

$$\frac{xy}{3+z} + \frac{zx}{3+y} + \frac{yz}{3+x} \leq \frac{1}{10}. \tag{1}$$

Let $a = 3 + x, b = 3 + y, c = 3 + z$.

Then clearly $a + b + c = 10$.

Inequality (1) is equivalent to

$$\frac{(a-3)(b-3)}{c} + \frac{(c-3)(a-3)}{b} + \frac{(b-3)(c-3)}{a} \leq \frac{1}{10},$$

i.e.

$$\frac{ab-3(a+b)+9}{c}+\frac{ca-3(c+a)+9}{b}+\frac{bc-3(b+c)+9}{a}\le\frac{1}{10}$$
$$\Leftrightarrow\quad \frac{ab+3c-21}{c}+\frac{ca+3b-21}{b}+\frac{bc+3a-21}{a}\le\frac{1}{10}$$
$$\Leftrightarrow\quad \frac{ab-21}{c}+\frac{ca-21}{b}+\frac{bc-21}{a}\le-\frac{89}{10}.$$

After clearing denominators, we obtain

$$21(a^3(b+c)+b^3(a+c)+c^3(b+a))+16(a^2bc+b^2ac+c^2ab)$$
$$\ge 58(a^2b^2+b^2c^2+c^2a^2)$$
$$\Leftrightarrow\quad (21ab-8c^2)(a-b)^2+(21bc-8a^2)(b-c)^2$$
$$+(21ca-8b^2)(c-a)^2\ge 0,$$

which is true since $a, b, c \in (3, 4)$, i.e.

$$21ab-8c^2\ge 21\cdot 3\cdot 3-8\cdot 4^2=61>0.$$

In the same way we find that $21bc-8a^2>0$ and $21ca-8b^2>0$. ■

128 Let $a, b, c \in \mathbb{R}^+$. Prove the inequality

$$(a+b+c)^6\ge 27(a^2+b^2+c^2)(ab+bc+ca)^2.$$

Solution Denote $x=a+b+c$, $y=ab+bc+ca$.
Then we have

$$x^6\ge 27(x^2-2y)y^2$$
$$\Leftrightarrow\quad x^6\ge 27x^2y^2-54y^3$$
$$\Leftrightarrow\quad (x^2-3y)(x^4+3x^2y-18y^2)\ge 0,$$

which is true, since

$$x^2=(a+b+c)^2\ge 3(ab+bc+ca)=3y,\qquad x^4\ge 9y^2\quad\text{and}$$
$$3x^2y\ge 3\cdot 3y\cdot y=9y^2,$$

i.e. we have

$$x^2-3y\ge 0\quad\text{and}\quad x^4+3x^2y-18y^2\ge 9y^2+9y^2-18y^2=0.$$ ■

129 Let $a, b, c \in [1, 2]$ be real numbers. Prove the inequality

$$a^3 + b^3 + c^3 \leq 5abc.$$

Solution Without loss of generality we may assume that $a \geq b \geq c$.

Then since $a, b, c \in [1, 2]$ we have

$$b^2 + b + 1 \leq a^2 + a + 1 \leq 2a + a + 1 \leq 5a \quad \text{and}$$

$$c^2 + c + 1 \leq a^2 + a + 1 \leq 5a \leq 5ab.$$

Because of the previous inequalities it follows that:

$$a^3 + 2 \leq 5a \quad \Leftrightarrow \quad (a-2)(a^2 + 2a - 1) \leq 0, \tag{1}$$

$$5a + b^3 \leq 5ab + 1 \quad \Leftrightarrow \quad (b-1)(b^2 + b + 1 - 5a) \leq 0, \tag{2}$$

$$5ab + c^3 \leq 5abc + 1 \quad \Leftrightarrow \quad (c-1)(c^2 + c + 1 - 5ab) \leq 0. \tag{3}$$

Adding (1), (2) and (3) gives us the desired inequality.

Equality holds iff $a = 2, b = c = 1$. ■

130 Let a, b, c be positive real numbers such that $ab + bc + ca = 3$. Prove the inequality

$$(a^7 - a^4 + 3)(b^5 - b^2 + 3)(c^4 - c + 3) \geq 27.$$

Solution For any real number x, the numbers $x - 1, x^2 - 1, x^3 - 1$ and $x^4 - 1$, are of the same sign.

Therefore

$$(x-1)(x^3-1) \geq 0, \qquad (x^2-1)(x^3-1) \geq 0 \quad \text{and} \quad (x^3-1)(x^4-1) \geq 0,$$

i.e.

$$c^4 - c^3 - c + 1 \geq 0, \tag{1}$$

$$b^5 - b^3 - b^2 + 1 \geq 0, \tag{2}$$

$$a^7 - a^4 - a^3 + 1 \geq 0. \tag{3}$$

By (1), (2) and (3) we have

$$a^7 - a^4 + 3 \geq a^3 + 2, \qquad b^5 - b^2 + 3 \geq b^3 + 2 \quad \text{and} \quad c^4 - c + 3 \geq c^3 + 2.$$

After multiplying these inequalities it follows that

$$(a^7 - a^4 + 3)(b^5 - b^2 + 3)(c^4 - c + 3) \geq (a^3 + 2)(b^3 + 2)(c^3 + 2). \tag{4}$$

Analogously as in Problem 126, we can prove that

$$(a^3+2)(b^3+2)(c^3+2) \geq (a+b+c)^3. \tag{5}$$

By the obvious inequality $(a+b+c)^2 \geq 3(ab+bc+ca)$, since $ab+bc+ca=3$ we deduce that

$$a+b+c \geq 3. \tag{6}$$

Finally from (4), (5) and (6) we obtain the required inequality.

Equality occurs iff $a=b=c=1$. ■

131 Let $a, b, c \in [1,2]$ be real numbers. Prove the inequality

$$(a+b+c)\left(\frac{1}{a}+\frac{1}{b}+\frac{1}{c}\right) \leq 10.$$

Solution The given inequality is equivalent to

$$\frac{a}{b}+\frac{b}{c}+\frac{c}{a}+\frac{b}{a}+\frac{c}{b}+\frac{a}{c} \leq 7. \tag{1}$$

Without loss of generality we may assume that $a \geq b \geq c$.

Then, since $(a-b)(b-c) \geq 0$ we deduce that

$$ab+bc \geq b^2+ac, \quad \text{i.e.} \quad \frac{a}{c}+1 \geq \frac{a}{b}+\frac{b}{c}.$$

Analogously as $ab+bc \geq b^2+ac$ we have $\frac{c}{a}+1 \geq \frac{c}{b}+\frac{b}{a}$.

Now we obtain

$$\frac{a}{b}+\frac{b}{c}+\frac{c}{b}+\frac{b}{a} \leq \frac{a}{c}+\frac{c}{a}+2.$$

So

$$\frac{a}{b}+\frac{b}{c}+\frac{c}{a}+\frac{b}{a}+\frac{c}{b}+\frac{a}{c} \leq 2+2\left(\frac{a}{c}+\frac{c}{a}\right). \tag{2}$$

Let $x=\frac{a}{c}$. Then $2 \geq x \geq 1$, i.e. we have that $(x-2)(x-1) \leq 0$, from which we deduce that

$$x+\frac{1}{x} \leq \frac{5}{2}. \tag{3}$$

Finally using (2) and (3) we obtain inequality (1).

Equality occurs iff $a=b=2, c=1$ or $a=2, b=c=1$. ■

132 Let $a, b, c \in \mathbb{R}^+$ such that $a+b+c=1$. Prove the inequality

$$10(a^3+b^3+c^3)-9(a^5+b^5+c^5) \geq 1.$$

Solution Denote $L = 10(a^3 + b^3 + c^3) - 9(a^5 + b^5 + c^5)$.

Let $x = a + b + c = 1$, $y = ab + bc + ca$, $z = abc$.

Then

$$\begin{aligned}10(a^3 + b^3 + c^3) &= 10((a + b + c)^3 - 3(a + b + c)(ab + bc + ca) + 3abc)\\ &= 10 - 30y + 30z\end{aligned}$$

and

$$\begin{aligned}9(a^5 + b^5 + c^5) &= 9(x^5 - 5x^3y + 5xy^2 + 5x^2z - 5yz)\\ &= 9(1 - 5y + 5y^2 + 5z - 5yz)\\ &= 9 - 45y + 45y^2 + 45z - 45yz.\end{aligned}$$

We have

$$L \geq 1 \quad \Leftrightarrow \quad 10 - 30y + 30z - 9 + 45y - 45y^2 - 45z + 45yz \geq 1,$$

i.e.

$$1 + 15y - 15z - 45y^2 + 45yz \geq 1,$$

i.e.

$$y - z - 3y(y - z) \geq 0 \quad \Leftrightarrow \quad (1 - 3y)(y - z) \geq 0. \tag{1}$$

Furthermore,

$$y = ab + bc + ca \leq \frac{(a + b + c)^2}{3} = \frac{1}{3}, \quad \text{i.e.} \quad 1 - 3y \geq 0$$

and

$$y = ab + bc + ca \geq 3\sqrt[3]{a^2b^2c^2} = 3\sqrt[3]{z^2} > z.$$

The last inequality is true since

$$z = abc \leq \left(\frac{a + b + c}{3}\right)^3 = 1 < 27.$$

From the previous two inequalities we get inequality (1), as desired. ■

133 Let $n \in \mathbb{N}$ and $x_1, x_2, \ldots, x_n \in (0, \pi)$. Find maximum value of the expression

$$\sin x_1 \cos x_2 + \sin x_2 \cos x_3 + \cdots + \sin x_n \cos x_1.$$

Solution It's clear that for all real numbers a, b we have $a^2 + b^2 \geq 2ab$. So we obtain

$$\begin{aligned}&\sin x_1 \cos x_2 + \sin x_2 \cos x_3 + \cdots + \sin x_n \cos x_1\\ &\quad\leq \frac{\sin^2 x_1 + \cos^2 x_2}{2} + \frac{\sin^2 x_2 + \cos^2 x_3}{2} + \cdots + \frac{\sin^2 x_n + \cos^2 x_1}{2} = \frac{n}{2}.\end{aligned}$$

Equality occurs iff $x_1 = x_2 = \cdots = x_n = \frac{\pi}{4}$. ■

134 Let $\alpha_i \in [\frac{\pi}{4}, \frac{5\pi}{4}]$, for $i = 1, 2, \ldots, n$. Prove the inequality

$$\left(\sin\alpha_1 + \sin\alpha_2 + \cdots + \sin\alpha_n + \frac{1}{4}\right)^2 \geq (\cos\alpha_1 + \cos\alpha_2 + \cdots + \cos\alpha_n).$$

Solution Let $S = \sin\alpha_1 + \sin\alpha_2 + \cdots + \sin\alpha_n$.
We have

$$\left(S + \frac{1}{4}\right)^2 = S^2 + \frac{S}{2} + \frac{1}{16} = S^2 - \frac{S}{2} + \frac{1}{16} + S = \left(S - \frac{1}{4}\right)^2 + S \geq S. \quad (1)$$

Since $\alpha_i \in [\frac{\pi}{4}, \frac{5\pi}{4}]$ we deduce that

$$\sin\alpha_i \geq \cos\alpha_i, \quad \text{for all } i = 1, 2, \ldots, n. \quad (2)$$

Using (1) and (2) we obtain the required inequality. ■

135 Let $a_1, a_2, \ldots, a_n; a_{n+1} = a_1, a_{n+2} = a_2$ be positive real numbers. Prove the inequality

$$\sum_{i=1}^{n} \frac{a_i - a_{i+2}}{a_{i+1} + a_{i+2}} \geq 0.$$

Solution Applying $AM \geq GM$ we have

$$\begin{aligned}\sum_{i=1}^{n} \frac{a_i + a_{i+1}}{a_{i+1} + a_{i+2}} &= \frac{a_1 + a_2}{a_2 + a_3} + \frac{a_2 + a_3}{a_3 + a_4} + \cdots + \frac{a_{n-1} + a_n}{a_n + a_1} + \frac{a_n + a_1}{a_1 + a_2} \\ &\geq n\sqrt[n]{\frac{a_1 + a_2}{a_2 + a_3} \cdot \frac{a_2 + a_3}{a_3 + a_4} \cdots \frac{a_{n-1} + a_n}{a_n + a_1} \cdot \frac{a_n + a_1}{a_1 + a_2}} = n. \quad (1)\end{aligned}$$

So

$$\begin{aligned}\sum_{i=1}^{n} \frac{a_{i+1}}{a_{i+1} + a_{i+2}} &= \sum_{i=1}^{n} \frac{a_{i+1} + a_{i+2}}{a_{i+1} + a_{i+2}} - \sum_{i=1}^{n} \frac{a_{i+2}}{a_{i+1} + a_{i+2}} \\ &= n - \sum_{i=1}^{n} \frac{a_{i+2}}{a_{i+1} + a_{i+2}} \\ &\overset{(1)}{\leq} \sum_{i=1}^{n} \frac{a_i + a_{i+1}}{a_{i+1} + a_{i+2}} - \sum_{i=1}^{n} \frac{a_{i+2}}{a_{i+1} + a_{i+2}},\end{aligned}$$

from where it follows that

$$\sum_{i=1}^{n} \frac{a_i - a_{i+2}}{a_{i+1} + a_{i+2}} \geq 0.$$

■

136 Let $n \geq 2, n \in \mathbb{N}$ and $x_1, x_2, \ldots, x_n$ be positive real numbers such that

$$\frac{1}{x_1+1998}+\frac{1}{x_2+1998}+\cdots+\frac{1}{x_n+1998}=\frac{1}{1998}.$$

Prove the inequality

$$\sqrt[n]{x_1 x_2 \cdots x_n} \geq 1998(n-1).$$

Solution After setting $\frac{1998}{x_i+1998}=a_i$, for $i=1,2,\ldots,n$, the identity

$$\frac{1}{x_1+1998}+\frac{1}{x_2+1998}+\cdots+\frac{1}{x_n+1998}=\frac{1}{1998}$$

becomes

$$a_1+a_2+\cdots+a_n=1.$$

We need to show that

$$\left(\frac{1}{a_1}-1\right)\left(\frac{1}{a_2}-1\right)\cdots\left(\frac{1}{a_n}-1\right) \geq (n-1)^n. \tag{1}$$

We have

$$\begin{aligned}\frac{1}{a_i}-1=\frac{1-a_i}{a_i}&=\frac{a_1+\cdots+a_{i-1}+a_{i+1}+\cdots+a_n}{a_i}\\ &\geq (n-1)\sqrt[n-1]{\frac{a_1\cdots a_{i-1}a_{i+1}\cdots a_n}{a_i^{n-1}}}.\end{aligned}$$

Multiplying these inequalities for $i=1,2,\ldots,n$ we obtain (1), as desired. ■

137 Let $a_1, a_2, \ldots, a_n \in \mathbb{R}^+$. Prove the inequality

$$\sum_{k=1}^{n} k a_k \leq \binom{n}{2}+\sum_{k=1}^{n} a_k^k.$$

Solution For $1 \leq k \leq n$ we have

$$a_k^k+(k-1)=a_k^k+\underbrace{1+1+\cdots+1}_{k-1} \geq k\sqrt[k]{a_k^k}=k a_k.$$

After adding these inequalities, for $1 \leq k \leq n$ we get

$$\sum_{k=1}^{n} k a_k \leq \sum_{k=1}^{n} a_k^k+\sum_{k=1}^{n}(k-1)=\sum_{k=1}^{n} a_k^k+\frac{n(n-1)}{2}=\sum_{k=1}^{n} a_k^k+\binom{n}{2}.$$ ■

138 Let $a_1, a_2, \ldots, a_n$ be positive real numbers such that $a_1 + a_2 + \cdots + a_n = n$. Prove that for every natural number k the following inequality holds

$$a_1^k + a_2^k + \cdots + a_n^k \geq a_1^{k-1} + a_2^{k-1} + \cdots + a_n^{k-1}.$$

Solution Using $AM \geq GM$ we get

$$(k-1)a_i^k + 1 = \underbrace{a_i^k + a_i^k + \cdots + a_i^k}_{k-1} + 1 \geq k\sqrt[k]{a_i^{k(k-1)}} = ka_i^{k-1}$$

and if we add these inequalities for $i = 1, 2, \ldots, n$ we obtain

$$(k-1)(a_1^k + a_2^k + \cdots + a_n^k) + n \geq k(a_1^{k-1} + a_2^{k-1} + \cdots + a_n^{k-1}). \qquad (1)$$

We'll show that

$$a_1^{k-1} + a_2^{k-1} + \cdots + a_n^{k-1} \geq n. \qquad (2)$$

One more application of $AM \geq GM$ gives us

$$a_i^{k-1} + (k-2) = a_i^{k-1} + \underbrace{1 + \cdots + 1}_{k-2} \geq (k-1)\sqrt[k-1]{a_i^{k-1}} = (k-1)a_i$$

and adding the previous inequalities for $i = 1, 2, \ldots, n$ we get

$$(a_1^{k-1} + a_2^{k-1} + \cdots + a_n^{k-1}) + n(k-2) \geq (k-1)(a_1 + a_2 + \cdots + a_n) = n(k-1),$$

from which we deduce

$$a_1^{k-1} + a_2^{k-1} + \cdots + a_n^{k-1} \geq n.$$

So we are done with (2).

Now from (1) and (2) we obtain

$$\begin{aligned}
&(k-1)(a_1^k + a_2^k + \cdots + a_n^k) + n \geq k(a_1^{k-1} + a_2^{k-1} + \cdots + a_n^{k-1}) \\
&\quad\Leftrightarrow\quad (k-1)(a_1^k + a_2^k + \cdots + a_n^k) + n \\
&\qquad\quad \geq (k-1)(a_1^{k-1} + a_2^{k-1} + \cdots + a_n^{k-1}) + (a_1^{k-1} + a_2^{k-1} + \cdots + a_n^{k-1}) \\
&\qquad\quad \geq (k-1)(a_1^{k-1} + a_2^{k-1} + \cdots + a_n^{k-1}) + n \\
&\quad\Leftrightarrow\quad a_1^k + a_2^k + \cdots + a_n^k \geq a_1^{k-1} + a_2^{k-1} + \cdots + a_n^{k-1},
\end{aligned}$$

as desired.

Equality holds iff $a_1 = a_2 = \cdots = a_n = 1$. ■

Remark The given inequality immediately follows by *Chebishev's inequality*.

139 Let a, b, c, d be positive real numbers. Prove the inequality

$$\left(\frac{a}{a+b}\right)^5+\left(\frac{b}{b+c}\right)^5+\left(\frac{c}{c+d}\right)^5+\left(\frac{d}{d+a}\right)^5\geq\frac{1}{8}.$$

Solution 1 Let $x=b/a$, $y=c/b$, $z=d/c$ and $t=a/d$.

Then it is clear that $xyzt=1$, and the given inequality becomes

$$A=\left(\frac{1}{1+x}\right)^5+\left(\frac{1}{1+y}\right)^5+\left(\frac{1}{1+z}\right)^5+\left(\frac{1}{1+t}\right)^5\geq\frac{1}{8}. \tag{1}$$

By the inequality $AM\geq GM$ we have

$$2\left(\frac{1}{1+x}\right)^5+\frac{3}{32}=\left(\frac{1}{1+x}\right)^5+\left(\frac{1}{1+x}\right)^5+\frac{1}{32}+\frac{1}{32}+\frac{1}{32}\geq\frac{5}{8}\left(\frac{1}{1+x}\right)^2,$$

i.e.

$$2\left(\frac{1}{1+x}\right)^5+\frac{3}{32}\geq\frac{5}{8}\left(\frac{1}{1+x}\right)^2.$$

So it follows that

$$2A+\frac{12}{32}\geq\frac{5}{8}\left(\left(\frac{1}{1+x}\right)^2+\left(\frac{1}{1+y}\right)^2+\left(\frac{1}{1+z}\right)^2+\left(\frac{1}{1+t}\right)^2\right). \tag{2}$$

We'll prove that for all positive real numbers x and y the following inequality holds

$$\frac{1}{(1+x)^2}+\frac{1}{(1+y)^2}\geq\frac{1}{1+xy}.$$

We have

$$\begin{aligned}\frac{1}{(1+x)^2}+\frac{1}{(1+y)^2}-\frac{1}{1+xy}&=\frac{xy(x^2+y^2)-x^2y^2-2xy+1}{(1+x)^2(1+y)^2(1+xy)}\\&=\frac{xy(x-y)^2+(xy-1)^2}{(1+x)^2(1+y)^2(1+xy)}\geq 0.\end{aligned}$$

Now according to the previous inequality and the condition $xyzt=1$, we deduce

$$\begin{aligned}&\left(\frac{1}{1+x}\right)^2+\left(\frac{1}{1+y}\right)^2+\left(\frac{1}{1+z}\right)^2+\left(\frac{1}{1+t}\right)^2\\&\quad\geq\frac{1}{1+xy}+\frac{1}{1+zt}=\frac{1}{1+xy}+\frac{1}{1+1/xy}=1.\end{aligned} \tag{3}$$

By (2) and (3) we get

$$2A+\frac{12}{32}\geq\frac{5}{8},\quad \text{i.e.}\quad A\geq\frac{1}{8}.$$

Equality occurs iff $x=y=z=t=1$, i.e. $a=b=c=d$. ■

140 Let $x_1, x_2, \ldots, x_n$ be positive real numbers not greater then 1. Prove the inequality

$$(1+x_1)^{\frac{1}{x_2}}(1+x_2)^{\frac{1}{x_3}}\cdots(1+x_n)^{\frac{1}{x_1}} \geq 2^n.$$

Solution From $0 < x_1, x_2, \ldots, x_n \leq 1$ it follows that

$$\frac{1}{x_1}, \frac{1}{x_2}, \ldots, \frac{1}{x_n} \geq 1.$$

By Corollary 4.7, (Chap. 4) we have that for every $x > -1$ and $\alpha \in [1, \infty)$, the following inequality

$$(1+x)^{\alpha} \geq 1 + x\alpha$$

holds.

Hence we get

$$(1+x_1)^{\frac{1}{x_2}}(1+x_2)^{\frac{1}{x_3}}\cdots(1+x_n)^{\frac{1}{x_1}} \geq \left(1+\frac{x_1}{x_2}\right)\left(1+\frac{x_2}{x_3}\right)\cdots\left(1+\frac{x_n}{x_1}\right). \quad (1)$$

Furthermore, applying $AM \geq GM$ we get

$$\left(1+\frac{x_1}{x_2}\right)\left(1+\frac{x_2}{x_3}\right)\cdots\left(1+\frac{x_n}{x_1}\right) \geq 2\sqrt{\frac{x_1}{x_2}}\cdot 2\sqrt{\frac{x_2}{x_3}}\cdot \cdots \cdot 2\sqrt{\frac{x_n}{x_1}} = 2^n. \quad (2)$$

By (1) and (2) we obtain

$$(1+x_1)^{\frac{1}{x_2}}(1+x_2)^{\frac{1}{x_3}}\cdots(1+x_n)^{\frac{1}{x_1}} \geq 2^n.$$

Equality occurs iff $x_1 = x_2 = \cdots = x_n = 1$. ■

141 Let $x_1, x_2, \ldots, x_n$ be non-negative real numbers such that $x_1 + x_2 + \cdots + x_n \leq \frac{1}{2}$. Prove the inequality

$$(1-x_1)(1-x_2)\cdots(1-x_n) \geq \frac{1}{2}.$$

Solution From $x_1 + x_2 + \cdots + x_n \leq \frac{1}{2}$ and the fact that $x_1, x_2, \ldots, x_n$ are non-negative we deduce that

$$0 \leq x_i \leq \frac{1}{2} < 1, \quad \text{i.e.} \quad -x_i > -1, \quad \text{for all } i = 1, 2, \ldots, n,$$

and it's clear that all $-x_i$ are of the same sign.

Applying *Bernoulli's inequality* we obtain

$$\begin{aligned}(1-x_1)(1-x_2)\cdots(1-x_n) &= (1+(-x_1))(1+(-x_2))\cdots(1+(-x_n)) \\ &\geq 1 + (-x_1 - x_2 - \cdots - x_n) \\ &= 1 - (x_1 + x_2 + \cdots + x_n) \geq 1 - \frac{1}{2} = \frac{1}{2}.\end{aligned}$$ ■

142 Let $a, b, c \in \mathbb{R}^+$ such that $abc = 1$. Prove the inequality

$$\frac{1}{a^3+b^3+1} + \frac{1}{b^3+c^3+1} + \frac{1}{c^3+a^3+1} \leq 1.$$

Solution We have

$$\frac{1}{a^3+b^3+1} = \frac{1}{(a+b)((a-b)^2+ab)+1} \leq \frac{1}{(a+b)ab+1},$$

and since $ab = \frac{1}{c}$ we deduce

$$\frac{1}{a^3+b^3+1} \leq \frac{1}{(a+b)ab+1} = \frac{c}{a+b+c}.$$

Similarly

$$\frac{1}{b^3+c^3+1} \leq \frac{a}{a+b+c} \quad \text{and} \quad \frac{1}{c^3+a^3+1} \leq \frac{b}{a+b+c}.$$

Adding the last three inequalities we obtain the required inequality.
Equality holds if and only if $a = b = c = 1$. ■

143 Let $0 \leq a, b, c \leq 1$. Prove the inequality

$$\frac{c}{7+a^3+b^3} + \frac{b}{7+c^3+a^3} + \frac{a}{7+b^3+c^3} \leq \frac{1}{3}.$$

Solution Since $0 \leq a, b, c \leq 1$ it follows that $0 \leq a^3, b^3, c^3 \leq 1$, so we have

$$\begin{aligned}
&\frac{c}{7+a^3+b^3} + \frac{b}{7+c^3+a^3} + \frac{a}{7+b^3+c^3} \\
&\quad \leq \frac{c}{6+a^3+b^3+c^3} + \frac{b}{6+c^3+a^3+b^3} + \frac{a}{6+b^3+c^3+a^3} \\
&\quad = \frac{a+b+c}{6+a^3+b^3+c^3}.
\end{aligned}$$

It suffices to prove that

$$3(a+b+c) \leq 6+a^3+b^3+c^3,$$

which is true since $t^3-3t+2 = (t-1)^2(t+2) \geq 0$, for $0 \leq t \leq 1$. ■

144 Let $a, b, c \in \mathbb{R}^+$ such that $abc = 1$. Prove the inequality

$$\frac{ab}{a^5+ab+b^5} + \frac{bc}{b^5+bc+c^5} + \frac{ca}{c^5+ca+a^5} \leq 1.$$

Solution Since

$$a^4 - a^3b - ab^3 + b^4 = a^3(a-b) - b^3(a-b) = (a-b)^2(a^2 - ab + b^2) \geq 0,$$

we have

$$a^5 + b^5 = (a+b)(a^4 - a^3b + a^2b^2 - ab^3 + b^4) \geq (a+b)a^2b^2.$$

So

$$\frac{ab}{a^5 + ab + b^5} \leq \frac{ab}{(a+b)a^2b^2 + ab} = \frac{abc^2}{(a+b)a^2b^2c^2 + abc^2} = \frac{c}{a+b+c}. \tag{1}$$

Analogously

$$\frac{bc}{b^5 + bc + c^5} \leq \frac{a}{a+b+c} \tag{2}$$

and

$$\frac{ca}{c^5 + ca + a^5} \leq \frac{b}{a+b+c}. \tag{3}$$

Adding (1), (2) and (3) gives us the required inequality. ■

145 Let $a, b, c \in \mathbb{R}^+$ such that $a + b + c = 3$. Prove the inequality

$$\frac{a^3}{a^2 + ab + b^2} + \frac{b^3}{b^2 + bc + c^2} + \frac{c^3}{c^2 + ca + a^2} \geq 1.$$

Solution We'll show that

$$A = \frac{a^3}{a^2 + ab + b^2} + \frac{b^3}{b^2 + bc + c^2} + \frac{c^3}{c^2 + ca + a^2} \geq \frac{a+b+c}{3}.$$

For every $x, y \in \mathbb{R}^+$ we have $\frac{x^3+y^3}{x^2+xy+y^2} \geq \frac{x+y}{3}$, in which equality occurs iff $x = y$.
(This inequality follows from the obvious inequality $2(x+y)(x-y)^2 \geq 0$.)
On the other hand, we have

$$A = \frac{a^3}{a^2 + ab + b^2} + \frac{b^3}{b^2 + bc + c^2} + \frac{c^3}{c^2 + ca + a^2} = \frac{b^3}{a^2 + ab + b^2} + \frac{c^3}{b^2 + bc + c^2} + \frac{a^3}{c^2 + ca + a^2},$$

so

$$2A = \frac{a^3 + b^3}{a^2 + ab + b^2} + \frac{b^3 + c^3}{b^2 + bc + c^2} + \frac{c^3 + a^3}{c^2 + ca + a^2} \geq \frac{a+b}{3} + \frac{b+c}{3} + \frac{c+a}{3},$$

i.e.

$$A \geq \frac{a+b+c}{3} = 1.$$

Equality occurs if and only if $a = b = c = 1/3$. ■

146 Let a, b, c be positive real numbers such that $a^2 + b^2 + c^2 = 3abc$. Prove the inequality

$$\frac{a}{b^2c^2} + \frac{b}{c^2a^2} + \frac{c}{a^2b^2} \geq \frac{9}{a+b+c}.$$

Solution The given inequality is equivalent to

$$(a^3 + b^3 + c^3)(a + b + c) \geq 9a^2b^2c^2.$$

Applying the *Cauchy–Schwarz inequality* we have

$$(a^3 + b^3 + c^3)(a + b + c) \geq (a^2 + b^2 + c^2)^2.$$

Since $a^2 + b^2 + c^2 = 3abc$ we obtain

$$(a^3 + b^3 + c^3)(a + b + c) \geq (a^2 + b^2 + c^2)^2 = (3abc)^2 = 9a^2b^2c^2.$$

Equality holds if and only if $a = b = c = 1$. ■

147 Let a, b, c, x, y, z be positive real number, and let $a + b = 3$. Prove the inequality

$$\frac{x}{ay+bz} + \frac{y}{az+bx} + \frac{z}{ax+by} \geq 1.$$

Solution We'll show that

$$\frac{x}{ay+bz} + \frac{y}{az+bx} + \frac{z}{ax+by} \geq \frac{3}{a+b},$$

and combining with $a + b = 3$ will give us the required inequality.

Applying the *Cauchy–Schwarz inequality* we have

$$\begin{aligned}
&\frac{x}{ay+bz} + \frac{y}{az+bx} + \frac{z}{ax+by} \\
&\quad = \frac{x^2}{axy+bxz} + \frac{y^2}{ayz+bxy} + \frac{z^2}{axz+byz} \geq \frac{(x+y+z)^2}{(a+b)(xy+yz+zx)} \\
&\quad \geq \frac{3}{a+b} = 1.
\end{aligned}$$

■

148 Let $x, y, z > 0$ be real numbers. Prove the inequality

$$\frac{x}{x+2y+3z}+\frac{y}{y+2z+3x}+\frac{z}{z+2x+3y}\geq\frac{1}{2}.$$

Solution The *Cauchy–Schwarz inequality* gives us

$$\begin{aligned}&\frac{x^2}{x^2+2xy+3xz}+\frac{y^2}{y^2+2yz+3xy}+\frac{z^2}{z^2+2xz+3yz}\\&\geq\frac{(x+y+z)^2}{x^2+y^2+z^2+5(xy+yz+zx)}.\end{aligned}$$

It suffices to prove that

$$2(x+y+z)^2\geq x^2+y^2+z^2+5(xy+yz+zx),$$

which is exactly $x^2+y^2+z^2\geq xy+yz+zx$, and clearly holds. ∎

149 Let $a, b, c, d \in \mathbb{R}^+$. Prove the inequality

$$\frac{c}{a+3b}+\frac{d}{b+3c}+\frac{a}{c+3d}+\frac{b}{d+3a}\geq 1.$$

Solution Let $L=\frac{c^2}{ac+3bc}+\frac{d^2}{bd+3cd}+\frac{a^2}{ca+3da}+\frac{b^2}{bd+3ab}$.

Applying the *Cauchy–Schwarz inequality* we get

$$\begin{aligned}&((ac+3bc)+(bd+3cd)+(ca+3da)+(bd+3ab))\cdot L\geq(a+b+c+d)^2\\ \Leftrightarrow\quad & L\geq\frac{(a+b+c+d)^2}{2ac+2bd+3bc+3cd+3ad+3ab}.\end{aligned}$$

It suffices to prove that

$$\begin{aligned}&(a+b+c+d)^2\geq 2ac+2bd+3bc+3cd+3ad+3ab\\ \Leftrightarrow\quad &(a-b)^2+(a-d)^2+(b-c)^2+(c-d)^2\geq 0,\end{aligned}$$

which is clearly true.

Equality holds iff $a=b=c=d$. ∎

150 Let a, b, c, d, e be positive real numbers. Prove the inequality

$$\frac{a}{b+c}+\frac{b}{c+d}+\frac{c}{d+e}+\frac{d}{e+a}+\frac{e}{a+b}\geq\frac{5}{2}.$$

Solution Applying the *Cauchy–Schwarz inequality* we have

$$\begin{aligned}
&\frac{a}{b+c}+\frac{b}{c+d}+\frac{c}{d+e}+\frac{d}{e+a}+\frac{e}{a+b} \\
&\quad=\frac{a^2}{ab+ac}+\frac{b^2}{bc+bd}+\frac{c^2}{cd+ce}+\frac{d^2}{de+ad}+\frac{e^2}{ae+be} \\
&\quad\geq\frac{(a+b+c+d+e)^2}{ab+ac+ad+ae+bc+bd+be+cd+ce+de}.
\end{aligned}$$

So it is suffices to show that

$$\frac{(a+b+c+d+e)^2}{ab+ac+ad+ae+bc+bd+be+cd+ce+de}\geq\frac{5}{2},$$

which clearly holds (Why?). ■

151 Prove that for all positive real numbers a, b, c the following inequality holds

$$\frac{a^3}{a^2+ab+b^2}+\frac{b^3}{b^2+bc+c^2}+\frac{c^3}{c^2+ca+a^2}\geq\frac{a^2+b^2+c^2}{a+b+c}.$$

Solution Applying the *Cauchy–Schwarz inequality* we have

$$\begin{aligned}
A&=\frac{a^3}{a^2+ab+b^2}+\frac{b^3}{b^2+bc+c^2}+\frac{c^3}{c^2+ca+a^2} \\
&=\frac{a^4}{a(a^2+ab+b^2)}+\frac{b^4}{b(b^2+bc+c^2)}+\frac{c^4}{c(c^2+ca+a^2)} \\
&\geq\frac{(a^2+b^2+c^2)^2}{(a(a^2+ab+b^2)+b(b^2+bc+c^2)+c(c^2+ca+a^2))}.
\end{aligned}$$

So it suffices to prove that

$$(a+b+c)(a^2+b^2+c^2)\geq a(a^2+ab+b^2)+b(b^2+bc+c^2)+c(c^2+ca+a^2),$$

which is true. ■

152 Let a, b, c be positive real numbers such that $ab+bc+ca=1$. Prove the inequality

$$\frac{1}{4a^2-bc+1}+\frac{1}{4b^2-ca+1}+\frac{1}{4c^2-ab+1}\geq\frac{3}{2}.$$

Solution Since $1-bc=ac+ab$, $1-ca=ab+bc$ and $1-ab=ac+bc$, the given inequality can be rewritten as

$$\frac{1}{a(4a+b+c)}+\frac{1}{b(4b+c+a)}+\frac{1}{c(4c+a+b)}\geq\frac{3}{2}.$$

By the *Cauchy–Schwarz inequality* we get

$$\left(\frac{1}{a(4a+b+c)}+\frac{1}{b(4b+c+a)}+\frac{1}{c(4c+a+b)}\right)$$
$$\times\left(\frac{4a+b+c}{a}+\frac{4b+c+a}{b}+\frac{4c+a+b}{c}\right)$$
$$\geq\left(\frac{1}{a}+\frac{1}{b}+\frac{1}{c}\right)^2=\frac{1}{a^2b^2c^2}.$$

So it suffices to prove that

$$\frac{2}{3a^2b^2c^2}\geq\frac{4a+b+c}{a}+\frac{4b+c+a}{b}+\frac{4c+a+b}{c}. \tag{1}$$

We have

$$\begin{aligned}\frac{4a+b+c}{a}+\frac{4b+c+a}{b}+\frac{4c+a+b}{c} &=9+\frac{a+b+c}{a}+\frac{b+c+a}{b}+\frac{c+a+b}{c}\\ &=9+(a+b+c)\left(\frac{1}{a}+\frac{1}{b}+\frac{1}{c}\right)\\ &=9+\frac{(a+b+c)(ab+bc+ca)}{abc}\\ &=9+\frac{a+b+c}{abc},\end{aligned}$$

so inequality (1) becomes

$$\frac{2}{3a^2b^2c^2}\geq 9+\frac{a+b+c}{abc}, \quad \text{i.e.} \quad 27a^2b^2c^2+3abc(a+b+c)\leq 2. \tag{2}$$

By $AM \geq GM$ we have

$$1=ab+bc+ca\geq 3\sqrt[3]{a^2b^2c^2}, \quad \text{i.e.} \quad 27a^2b^2c^2\leq 1. \tag{3}$$

By the well-known inequality $(x+y+z)^2\geq 3(xy+yz+zx)$ we get

$$3abc(a+b+c)\leq(ab+bc+ca)^2=1. \tag{4}$$

Finally by (3) and (4) we get inequality (2), as required.

Equality occurs iff $a=b=c=\frac{1}{\sqrt{3}}$. ■

153 Let a,b,c be positive real numbers such that

$$\frac{1}{a^2+b^2+1}+\frac{1}{b^2+c^2+1}+\frac{1}{c^2+a^2+1}\geq 1.$$

Prove the inequality

$$ab + bc + ca \le 3.$$

Solution Using the *Cauchy–Schwarz inequality* gives us

$$(a^2 + b^2 + 1)(1 + 1 + c^2) \ge (a + b + c)^2, \quad \text{i.e.} \quad \frac{1}{a^2 + b^2 + 1} \le \frac{2 + c^2}{(a + b + c)^2}.$$

Analogous we obtain

$$\frac{1}{b^2 + c^2 + 1} \le \frac{2 + a^2}{(a + b + c)^2} \quad \text{and} \quad \frac{1}{c^2 + a^2 + 1} \le \frac{2 + b^2}{(a + b + c)^2}.$$

So we have

$$1 \le \frac{1}{a^2 + b^2 + 1} + \frac{1}{b^2 + c^2 + 1} + \frac{1}{c^2 + a^2 + 1} \le \frac{6 + a^2 + b^2 + c^2}{(a + b + c)^2},$$

i.e.

$$6 + a^2 + b^2 + c^2 \ge (a + b + c)^2, \quad \text{i.e.} \quad ab + bc + ca \le 3. \qquad \blacksquare$$

154 Let a, b, c be positive real numbers such that $ab + bc + ca = 1/3$. Prove the inequality

$$\frac{a}{a^2 - bc + 1} + \frac{b}{b^2 - ca + 1} + \frac{c}{c^2 - ab + 1} \ge \frac{1}{a + b + c}.$$

Solution Applying the *Cauchy–Schwarz inequality* we have

$$\begin{aligned} &\frac{a}{a^2 - bc + 1} + \frac{b}{b^2 - ca + 1} + \frac{c}{c^2 - ab + 1} \\ &\quad = \frac{a^2}{a^3 - abc + a} + \frac{b^2}{b^3 - abc + b} + \frac{c^2}{c^3 - abc + c} \\ &\quad \ge \frac{(a + b + c)^2}{a^3 + b^3 + c^3 + a + b + c - 3abc}. \end{aligned}$$

Furthermore, since

$$\begin{aligned} a^3 + b^3 + c^3 - 3abc &= (a + b + c)(a^2 + b^2 + c^2 - ab - bc - ca) \\ &= (a + b + c)(a^2 + b^2 + c^2 - 1/3), \end{aligned}$$

we obtain

$$\frac{(a+b+c)^2}{a^3+b^3+c^3+a+b+c-3abc} = \frac{(a+b+c)^2}{(a+b+c)(a^2+b^2+c^2+1-1/3)}$$
$$= \frac{a+b+c}{a^2+b^2+c^2+2/3}$$
$$= \frac{a+b+c}{a^2+b^2+c^2+2(ab+bc+ca)}$$
$$= \frac{1}{a+b+c},$$

as required. ■

155 Let a, b, c be positive real numbers. Prove the inequality

$$\frac{a^3}{a^3+b^3+abc} + \frac{b^3}{b^3+c^3+abc} + \frac{c^3}{c^3+a^3+abc} \geq 1.$$

Solution Let $x = \frac{b}{a}$, $y = \frac{c}{b}$, $z = \frac{a}{c}$. Then clearly $xyz = 1$.
Therefore

$$\frac{a^3}{a^3+b^3+abc} = \frac{1}{1+x^3+\frac{x}{z}} = \frac{1}{1+x^3+x^2y} = \frac{xyz}{xyz+x^3+x^2y}$$
$$= \frac{yz}{yz+x^2+xy}.$$

Similarly we deduce

$$\frac{b^3}{b^3+c^3+abc} = \frac{xz}{xz+y^2+zy} \quad \text{and} \quad \frac{c^3}{c^3+a^3+abc} = \frac{xy}{xy+z^2+xz}.$$

So it suffices to prove that

$$\frac{yz}{yz+x^2+xy} + \frac{xz}{xz+y^2+zy} + \frac{xy}{xy+z^2+xz} \geq 1.$$

According to the *Cauchy–Schwarz inequality* (Corollary 4.3, Chap. 4) we have

$$\frac{yz}{yz+x^2+xy} + \frac{xz}{xz+y^2+zy} + \frac{xy}{xy+z^2+xz}$$
$$\geq \frac{(xy+yz+zx)^2}{yz(yz+x^2+xy)+xz(xz+y^2+zy)+xy(xy+z^2+xz)}.$$

We need to prove that

$$(xy+yz+zx)^2 \geq yz(yz+x^2+xy)+xz(xz+y^2+zy)+xy(xy+z^2+xz),$$

which is in fact an equality.

Equality holds iff $x = y = z$, i.e. $a = b = c$. ■

156 Let a, b, c be positive real numbers such that $a^2 + b^2 + c^2 = 3$. Prove the inequality

$$\frac{a}{a^2+2b+3} + \frac{b}{b^2+2c+3} + \frac{c}{c^2+2a+3} \le \frac{1}{2}.$$

Solution Clearly $x^2 + 1 \ge 2x$, for every real x, and therefore

$$\frac{a}{a^2+2b+3} + \frac{b}{b^2+2c+3} + \frac{c}{c^2+2a+3} \le \frac{a}{2(a+b+1)} + \frac{b}{2(b+c+1)} + \frac{c}{2(c+a+1)}.$$

So it remains to prove that

$$\frac{a}{a+b+1} + \frac{b}{b+c+1} + \frac{c}{c+a+1} \le 1. \tag{1}$$

Inequality (1) is equivalent to

$$\frac{b+1}{a+b+1} + \frac{c+1}{b+c+1} + \frac{a+1}{c+a+1} \ge 2.$$

According to the *Cauchy–Schwarz inequality* (Corollary 4.3) we have

$$\frac{b+1}{a+b+1} + \frac{c+1}{b+c+1} + \frac{a+1}{c+a+1} \ge \frac{(a+b+c+3)^2}{(b+1)(a+b+1)+(c+1)(b+c+1)+(a+1)(c+a+1)} = 2.$$

Equality holds iff $a = b = c = 1$. ■

157 Let $a, b, c, d > 1$ be real numbers. Prove the inequality

$$\sqrt{a-1} + \sqrt{b-1} + \sqrt{c-1} + \sqrt{d-1} \le \sqrt{(ab+1)(cd+1)}.$$

Solution We'll prove that for every $x, y \in \mathbb{R}^+$ we have $\sqrt{x-1} + \sqrt{y-1} \le \sqrt{xy}$.

Applying the *Cauchy–Schwarz inequality* for $a_1 = \sqrt{x-1}, a_2 = 1; b_1 = 1, b_2 = \sqrt{y-1}$ gives us

$$(\sqrt{x-1} + \sqrt{y-1})^2 \le xy, \quad \text{i.e.} \quad \sqrt{x-1} + \sqrt{y-1} \le \sqrt{xy}.$$

Now we easily deduce that

$$\sqrt{a-1} + \sqrt{b-1} + \sqrt{c-1} + \sqrt{d-1} \le \sqrt{ab} + \sqrt{cd} \le \sqrt{(ab+1)(cd+1)}.$$ ■

158 Let $a_1, a_2, \ldots, a_n \in \mathbb{R}^+$ such that $a_1 a_2 \cdots a_n = 1$. Prove the inequality

$$\sqrt{a_1} + \sqrt{a_2} + \cdots + \sqrt{a_n} \le a_1 + a_2 + \cdots + a_n.$$

Solution Applying $AM \geq GM$ we obtain

$$\frac{\sqrt{a_1}+\sqrt{a_2}+\cdots+\sqrt{a_n}}{n} \geq \sqrt[n]{\sqrt{a_1}\sqrt{a_2}\cdots\sqrt{a_n}} = 1$$

i.e.

$$\sqrt{a_1}+\sqrt{a_2}+\cdots+\sqrt{a_n} \geq n. \tag{1}$$

Now we'll use the *Cauchy–Schwarz inequality*.

We have

$$(\sqrt{a_1}+\sqrt{a_2}+\cdots+\sqrt{a_n})^2 \leq (a_1+a_2+\cdots+a_n)(1+1+\cdots+1),$$

i.e.

$$(\sqrt{a_1}+\sqrt{a_2}+\cdots+\sqrt{a_n})^2 \leq n(a_1+a_2+\cdots+a_n). \tag{2}$$

Using (1) and (2) gives us

$$(\sqrt{a_1}+\sqrt{a_2}+\cdots+\sqrt{a_n})^2 \leq n(a_1+a_2+\cdots+a_n) \leq (\sqrt{a_1}+\sqrt{a_2}+\cdots+\sqrt{a_n}) \times (a_1+a_2+\cdots+a_n)$$

i.e.

$$\sqrt{a_1}+\sqrt{a_2}+\cdots+\sqrt{a_n} \leq a_1+a_2+\cdots+a_n,$$

as required. ■

159 Let a, b, c be positive real numbers such that $a+b+c=1$. Prove the inequality

$$a\sqrt{b}+b\sqrt{c}+c\sqrt{a} \leq \frac{1}{\sqrt{3}}.$$

Solution Applying the *Cauchy–Schwarz inequality* we have

$$(a\sqrt{b}+b\sqrt{c}+c\sqrt{a})^2 \leq (a^2+b^2+c^2)(a+b+c) = a^2+b^2+c^2. \tag{1}$$

One more use of the *Cauchy–Schwarz inequality* for

$$A_1=\sqrt{a}, \quad A_2=\sqrt{b}, \quad A_3=\sqrt{c} \quad \text{and}$$
$$B_1=\sqrt{ab}, \quad B_2=\sqrt{bc}, \quad B_3=\sqrt{ca}$$

gives us

$$(a\sqrt{b}+b\sqrt{c}+c\sqrt{a})^2 \leq (a+b+c)(ab+bc+ca) = ab+bc+ca,$$

i.e.

$$2(a\sqrt{b}+b\sqrt{c}+c\sqrt{a})^2 \leq 2(ab+bc+ca). \tag{2}$$

By adding (1) and (2) we get

$$3(a\sqrt{b}+b\sqrt{c}+c\sqrt{a})^2 \leq a^2+b^2+c^2+2(ab+bc+ca),$$

i.e.

$$3(a\sqrt{b}+b\sqrt{c}+c\sqrt{a})^2 \leq (a+b+c)^2 = 1,$$

i.e.

$$a\sqrt{b}+b\sqrt{c}+c\sqrt{a} \leq \frac{1}{\sqrt{3}}.$$

■

160 Let $a, b, c \in (0, 1)$ be real numbers. Prove the inequality

$$\sqrt{abc}+\sqrt{(1-a)(1-b)(1-c)} < 1.$$

Solution 1 For $x \in (0, 1)$ we have $\sqrt{x} < \sqrt[3]{x}$.
So

$$\sqrt{abc} < \sqrt[3]{abc} \quad \text{and} \quad \sqrt{(1-a)(1-b)(1-c)} < \sqrt[3]{(1-a)(1-b)(1-c)}. \tag{1}$$

Using (1) and $AM \geq GM$ gives us

$$\begin{aligned}\sqrt{abc}+\sqrt{(1-a)(1-b)(1-c)} &< \sqrt[3]{abc}+\sqrt[3]{(1-a)(1-b)(1-c)}\\ &\leq \frac{a+b+c}{3}+\frac{1-a+1-b+1-c}{3}=1.\end{aligned}$$

■

Solution 2 Since $a, b, c \in (0, 1)$ we obtain

$$\sqrt{abc}+\sqrt{(1-a)(1-b)(1-c)} < \sqrt{b}\sqrt{c}+\sqrt{1-b}\sqrt{1-c}. \tag{1}$$

Using the *Cauchy–Schwarz inequality* we have

$$\sqrt{b}\sqrt{c}+\sqrt{1-b}\sqrt{1-c} \leq \sqrt{(b+1-b)^2(c+1-c)^2} = 1. \tag{2}$$

From (1) and (2), we obtain the required inequality. ■

161 Let a, b, c be positive real numbers such that $a+b+c=3$. Prove the inequality

$$\frac{a^3+2}{b+2}+\frac{b^3+2}{c+2}+\frac{c^3+2}{a+2} \geq 3.$$

Solution By $AM \geq GM$ we have

$$\frac{a^3+2}{b+2} = \frac{a^3+1+1}{b+2} \geq \frac{3\sqrt[3]{a^3\cdot 1\cdot 1}}{b+2} = \frac{3a}{b+2}.$$

Similarly we get

$$\frac{b^3+2}{c+2} \geq \frac{3b}{c+2} \quad \text{and} \quad \frac{c^3+2}{a+2} \geq \frac{3c}{a+2}.$$

Therefore

$$\frac{a^3+2}{b+2} + \frac{b^3+2}{c+2} + \frac{c^3+2}{a+2} \geq 3\left(\frac{a}{b+2} + \frac{b}{c+2} + \frac{c}{a+2}\right). \tag{1}$$

Applying the *Cauchy–Schwarz inequality* (Corollary 4.3) we obtain

$$\begin{aligned}\frac{a}{b+2} + \frac{b}{c+2} + \frac{c}{a+2} &= \frac{a^2}{a(b+2)} + \frac{b^2}{b(c+2)} + \frac{c^2}{c(a+2)} \\ &\geq \frac{(a+b+c)^2}{a(b+2)+b(c+2)+c(a+2)} \\ &= \frac{(a+b+c)^2}{ab+bc+ca+2(a+b+c)}. \end{aligned} \tag{2}$$

Since $(a+b+c)^2 \geq 3(ab+bc+ca)$ we deduce that

$$\frac{1}{ab+bc+ca} \geq \frac{3}{(a+b+c)^2}. \tag{3}$$

From (2) and (3) we get

$$\begin{aligned}\frac{a}{b+2} + \frac{b}{c+2} + \frac{c}{a+2} &\geq \frac{(a+b+c)^2}{ab+bc+ca+2(a+b+c)} \\ &\geq \frac{(a+b+c)^2}{(a+b+c)^2/3+2(a+b+c)} \\ &= \frac{3(a+b+c)^2}{(a+b+c)^2+6(a+b+c)} = \frac{3(a+b+c)}{(a+b+c)+6}. \end{aligned} \tag{4}$$

Finally by (1), (4) and since $a+b+c=3$ we obtain

$$A \geq 3\left(\frac{a}{b+2} + \frac{b}{c+2} + \frac{c}{a+2}\right) \geq \frac{9(a+b+c)}{(a+b+c)+6} = \frac{27}{9} = 3,$$

as required. Equality occurs iff $a=b=c=1$. ■

162 Let a, b, c be positive real numbers such that $a^2+b^2+c^2=3$. Prove the inequality

$$\frac{1}{2-a} + \frac{1}{2-b} + \frac{1}{2-c} \geq 3.$$

Solution Rewrite the given inequality as follows

$$\frac{a}{2-a}+\frac{b}{2-b}+\frac{c}{2-c}\geq 3,$$

i.e.

$$\frac{a^2}{2a-a^2}+\frac{b^2}{2b-b^2}+\frac{c^2}{2c-c^2}\geq 3.$$

Clearly $a,b,c\in(0,\sqrt{3})$, so $2a-a^2, 2b-b^2, 2c-c^2>0$.

Now by the *Cauchy–Schwarz inequality* (Corollary 4.3) we obtain

$$\frac{a^2}{2a-a^2}+\frac{b^2}{2b-b^2}+\frac{c^2}{2c-c^2}\geq \frac{(a+b+c)^2}{2(a+b+c)-(a^2+b^2+c^2)}$$
$$=\frac{9}{2(a+b+c)-3}.$$

So it remains to prove that

$$\frac{(a+b+c)^2}{2(a+b+c)-3}\geq 3,$$

which is equivalent to $(a+b+c-3)^2\geq 0$, and clearly holds.

Equality holds iff $a=b=c=1$. ∎

163 Let a,b,c be positive real numbers such that $abc=8$. Prove the inequality

$$\frac{a-2}{a+1}+\frac{b-2}{b+1}+\frac{c-2}{c+1}\leq 0.$$

Solution Rewrite the given inequality as follows

$$\frac{a+1-3}{a+1}+\frac{b+1-3}{b+1}+\frac{c+1-3}{c+1}\leq 0$$

or

$$\frac{1}{a+1}+\frac{1}{b+1}+\frac{1}{c+1}\geq 1.$$

Let $a=\frac{2x}{y}, b=\frac{2y}{z}, c=\frac{2z}{x}$.

Then

$$\frac{1}{a+1}+\frac{1}{b+1}+\frac{1}{c+1}=\frac{1}{\frac{2x}{y}+1}+\frac{1}{\frac{2y}{z}+1}+\frac{1}{\frac{2z}{x}+1}$$
$$=\frac{y}{2x+y}+\frac{z}{2y+z}+\frac{x}{2z+x}$$

$$= \frac{y^2}{2xy + y^2} + \frac{z^2}{2yz + z^2} + \frac{x^2}{2zx + x^2}$$

$$\geq \frac{(x + y + z)^2}{2xy + y^2 + 2yz + z^2 + 2zx + x^2} = 1.$$

In the last step we used the *Cauchy–Schwarz inequality* (Corollary 4.3). ■

164 Let $a, b, c \in \mathbb{R}^+$ such that $a^2 + b^2 + c^2 = 1$. Prove the inequality

$$a + b + c - 2abc \leq \sqrt{2}.$$

Solution Since $a^2 + b^2 + c^2 = 1$ and $a^2 \geq 0$ it follows that $b^2 + c^2 \leq 1$, i.e. $2bc \leq 1$. Applying the *Cauchy–Schwarz inequality* we have

$$a + b + c - 2abc = a(1 - 2bc) + (b + c) \cdot 1 \leq \sqrt{a^2 + (b + c)^2}\sqrt{(1 - 2bc)^2 + 1}$$

$$= \sqrt{(a^2 + b^2 + c^2 + 2bc)(2 - 4bc + 4b^2c^2)}$$

$$= \sqrt{(1 + 2bc)(2 - 4bc + 4b^2c^2)}.$$

So it suffices to show that

$$(1 + 2bc)(2 - 4bc + 4b^2c^2) \leq 2.$$

We have

$$2 - (1 + 2bc)(2 - 4bc + 4b^2c^2) = 4b^2c^2(1 - 2bc) \geq 0.$$ ■

165 Let $x, y, z \in \mathbb{R}^+$ such that $x^2 + y^2 + z^2 = 2$. Prove the inequality

$$x + y + z \leq 2 + xyz.$$

Solution 1 Let $x = a\sqrt{2}, y = b\sqrt{2}, z = c\sqrt{2}$. Then $a^2 + b^2 + c^2 = 1$ and the given inequality becomes $a + b + c - 2abc \leq \sqrt{2}$, which is true (Problem 127). ■

Solution 2 The given inequality becomes

$$x(1 - yz) + y + z \leq 2.$$

Using the *Cauchy–Schwarz inequality* we get

$$(x(1 - yz) + (y + z) \cdot 1)^2 \leq (x^2 + (y + z)^2)((1 - yz)^2 + 1^2)$$

$$\Leftrightarrow \quad (x + y + z - xyz)^2 \leq (x^2 + y^2 + z^2 + 2yz)(2 - 2yz + y^2z^2)$$

$$\Leftrightarrow \quad (x + y + z - xyz)^2 \leq 2(1 + yz)(2 - 2yz + y^2z^2).$$

So it suffices to show that

$$2(1+yz)(2-2yz+y^2z^2)\le 4,$$

i.e.

$$(1+yz)(2-2yz+y^2z^2)\le 2 \quad \Leftrightarrow \quad y^3z^3\le y^2z^2,$$

i.e.

$$yz\le 1.$$

The last inequality is true since $2yz\le y^2+z^2\le x^2+y^2+z^2=2$. ∎

166 Let $x, y, z>-1$ be real numbers. Prove the inequality

$$\frac{1+x^2}{1+y+z^2}+\frac{1+y^2}{1+z+x^2}+\frac{1+z^2}{1+x+y^2}\ge 2.$$

Solution Notice that $\frac{1+y^2}{2}\ge y$ and $1+y+z^2>0$.

So

$$\frac{1+x^2}{1+y+z^2}\ge\frac{1+x^2}{1+\frac{1+y^2}{2}+z^2}=\frac{2(1+x^2)}{2(1+z^2)+1+y^2}.$$

Analogously

$$\frac{1+y^2}{1+z+x^2}\ge\frac{2(1+y^2)}{2(1+x^2)+1+z^2} \quad \text{and} \quad \frac{1+z^2}{1+x+y^2}\ge\frac{2(1+z^2)}{2(1+y^2)+1+x^2}.$$

It suffices to show that

$$\frac{2(1+x^2)}{2(1+z^2)+1+y^2}+\frac{2(1+y^2)}{2(1+x^2)+1+z^2}+\frac{2(1+z^2)}{2(1+y^2)+1+x^2}\ge 2.$$

Let $1+x^2=a$, $1+y^2=b$, $1+z^2=c$, i.e. we need to show that

$$\frac{a}{2c+b}+\frac{b}{2a+c}+\frac{c}{2b+a}\ge 1.$$

Applying the *Cauchy–Schwarz inequality* we obtain

$$3\left(\frac{a^2}{2ca+ab}+\frac{b^2}{2ab+bc}+\frac{c^2}{2bc+ca}\right)(ab+bc+ca)\ge(a+b+c)^2$$

i.e.

$$\frac{a}{2c+b}+\frac{b}{2a+c}+\frac{c}{2b+a}\ge\frac{(a+b+c)^2}{3(ab+bc+ca)}\ge 1,$$

as required. ∎

167 Let a, b, c, d be positive real numbers such that $abcd = 1$. Prove the inequality

$$(1+a^2)(1+b^2)(1+c^2)(1+d^2) \geq (a+b+c+d)^2.$$

Solution Since $abcd = 1$, there are two numbers x, y among a, b, c, d, such that $x, y \geq 1$ or $x, y \leq 1$. Without loss of generality we may suppose that they are b and d. Then clearly $(b-1)(d-1) \geq 0$, i.e. $bd + 1 \geq b + d$.

According to the *Cauchy–Schwarz inequality* and the previous note, we obtain

$$\begin{aligned}(1+a^2)(1+b^2)(1+c^2)(1+d^2) &= (1+a^2+b^2+a^2b^2)(c^2+1+d^2+c^2d^2)\\ &\geq (c+a+bd+1)^2 \geq (a+b+c+d)^2.\end{aligned}$$

Equality holds iff $a = b = c = d = 1$. ∎

168 Let $a, b, c, d \in \mathbb{R}^+$ such that $\frac{1}{a} + \frac{1}{b} + \frac{1}{c} + \frac{1}{d} = 4$. Prove the inequality

$$\sqrt[3]{\frac{a^3+b^3}{2}} + \sqrt[3]{\frac{b^3+c^3}{2}} + \sqrt[3]{\frac{c^3+d^3}{2}} + \sqrt[3]{\frac{d^3+a^3}{2}} \leq 2(a+b+c+d) - 4.$$

Solution

Lemma 21.2 *If* $x, y \in \mathbb{R}^+$ *then* $\sqrt[3]{\frac{x^3+y^3}{2}} \leq \frac{x^2+y^2}{x+y}$.

Proof The given inequality is equivalent to $(x-y)^4(x^2+xy+y^2) \geq 0$. □

So it follows that

$$\begin{aligned}&\sqrt[3]{\frac{a^3+b^3}{2}} + \sqrt[3]{\frac{b^3+c^3}{2}} + \sqrt[3]{\frac{c^3+d^3}{2}} + \sqrt[3]{\frac{d^3+a^3}{2}}\\ &\quad \leq \frac{a^2+b^2}{a+b} + \frac{b^2+c^2}{b+c} + \frac{c^2+d^2}{c+d} + \frac{d^2+a^2}{d+a}.\end{aligned}$$

Furthermore, we have

$$(a+b) - \frac{a^2+b^2}{a+b} = \frac{2ab}{a+b}.$$

So

$$L \leq (a+b) - \frac{2ab}{a+b} + (b+c) - \frac{2bc}{b+c} + (c+d) - \frac{2cd}{c+d} + (d+a) - \frac{2da}{d+a},$$

and it is sufficient to prove that

$$\frac{ab}{a+b} + \frac{bc}{b+c} + \frac{cd}{c+d} + \frac{da}{d+a} \geq 2.$$

Applying the *Cauchy–Schwarz inequality* we obtain

$$\left(\frac{ab}{a+b}+\frac{bc}{b+c}+\frac{cd}{c+d}+\frac{da}{d+a}\right)\cdot\left(2\left(\frac{1}{a}+\frac{1}{b}+\frac{1}{c}+\frac{1}{d}\right)\right)\geq 4^2,$$

i.e.

$$\frac{ab}{a+b}+\frac{bc}{b+c}+\frac{cd}{c+d}+\frac{da}{d+a}\geq\frac{16}{8}=2,$$

as required. ■

169 Let $x, y, z \in [-1, 1]$ be real numbers such that $x+y+z+xyz=0$. Prove the inequality

$$\sqrt{x+1}+\sqrt{y+1}+\sqrt{z+1}\leq 3.$$

Solution Applying the *Cauchy–Schwarz inequality* we have

$$\sqrt{x+1}+\sqrt{y+1}+\sqrt{z+1}\leq\sqrt{3(x+y+z+3)}.$$

If $x+y+z\leq 0$ then $\sqrt{x+1}+\sqrt{y+1}+\sqrt{z+1}\leq 3$, and the given inequality clearly holds.

So let us assume that $x+y+z>0$. Then we have $xyz=-(x+y+z)<0$. Without loss of generality we may assume that $z<0$ and then it's clear that $x, y\in(0, 1]$.

Applying once more, the *Cauchy–Schwarz inequality* we obtain

$$\sqrt{x+1}+\sqrt{y+1}+\sqrt{z+1}\leq\sqrt{2x+2y+4}+\sqrt{z+1}.$$

So it suffices to show that

$$\sqrt{2x+2y+4}+\sqrt{z+1}\leq 3.$$

We have

$$\begin{aligned}
\sqrt{2x+2y+4}+\sqrt{z+1}\leq 3 \quad&\Leftrightarrow\quad \sqrt{2x+2y+4}-2\leq 1-\sqrt{z+1}\\
\Leftrightarrow\quad&\frac{2(x+y)}{\sqrt{2x+2y+4}+2}\leq\frac{-z}{\sqrt{z+1}+1}\\
\Leftrightarrow\quad&\frac{-2z(1+xy)}{\sqrt{2x+2y+4}+2}\leq\frac{-z}{\sqrt{z+1}+1}\\
\Leftrightarrow\quad&2(1+xy)(1+\sqrt{1+z})\leq\sqrt{2x+2y+4}+2\\
\Leftrightarrow\quad&2xy+2(1+xy)\sqrt{1+z}\leq\sqrt{2x+2y+4}.
\end{aligned}\tag{1}$$

We can easily deduce that $1+z=\frac{(1-x)(1-y)}{1+xy}$, and then inequality (1) is equivalent to

$$xy+\sqrt{(1-x)(1-y)(1+xy)}\leq\sqrt{1+\frac{x+y}{2}}.$$

Finally, using the *Cauchy–Schwarz inequality* we obtain

$$\begin{aligned} xy+\sqrt{(1-x)(1-y)(1+xy)} &= \sqrt{x}\sqrt{xy^2}+\sqrt{1-x}\sqrt{1+xy-y-xy^2} \\ &\le \sqrt{(x+1-x)(xy^2+1+xy-y-xy^2)} \\ &= \sqrt{1+y(1-x)} \le 1 \le \sqrt{1+\frac{x+y}{2}}, \end{aligned}$$

as desired. ■

170 Let $a, b, c > 0$ be positive real numbers such that $a+b+c = abc$. Prove the inequality

$$ab+bc+ca \ge 3+\sqrt{a^2+1}+\sqrt{b^2+1}+\sqrt{c^2+1}.$$

Solution First we'll show that

$$a^2b^2+b^2c^2+c^2a^2 \ge a^2b^2c^2. \quad (1)$$

We have

$$(ab)^2+(bc)^2+(ca)^2 \ge (ab)(bc)+(bc)(ca)+(ca)(ab) = abc(a+b+c)$$

i.e.

$$\begin{aligned} &\frac{1}{a^2}+\frac{1}{b^2}+\frac{1}{c^2} = \frac{(ab)^2+(bc)^2+(ca)^2}{(abc)^2} \ge \frac{a+b+c}{abc} = 1 \\ \Leftrightarrow \quad & a^2b^2+b^2c^2+c^2a^2 \ge a^2b^2c^2. \end{aligned}$$

Furthermore

$$\begin{aligned} (ab+bc+ca)^2 &= a^2b^2+b^2c^2+c^2a^2+2abc(a+b+c) \\ &\overset{(1)}{\ge} a^2b^2c^2+2abc(a+b+c) = 3(a+b+c)^2. \end{aligned} \quad (2)$$

So

$$\begin{aligned} (ab+bc+ca-3)^2 &= (ab+bc+ca)^2-6(ab+bc+ca)+9 \\ &\overset{(2)}{\ge} 3(a+b+c)^2-6(ab+bc+ca)+9 \\ &= 3(a^2+b^2+c^2)+9, \end{aligned}$$

i.e.

$$ab+bc+ca \ge 3+\sqrt{3(a^2+b^2+c^2)+9}. \quad (3)$$

Applying the *Cauchy–Schwarz inequality* we have

$$3(a^2+b^2+c^2)+9=3((a^2+1)+(b^2+1)+(c^2+1))$$
$$\geq (\sqrt{a^2+1}+\sqrt{b^2+1}+\sqrt{c^2+1})^2,$$

i.e.

$$\sqrt{3(a^2+b^2+c^2)+9}\geq \sqrt{a^2+1}+\sqrt{b^2+1}+\sqrt{c^2+1}. \qquad (4)$$

Using (3) and (4) we obtain

$$ab+bc+ca\geq 3+\sqrt{3(a^2+b^2+c^2)+9}\geq 3+\sqrt{a^2+1}+\sqrt{b^2+1}+\sqrt{c^2+1},$$

as required. ■

171 Let a,b,c,x,y,z be positive real numbers such that $ax+by+cz=xyz$. Prove the inequality

$$\sqrt{a+b}+\sqrt{b+c}+\sqrt{c+a}<x+y+z.$$

Solution We have $\frac{a}{yz}+\frac{b}{xz}+\frac{c}{xy}=1$.

Let

$$u=\frac{a}{yz},\qquad v=\frac{b}{xz},\qquad w=\frac{c}{xy}.$$

We need to show that

$$\sqrt{z(yu+xv)}+\sqrt{x(zv+yw)}+\sqrt{y(xw+zu)}<x+y+z,$$

where $u+v+w=1$.

Applying the *Cauchy–Schwarz inequality* we obtain

$$(\sqrt{z(yu+xv)}+\sqrt{x(zv+yw)}+\sqrt{y(xw+zu)})^2$$
$$\leq (x+y+z)(yu+xv+zv+yw+xw+zu).$$

Also we have

$$yu+xv+zv+yw+xw+zu=x(1-u)+y(1-v)+z(1-w)$$
$$=x+y+z-(xu+yv+zw)<x+y+z.$$

Now we obtain

$$(\sqrt{z(yu+xv)}+\sqrt{x(zv+yw)}+\sqrt{y(xw+zu)})^2<(x+y+z)^2,$$

i.e.

$$\sqrt{z(yu+xv)}+\sqrt{x(zv+yw)}+\sqrt{y(xw+zu)}<x+y+z.$$ ■

172 Let a, b, c be non-negative real numbers such that $a^2+b^2+c^2=1$. Prove the inequality

$$\frac{a}{b^2+1}+\frac{b}{c^2+1}+\frac{c}{a^2+1}\geq\frac{3}{4}(a\sqrt{a}+b\sqrt{b}+c\sqrt{c})^2.$$

Solution We'll use the *Cauchy–Schwarz inequality*, i.e.

$$(a_1^2+a_2^2+a_3^2)(b_1^2+b_2^2+b_3^2)\geq(a_1b_1+a_2b_2+a_3b_3)^2. \tag{1}$$

Let

$$a_1=\sqrt{a^2(b^2+1)},\qquad a_2=\sqrt{b^2(c^2+1)},\qquad a_3=\sqrt{c^2(a^2+1)}\quad\text{and}$$

$$b_1=\sqrt{\frac{a}{b^2+1}},\qquad b_2=\sqrt{\frac{b}{c^2+1}},\qquad b_3=\sqrt{\frac{c}{a^2+1}}.$$

Then using (1) we get

$$(a^2(b^2+1)+b^2(c^2+1)+c^2(a^2+1))\left(\frac{a}{b^2+1}+\frac{b}{c^2+1}+\frac{c}{a^2+1}\right)$$
$$\geq(a\sqrt{a}+b\sqrt{b}+c\sqrt{c})^2,$$

i.e.

$$\frac{a}{b^2+1}+\frac{b}{c^2+1}+\frac{c}{a^2+1}\geq\frac{(a\sqrt{a}+b\sqrt{b}+c\sqrt{c})^2}{a^2(b^2+1)+b^2(c^2+1)+c^2(a^2+1)}.$$

So it suffices to show that

$$a^2(b^2+1)+b^2(c^2+1)+c^2(a^2+1)\leq\frac{4}{3}.$$

From the obvious inequality $(a^2-b^2)^2+(b^2-c^2)^2+(c^2-a^2)^2\geq 0$ we deduce that

$$a^2b^2+b^2c^2+c^2a^2\leq a^4+b^4+c^4. \tag{2}$$

Now we have

$$a^2(b^2+1)+b^2(c^2+1)+c^2(a^2+1)$$
$$=a^2+b^2+c^2+a^2b^2+b^2c^2+c^2a^2$$
$$=a^2+b^2+c^2+\frac{3(a^2b^2+b^2c^2+c^2a^2)}{3}$$
$$\overset{(2)}{\leq}a^2+b^2+c^2+\frac{2(a^2b^2+b^2c^2+c^2a^2)+a^4+b^4+c^4}{3}$$
$$=a^2+b^2+c^2+\frac{(a^2+b^2+c^2)^2}{3}=1+\frac{1}{3}=\frac{4}{3},$$

as required.

Equality occurs iff $a=b=c=1/\sqrt{3}$. ■

173 Let a, b, c be positive real numbers. Prove the inequality

$$\frac{a}{(b+c)^2}+\frac{b}{(c+a)^2}+\frac{c}{(a+b)^2}\geq\frac{9}{4(a+b+c)}.$$

Solution Applying the *Cauchy–Schwarz inequality* gives us

$$(a+b+c)\left(\frac{a}{(b+c)^2}+\frac{b}{(c+a)^2}+\frac{c}{(a+b)^2}\right)\geq\left(\frac{a}{b+c}+\frac{b}{c+a}+\frac{c}{a+b}\right)^2. \tag{1}$$

Recalling *Nesbitt's inequality* we have

$$\frac{a}{b+c}+\frac{b}{c+a}+\frac{c}{a+b}\geq\frac{3}{2}. \tag{2}$$

From (1) and (2) we obtain the required inequality. ■

174 Let $x \geq y \geq z > 0$ be real numbers. Prove the inequality

$$\frac{x^2y}{z}+\frac{y^2z}{x}+\frac{z^2x}{y}\geq x^2+y^2+z^2.$$

Solution Applying the *Cauchy–Schwarz inequality* we obtain

$$\left(\frac{x^2y}{z}+\frac{y^2z}{x}+\frac{z^2x}{y}\right)\left(\frac{x^2z}{y}+\frac{y^2x}{z}+\frac{z^2y}{x}\right)\geq(x^2+y^2+z^2)^2. \tag{1}$$

We'll prove that

$$\frac{x^2y}{z}+\frac{y^2z}{x}+\frac{z^2x}{y}\geq\frac{x^2z}{y}+\frac{y^2x}{z}+\frac{z^2y}{x}. \tag{2}$$

From

$$\left(\frac{x^2y}{z}+\frac{y^2z}{x}+\frac{z^2x}{y}\right)-\left(\frac{x^2z}{y}+\frac{y^2x}{z}+\frac{z^2y}{x}\right)$$
$$=\frac{(xy+yz+zx)(x-y)(x-z)(y-z)}{xyz}\geq 0,$$

we deduce that

$$\frac{x^2y}{z}+\frac{y^2z}{x}+\frac{z^2x}{y}\geq\frac{x^2z}{y}+\frac{y^2x}{z}+\frac{z^2y}{x}.$$

Combining (1) and (2) give us

$$\left(\frac{x^2y}{z}+\frac{y^2z}{x}+\frac{z^2x}{y}\right)^2 \geq \left(\frac{x^2y}{z}+\frac{y^2z}{x}+\frac{z^2x}{y}\right)\left(\frac{x^2z}{y}+\frac{y^2x}{z}+\frac{z^2y}{x}\right)$$
$$\geq (x^2+y^2+z^2)^2,$$

i.e.

$$\frac{x^2y}{z}+\frac{y^2z}{x}+\frac{z^2x}{y} \geq x^2+y^2+z^2.$$

Equality occurs if and only if $x=y=z$. ■

175 Let a, b, c be positive real numbers such that $abc=1$. Prove the inequality

$$\frac{1}{2+a}+\frac{1}{2+b}+\frac{1}{2+c} \leq 1.$$

Solution The given inequality can be rewritten as

$$1-\frac{2}{2+a}+1-\frac{2}{2+b}+1-\frac{2}{2+c} \geq 1,$$

which is equivalent with

$$\frac{a}{2+a}+\frac{b}{2+b}+\frac{c}{2+c} \geq 1. \tag{1}$$

Let $a=\frac{x}{y}, b=\frac{y}{z}, c=\frac{z}{x}$.

Inequality (1) becomes

$$\frac{x}{x+2y}+\frac{y}{y+2z}+\frac{z}{z+2x} \geq 1. \tag{2}$$

Applying the *Cauchy–Schwarz inequality* we have

$$\frac{x}{x+2y}+\frac{y}{y+2z}+\frac{z}{z+2x} = \frac{x^2}{x^2+2xy}+\frac{y^2}{y^2+2yz}+\frac{z^2}{z^2+2zx}$$
$$\geq \frac{(x+y+z)^2}{x^2+y^2+z^2+2xy+2yz+2zx} = 1.$$

So we have proved (2) and we are done.

Equality occurs iff $x=y=z$, i.e. $a=b=c=1$. ■

176 Let a, b, c be positive real numbers such that $abc \geq 1$. Prove the inequality

$$\frac{1}{a^4+b^3+c^2}+\frac{1}{b^4+c^3+a^2}+\frac{1}{c^4+a^3+b^2} \leq 1.$$

Solution By the *Cauchy–Schwarz inequality* we have

$$\frac{1}{a^4+b^3+c^2}=\frac{1+b+c^2}{(a^4+b^3+c^2)(1+b+c^2)}\le\frac{1+b+c^2}{(a^2+b^2+c^2)^2}.$$

Similarly we get

$$\frac{1}{b^4+c^3+a^2}\le\frac{1+c+a^2}{(a^2+b^2+c^2)^2}\quad\text{and}\quad\frac{1}{c^4+a^3+b^2}\le\frac{1+a+b^2}{(a^2+b^2+c^2)^2}.$$

It follows that

$$\frac{1}{a^4+b^3+c^2}+\frac{1}{b^4+c^3+a^2}+\frac{1}{c^4+a^3+b^2}\le\frac{a^2+b^2+c^2+a+b+c+3}{(a^2+b^2+c^2)^2}.$$

So it remains to prove that

$$\frac{a^2+b^2+c^2+a+b+c+3}{(a^2+b^2+c^2)^2}\le 1.$$

By $AM \ge GM$ we have $a+b+c\ge 3$ and $a^2+b^2+c^2\ge 3$.

Consider the well-known inequality $3(a^2+b^2+c^2)\ge(a+b+c)^2$.

Then we obtain

$$\begin{aligned}\frac{a^2+b^2+c^2+a+b+c+3}{(a^2+b^2+c^2)^2}&\le\frac{a^2+b^2+c^2+\frac{(a+b+c)^2}{3}+\frac{(a+b+c)^2}{3}}{(a^2+b^2+c^2)^2}\\&\le\frac{a^2+b^2+c^2+(a^2+b^2+c^2)+(a^2+b^2+c^2)}{(a^2+b^2+c^2)^2}\\&=\frac{3}{a^2+b^2+c^2}\le 1,\end{aligned}$$

as required. ∎

177 Let a, b, c, d be positive real numbers such that $abcd = 1$. Prove the inequality

$$\frac{1}{a(1+b)}+\frac{1}{b(1+c)}+\frac{1}{c(1+d)}+\frac{1}{d(1+a)}\ge 2.$$

Solution With the substitutions $a=\frac{x}{y}, b=\frac{t}{x}, c=\frac{z}{t}, d=\frac{y}{z}$, the given inequality becomes

$$\frac{x}{z+t}+\frac{y}{x+t}+\frac{z}{x+y}+\frac{t}{z+y}\ge 2.$$

By the *Cauchy–Schwarz inequality* we have

$$\frac{x}{z+t}+\frac{y}{x+t}+\frac{z}{x+y}+\frac{t}{z+y}=\frac{x^2}{xz+xt}+\frac{y^2}{yx+yt}+\frac{z^2}{zx+zy}+\frac{t^2}{tz+ty}$$
$$\geq \frac{(x+y+z+t)^2}{2xz+2yt+xt+yx+zy+tz}.$$

Hence it suffices to prove that

$$\frac{(x+y+z+t)^2}{2xz+2yt+xt+yx+zy+tz}\geq 2,$$

which is equivalent to

$$(x-z)^2+(y-t)^2\geq 0.$$

Equality occurs iff $x=z$, $y=t$, i.e. $a=c=1/b=1/d$. ■

178 Let a,b,c be non-negative real numbers such that $a+b+c=1$. Prove the inequality

$$\frac{ab}{c+1}+\frac{bc}{a+1}+\frac{ca}{b+1}\leq \frac{1}{4}.$$

Solution If one of a,b,c is equal to zero then it is easy to show that the given inequality is true. Equality in this case occurs iff one of a,b,c is zero, and the other two numbers are equal to $1/2$.

Because of this we can assume that $a,b,c\in\mathbb{R}^+$.

From $a+b+c=1$ it follows that at least one of the numbers a,b,c is less then $4/9$. In the opposite case, if all of them are greater then $4/9$, we will have

$$a+b+c>3\cdot\frac{4}{9}=\frac{4}{3}>1,$$

a contradiction.

So we can assume that

$$c<4/9. \tag{1}$$

Let $A=\frac{ab}{c+1}+\frac{bc}{a+1}+\frac{ca}{b+1}$.

Then

$$A=abc\left(\frac{1}{a}+\frac{1}{b}+\frac{1}{c}-\frac{1}{a+1}-\frac{1}{b+1}-\frac{1}{c+1}\right). \tag{2}$$

Since

$$\frac{1}{a+1}+\frac{1}{b+1}+\frac{1}{c+1}=\frac{1}{4}((a+1)+(b+1)+(c+1))\left(\frac{1}{a+1}+\frac{1}{b+1}+\frac{1}{c+1}\right),$$

applying the *Cauchy–Schwarz inequality* we have

$$\begin{aligned}
&\frac{1}{a+1}+\frac{1}{b+1}+\frac{1}{c+1}\\
&\quad=\frac{1}{4}((a+1)+(b+1)+(c+1))\left(\frac{1}{a+1}+\frac{1}{b+1}+\frac{1}{c+1}\right)\\
&\quad\geq\frac{1}{4}(1+1+1)^2=\frac{9}{4}.
\end{aligned}\tag{3}$$

Now using (2) and (3) we obtain

$$\begin{aligned}
A&=abc\left(\frac{1}{a}+\frac{1}{b}+\frac{1}{c}-\left(\frac{1}{a+1}+\frac{1}{b+1}+\frac{1}{c+1}\right)\right)\\
&\leq abc\left(\frac{1}{a}+\frac{1}{b}+\frac{1}{c}-\frac{9}{4}\right)=ab+ac+bc-\frac{9abc}{4}.
\end{aligned}\tag{4}$$

On the other hand, we have

$$(1-c)^2=(a+b)^2\geq 4ab,$$

i.e.

$$ab\leq\frac{(1-c)^2}{4},\tag{5}$$

and using (4) we get

$$\begin{aligned}
A-\frac{1}{4}&\leq ab+ac+bc-\frac{9abc}{4}-\frac{1}{4}=ab+c(a+b)-\frac{9abc}{4}-\frac{1}{4}\\
&=ab\left(1-\frac{9c}{4}\right)+c(1-c)-\frac{1}{4}\\
&\overset{(1),(5)}{\leq}\frac{(1-c)^2}{4}\left(1-\frac{9c}{4}\right)+c(1-c)-\frac{1}{4}\\
&=\frac{1}{16}(-9c^3+6c^2-c)=\frac{-c}{16}(9c^2-6c+1)=\frac{-c(3c-1)^2}{16}\leq 0,
\end{aligned}$$

as required.

Equality occurs iff $a=b=c=1/3$. ■

179 Let a,b,c be positive real numbers such that $abc=1$. Prove the inequality

$$\frac{1}{(a+1)^2(b+c)}+\frac{1}{(b+1)^2(c+a)}+\frac{1}{(c+1)^2(a+b)}\leq\frac{3}{8}.$$

Solution Let $a = x^2, b = y^2, c = z^2$. The given inequality becomes

$$\frac{1}{(x^2+1)^2(y^2+z^2)} + \frac{1}{(y^2+1)^2(z^2+x^2)} + \frac{1}{(z^2+1)^2(x^2+y^2)} \le \frac{3}{8}.$$

By the *Cauchy–Schwarz inequality* we have

$$\sqrt{(x^2+1)(y^2+z^2)} \ge xy + z = \frac{1}{z} + z = \frac{z^2+1}{z}$$

and

$$\sqrt{(x^2+1)(z^2+y^2)} \ge xz + y = \frac{1}{y} + y = \frac{y^2+1}{y}.$$

Multiplying these two inequalities we get

$$(x^2+1)(z^2+y^2) \ge \frac{(y^2+1)(z^2+1)}{yz}$$

i.e.

$$(x^2+1)^2(z^2+y^2) \ge \frac{(x^2+1)(y^2+1)(z^2+1)}{yz}.$$

Hence

$$\frac{1}{(x^2+1)^2(y^2+z^2)} \le \frac{yz}{(x^2+1)(y^2+1)(z^2+1)}.$$

Similarly we obtain

$$\frac{1}{(y^2+1)^2(z^2+x^2)} \le \frac{zx}{(x^2+1)(y^2+1)(z^2+1)}$$

and

$$\frac{1}{(z^2+1)^2(x^2+y^2)} \le \frac{xy}{(x^2+1)(y^2+1)(z^2+1)}.$$

We have

$$\frac{1}{(x^2+1)^2(y^2+z^2)} + \frac{1}{(y^2+1)^2(z^2+x^2)} + \frac{1}{(z^2+1)^2(x^2+y^2)}$$
$$\le \frac{xy+yz+zx}{(x^2+1)(y^2+1)(z^2+1)},$$

and it suffices to prove that

$$\frac{xy+yz+zx}{(x^2+1)(y^2+1)(z^2+1)} \le \frac{3}{8}$$

i.e.

$$(x^2+1)(y^2+1)(z^2+1) \geq \frac{8}{3}(xy+yz+zx).$$

By the *Cauchy–Schwarz inequality* we have

$$\sqrt{(x^2+1)(1+y^2)} \geq x+y, \qquad \sqrt{(z^2+1)(1+x^2)} \geq z+x \quad \text{and}$$
$$\sqrt{(y^2+1)(1+z^2)} \geq y+z.$$

Multiplying these three inequalities gives us

$$(x^2+1)(y^2+1)(z^2+1) \geq (x+y)(y+z)(z+x). \tag{1}$$

By the well-known inequality

$$(x+y)(y+z)(z+x) \geq \frac{8}{9}(x+y+z)(xy+yz+zx),$$

and the $AM \geq GM$ we obtain

$$(x+y)(y+z)(z+x) \geq \frac{8}{9}(x+y+z)(xy+yz+zx) \geq \frac{8}{3}(xy+yz+zx). \tag{2}$$

By (1) and (2) we obtain

$$(x^2+1)(y^2+1)(z^2+1) \geq (x+y)(y+z)(z+x) \geq \frac{8}{3}(xy+yz+zx),$$

as required.

Equality occurs iff $x=y=z=1$ i.e. $a=b=c=1$. ■

180 Let x, y, z be positive real numbers. Prove the inequality

$$xy(x+y-z)+yz(y+z-x)+zx(z+x-y) \geq \sqrt{3(x^3y^3+y^3z^3+z^3x^3)}.$$

Solution Notice that

$$xy(x+y-z)+yz(y+z-x)+zx(z+x-y)$$
$$= \frac{x(y^3+z^3)}{y+z} + \frac{y(z^3+x^3)}{z+x} + \frac{z(x^3+y^3)}{x+y}.$$

Let $a=x^3, b=y^3, c=z^3$.

Using Corollary 4.5 (Chap. 4) and the previous identity we obtain

$$\begin{aligned}
&xy(x+y-z)+yz(y+z-x)+zx(z+x-y)\\
&=\frac{x(y^3+z^3)}{y+z}+\frac{y(z^3+x^3)}{z+x}+\frac{z(x^3+y^3)}{x+y}\\
&=\frac{x}{y+z}(b+c)+\frac{y}{z+x}(c+a)+\frac{z}{x+y}(a+b)\\
&\ge\sqrt{3(ab+bc+ca)}=\sqrt{3(x^3y^3+y^3z^3+z^3x^3)},
\end{aligned}$$

as required. ■

181 Let a, b, c be positive real numbers. Prove the inequality

$$\frac{ab(a^3+b^3)}{a^2+b^2}+\frac{bc(b^3+c^3)}{b^2+c^2}+\frac{ca(c^3+a^3)}{c^2+a^2}\ge\sqrt{3abc(a^3+b^3+c^3)}.$$

Solution Let

$$x=\frac{1}{c^2},\qquad y=\frac{1}{b^2},\qquad z=\frac{1}{a^2}\quad\text{and}$$

$$A=\frac{a^2b^2}{c},\qquad B=\frac{b^2c^2}{a},\qquad C=\frac{a^2c^2}{b}.$$

We have

$$\frac{x}{y+z}(B+C)=\frac{ab(a^3+b^3)}{a^2+b^2},\qquad \frac{y}{z+x}(C+A)=\frac{bc(b^3+c^3)}{b^2+c^2}$$

and

$$\frac{z}{x+y}(A+B)=\frac{ca(c^3+a^3)}{c^2+a^2}.$$

Using Corollary 4.5 (Chap. 4) and the previous identities we obtain

$$\begin{aligned}
&\frac{ab(a^3+b^3)}{a^2+b^2}+\frac{bc(b^3+c^3)}{b^2+c^2}+\frac{ca(c^3+a^3)}{c^2+a^2}\\
&=\frac{x}{y+z}(B+C)+\frac{y}{z+x}(C+A)+\frac{z}{x+y}(A+B)\\
&\ge\sqrt{3(AB+BC+CA)}=\sqrt{3abc(a^3+b^3+c^3)}.
\end{aligned}$$ ■

182 Let a, b, c be positive real numbers. Prove the inequality.

$$ab\frac{a+c}{b+c}+bc\frac{b+a}{c+a}+ca\frac{c+b}{a+b}\ge\sqrt{3abc(a+b+c)}.$$

Solution Let $x=\frac{1}{bc}, y=\frac{1}{ac}, z=\frac{1}{ab}$ and $A=ac, B=ab, C=bc$.

We have

$$\frac{x}{y+z}(B+C)=ab\frac{a+c}{b+c}, \qquad \frac{y}{z+x}(C+A)=bc\frac{b+a}{c+a} \quad \text{and}$$

$$\frac{z}{x+y}(A+B)=ca\frac{c+b}{a+b}.$$

Using Corollary 4.5 (Chap. 4) and the previous identities we obtain

$$\begin{aligned} &ab\frac{a+c}{b+c}+bc\frac{b+a}{c+a}+ca\frac{c+b}{a+b} \\ &= \frac{x}{y+z}(B+C)+\frac{y}{z+x}(C+A)+\frac{z}{x+y}(A+B) \\ &\geq \sqrt{3(AB+BC+CA)}=\sqrt{3abc(a+b+c)}. \end{aligned}$$

■

183 Let a,b,c and x,y,z be positive real numbers. Prove the inequality

$$a(y+z)+b(z+x)+c(x+y)\geq 2\sqrt{(xy+yz+zx)(ab+bc+ca)}.$$

Solution Since the given inequality is homogenous we may assume that $x+y+z=1$.

Now the given inequality can be written as follows

$$2\sqrt{(xy+yz+zx)(ab+bc+ca)}+ax+by+cz\leq a+b+c.$$

Applying the *Cauch–Schwarz inequality* twice we have

$$\begin{aligned} &ax+by+cz+2\sqrt{(xy+yz+zx)(ab+bc+ca)} \\ &\leq \sqrt{a^2+b^2+c^2}\cdot\sqrt{x^2+y^2+z^2}+\sqrt{2(xy+yz+zx)}\cdot\sqrt{2(ab+bc+ca)} \\ &\leq \sqrt{a^2+b^2+c^2+2(ab+bc+ca)}\cdot\sqrt{x^2+y^2+z^2+2(xy+yz+zx)} \\ &= a+b+c. \end{aligned}$$

■

184 Let a,b,c be positive real numbers such that $abc\geq 1$. Prove the inequality

$$a^3+b^3+c^3\geq ab+bc+ca.$$

Solution By *Chebishev's inequality* it is easy to obtain

$$3(a^3+b^3+c^3)\geq (a+b+c)(a^2+b^2+c^2). \tag{1}$$

Now by $AM\geq GM$ we have

$$a+b+c\geq 3\sqrt[3]{abc}\geq 3$$

and clearly

$$a^2+b^2+c^2 \geq ab+bc+ca.$$

So by (1) we obtain

$$a^3+b^3+c^3 \geq \frac{(a+b+c)(a^2+b^2+c^2)}{3} \geq \frac{3(ab+bc+ca)}{3} = ab+bc+ca.$$

■

185 Let $a, b, c > 0$ be real numbers such that $a^{2/3}+b^{2/3}+c^{2/3}=3$. Prove the inequality

$$a^2+b^2+c^2 \geq a^{4/3}+b^{4/3}+c^{4/3}.$$

Solution After setting $a^{1/3}=x, b^{1/3}=y, c^{1/3}=z$ the initial condition becomes

$$x^2+y^2+z^2=3, \tag{1}$$

and the given inequality is equivalent to

$$x^6+y^6+z^6 \geq x^4+y^4+z^4.$$

Assume that $x^2 \leq y^2 \leq z^2$. Then it is clear that $x^4 \leq y^4 \leq z^4$.

Applying *Chebishev's inequality* we get

$$(x^2+y^2+z^2)(x^4+y^4+z^4) \leq 3(x^6+y^6+z^6),$$

and using (1) we obtain $x^6+y^6+z^6 \geq x^4+y^4+z^4$, as required. ■

186 Let a, b, c be positive real numbers such that $a+b+c=3$. Prove the inequality

$$\frac{1}{c^2+a+b}+\frac{1}{a^2+b+c}+\frac{1}{b^2+c+a} \leq 1.$$

Solution Observe that

$$\frac{1}{a^2+b+c}-\frac{1}{3}=\frac{1}{a^2-a+3}-\frac{1}{3}=\frac{a(1-a)}{3(a^2-a+3)}.$$

Analogously

$$\frac{1}{b^2+c+a}-\frac{1}{3}=\frac{b(1-b)}{3(b^2-b+3)} \quad \text{and} \quad \frac{1}{c^2+a+b}-\frac{1}{3}=\frac{c(1-c)}{3(c^2-c+3)}.$$

Now the given inequality is equivalent to

$$\frac{a(a-1)}{a^2-a+3}+\frac{b(b-1)}{b^2-b+3}+\frac{c(c-1)}{c^2-c+3} \geq 0$$

i.e.

$$\frac{a-1}{a-1+3/a}+\frac{b-1}{b-1+3/b}+\frac{c-1}{c-1+3/c} \geq 0.$$

Without loss of generality we may assume that $a \geq b \geq c$.

Then clearly $a-1 \geq b-1 \geq c-1$ and since $a+b+c=3$ it follows that $ab, bc, ca \leq 3$. Now we can easily show that

$$\frac{1}{a-1+3/a} \geq \frac{1}{b-1+3/b} \geq \frac{1}{c-1+3/c}.$$

Applying *Chebishev's inequality* we obtain

$$(a-1+b-1+c-1)\left(\frac{1}{a-1+3/a}+\frac{1}{b-1+3/b}+\frac{1}{c-1+3/c}\right) \leq 3A$$

i.e.

$$A \geq 0.$$

Equality occurs iff $a=b=c=1$. ■

187 Let $a, b, c \in \mathbb{R}^+$. Prove the inequality

$$\frac{2a^2}{b+c}+\frac{2b^2}{c+a}+\frac{2c^2}{a+b} \geq a+b+c.$$

Solution Without loss of generality we can assume that $a \geq b \geq c$.
Then clearly

$$\frac{1}{b+c} \geq \frac{1}{c+a} \geq \frac{1}{a+b}.$$

By *Chebishev's inequality* we have

$$\frac{2a^2}{b+c}+\frac{2b^2}{c+a}+\frac{2c^2}{a+b} \geq \frac{2}{3}(a^2+b^2+c^2)\left(\frac{1}{b+c}+\frac{1}{c+a}+\frac{1}{a+b}\right). \quad (1)$$

Applying $QM \geq AM$ we deduce

$$\sqrt{\frac{a^2+b^2+c^2}{3}} \geq \frac{a+b+c}{3}, \quad \text{i.e.} \quad \frac{a^2+b^2+c^2}{3} \geq \left(\frac{a+b+c}{3}\right)^2.$$

By (1) and the previous inequality it follows that

$$\frac{2a^2}{b+c}+\frac{2b^2}{c+a}+\frac{2c^2}{a+b} \geq \frac{2}{9}(a+b+c)^2\left(\frac{1}{b+c}+\frac{1}{c+a}+\frac{1}{a+b}\right). \quad (2)$$

Applying $AM \geq HM$ we deduce

$$\frac{1}{b+c}+\frac{1}{c+a}+\frac{1}{a+b} \geq \frac{9}{(b+c)+(c+a)+(a+b)} = \frac{9}{2(a+b+c)}.$$

Finally from the previous inequality and (2), we get required result. ■

188 Let a, b, c be positive real numbers such that $abc = 2$. Prove the inequality.

$$a^3+b^3+c^3 \geq a\sqrt{b+c}+b\sqrt{c+a}+c\sqrt{a+b}.$$

Solution Applying the *Cauchy–Schwarz inequality* we get

$$a\sqrt{b+c}+b\sqrt{c+a}+c\sqrt{a+b} \leq \sqrt{2(a^2+b^2+c^2)(a+b+c)}. \quad (1)$$

Using *Chebishev's inequality* we get

$$\sqrt{2(a^2+b^2+c^2)(a+b+c)} \leq \sqrt{6(a^3+b^3+c^3)}. \quad (2)$$

Also from $AM \geq GM$ we have

$$a^3+b^3+c^3 \geq 3abc = 6. \quad (3)$$

Combining (1), (2) and (3) we have

$$a^3+b^3+c^3 \geq \sqrt{6(a^3+b^3+c^3)} \geq a\sqrt{b+c}+b\sqrt{c+a}+c\sqrt{a+b}.$$

Equality holds iff $a = b = c = \sqrt[3]{2}$. ■

189 Let $a_1, a_2, \dots, a_n$ be positive real numbers. Prove the inequality

$$\frac{1}{\frac{1}{1+a_1}+\frac{1}{1+a_2}+\cdots+\frac{1}{1+a_n}} - \frac{1}{\frac{1}{a_1}+\frac{1}{a_2}+\cdots+\frac{1}{a_n}} \geq \frac{1}{n}.$$

Solution We can assume that $a_1 \geq a_2 \geq \cdots \geq a_n$.

If we take $x_i = \frac{1}{a_i}$, $y_i = \frac{1}{a_i+1}$ for $i = 1, 2, \dots, n$ then

$$\frac{1}{a_1} \leq \frac{1}{a_2} \leq \cdots \leq \frac{1}{a_n} \quad \text{and} \quad \frac{1}{a_1+1} \leq \frac{1}{a_2+1} \leq \cdots \leq \frac{1}{a_n+1}.$$

Also we have that

$$x_i y_i = \frac{1}{a_i(a_i+1)} = \frac{1}{a_i} - \frac{1}{a_i+1} = x_i - y_i.$$

So we can use *Chebishev's inequality*, i.e. we have

$$\sum_{i=1}^{n} \frac{1}{a_i} \cdot \sum_{i=1}^{n} \frac{1}{a_i+1} \leq n \cdot \sum_{i=1}^{n} \frac{1}{a_i(a_i+1)} = n \cdot \sum_{i=1}^{n} \frac{1}{a_i} - \frac{1}{a_i+1}$$

$$= n \cdot \left(\sum_{i=1}^{n} \frac{1}{a_i} - \sum_{i=1}^{n} \frac{1}{a_i+1} \right).$$

Now we easily obtain

$$\frac{1}{\frac{1}{1+a_1}+\frac{1}{1+a_2}+\cdots+\frac{1}{1+a_n}}-\frac{1}{\frac{1}{a_1}+\frac{1}{a_2}+\cdots+\frac{1}{a_n}}\geq\frac{1}{n}.$$

Equality holds iff $a_1=a_2=\cdots=a_n$. ■

190 Let $a,b,c,d\in\mathbb{R}^+$ such that $ab+bc+cd+da=1$. Prove the inequality

$$\frac{a^3}{b+c+d}+\frac{b^3}{a+c+d}+\frac{c^3}{b+d+a}+\frac{d^3}{b+c+a}\geq\frac{1}{3}.$$

Solution Let $a+b+c+d=s$. Then the given inequality is equivalent to

$$A=\frac{a^3}{s-a}+\frac{b^3}{s-b}+\frac{c^3}{s-c}+\frac{d^3}{s-d}\geq\frac{1}{3}. \tag{1}$$

Let us assume $a\geq b\geq c\geq d$. Then

$$a^3\geq b^3\geq c^3\geq d^3 \quad\text{and}\quad \frac{1}{s-a}\geq\frac{1}{s-b}\geq\frac{1}{s-c}\geq\frac{1}{s-d}.$$

Applying *Chebishev's inequality* we get

$$\begin{aligned}&(a^3+b^3+c^3+d^3)\left(\frac{1}{s-a}+\frac{1}{s-b}+\frac{1}{s-c}+\frac{1}{s-d}\right)\\&\quad\leq4\left(\frac{a^3}{s-a}+\frac{b^3}{s-b}+\frac{c^3}{s-c}+\frac{d^3}{s-d}\right)\end{aligned}$$

i.e.

$$4A\geq(a^3+b^3+c^3+d^3)\left(\frac{1}{s-a}+\frac{1}{s-b}+\frac{1}{s-c}+\frac{1}{s-d}\right). \tag{2}$$

Since $a\geq b\geq c\geq d$ it follows that $a^2\geq b^2\geq c^2\geq d^2$, and one more application of *Chebishev's inequality* gives us

$$(a^2+b^2+c^2+d^2)(a+b+c+d)\leq4(a^3+b^3+c^3+d^3),$$

i.e.

$$a^3+b^3+c^3+d^3\geq\frac{(a^2+b^2+c^2+d^2)(a+b+c+d)}{4}. \tag{3}$$

Furthermore

$$\begin{aligned}a^2+b^2+c^2+d^2&=\frac{a^2+b^2}{2}+\frac{b^2+c^2}{2}+\frac{c^2+d^2}{2}+\frac{d^2+a^2}{2}\\&\geq ab+bc+cd+da=1.\end{aligned}$$

So in (3) we deduce

$$a^3+b^3+c^3+d^3 \geq \frac{a+b+c+d}{4} \tag{4}$$

and clearly we have

$$a+b+c+d = \frac{(s-a)+(s-b)+(s-c)+(s-d)}{3}. \tag{5}$$

Now from (4) and (5) we obtain

$$a^3+b^3+c^3+d^3 \geq \frac{(s-a)+(s-b)+(s-c)+(s-d)}{12}. \tag{6}$$

Using (2), (6) and $AM \geq HM$ we have

$$\begin{aligned} 4A &\geq \left(\frac{(s-a)+(s-b)+(s-c)+(s-d)}{12}\right)\left(\frac{1}{s-a}+\frac{1}{s-b}+\frac{1}{s-c}+\frac{1}{s-d}\right) \\ &\geq \frac{16}{12} = \frac{4}{3}, \end{aligned}$$

i.e. it follows that $A \geq \frac{1}{3}$, as required. ■

191 Let α, x, y, z be positive real numbers such that $xyz = 1$ and $\alpha \geq 1$. Prove the inequality

$$\frac{x^\alpha}{y+z} + \frac{y^\alpha}{z+x} + \frac{z^\alpha}{x+y} \geq \frac{3}{2}.$$

Solution Without loss of generality we may assume that $x \geq y \geq z$.
Then

$$\frac{x}{y+z} \geq \frac{y}{z+x} \geq \frac{z}{x+y} \quad \text{and} \quad x^{\alpha-1} \geq y^{\alpha-1} \geq z^{\alpha-1}.$$

Applying *Chebishev's inequality* we have

$$(x^{\alpha-1}+y^{\alpha-1}+z^{\alpha-1})\left(\frac{x}{y+z}+\frac{y}{z+x}+\frac{z}{x+y}\right) \leq 3\left(\frac{x^\alpha}{y+z}+\frac{y^\alpha}{z+x}+\frac{z^\alpha}{x+y}\right). \tag{1}$$

Recalling $AM \geq GM$ we get

$$x^{\alpha-1}+y^{\alpha-1}+z^{\alpha-1} \geq 3\sqrt[3]{(xyz)^{\alpha-1}} = 3. \tag{2}$$

Nesbitt's inequality gives us

$$\frac{x}{y+z}+\frac{y}{z+x}+\frac{z}{x+y} \geq \frac{3}{2}. \tag{3}$$

Finally using (1), (2) and (3) we obtain

$$3\left(\frac{x^\alpha}{y+z}+\frac{y^\alpha}{z+x}+\frac{z^\alpha}{x+y}\right)\geq (x^{\alpha-1}+y^{\alpha-1}+z^{\alpha-1})\left(\frac{x}{y+z}+\frac{y}{z+x}+\frac{z}{x+y}\right)$$
$$=3\cdot\frac{3}{2}$$

i.e.

$$\frac{x^\alpha}{y+z}+\frac{y^\alpha}{z+x}+\frac{z^\alpha}{x+y}\geq\frac{3}{2}.$$ ■

192 Let $x_1, x_2, \ldots, x_n$ be positive real numbers such that

$$\frac{1}{1+x_1}+\frac{1}{1+x_2}+\cdots+\frac{1}{1+x_n}=1.$$

Prove the inequality

$$\frac{\sqrt{x_1}+\sqrt{x_2}+\cdots+\sqrt{x_n}}{n-1}\geq\frac{1}{\sqrt{x_1}}+\frac{1}{\sqrt{x_2}}+\cdots+\frac{1}{\sqrt{x_n}}.$$

Solution Let $\frac{1}{1+x_i}=a_i$, for $i=1,2,\ldots,n$.

Clearly $\sum_{i=1}^n a_i=1$ and the given inequality becomes

$$\sum_{i=1}^n\sqrt{\frac{1-a_i}{a_i}}\geq(n-1)\sum_{i=1}^n\sqrt{\frac{a_i}{1-a_i}}\quad\Leftrightarrow\quad\sum_{i=1}^n\sqrt{\frac{1}{a_i(1-a_i)}}\geq n\sum_{i=1}^n\sqrt{\frac{a_i}{1-a_i}}$$
$$\Leftrightarrow\quad n\sum_{i=1}^n\sqrt{\frac{a_i}{1-a_i}}\leq\left(\sum_{i=1}^n a_i\right)\left(\sum_{i=1}^n\sqrt{\frac{1}{a_i(1-a_i)}}\right).$$

The last inequality is true according to *Chebishev's inequality* applied to the sequences

$$(a_1,a_2,\ldots,a_n)\quad\text{and}\quad\left(\frac{1}{\sqrt{a_1(1-a_1)}},\frac{1}{\sqrt{a_2(1-a_2)}},\ldots,\frac{1}{\sqrt{a_n(1-a_n)}}\right).$$ ■

193 Let $x_1, x_2, \ldots, x_n > 0$ be real numbers. Prove the inequality

$$x_1^{x_1}x_2^{x_2}\cdots x_n^{x_n}\geq(x_1x_2\cdots x_n)^{\frac{x_1+x_2+\cdots+x_n}{n}}.$$

Solution If we take the logarithm of both sides the given inequality becomes:

$$x_1\ln x_1+x_2\ln x_2+\cdots+x_n\ln x_n\geq\frac{x_1+x_2+\cdots+x_n}{n}(\ln x_1+\ln x_2+\cdots+\ln x_n).\tag{1}$$

We may assume that $x_1\geq x_2\geq\cdots\geq x_n$, then $\ln x_1\geq\ln x_2\geq\cdots\geq\ln x_n$.

Applying *Chebishev's inequality* we get

$$(x_1+x_2+\cdots+x_n)(\ln x_1+\ln x_2+\cdots+\ln x_n)\leq n(x_1\ln x_1+x_2\ln x_2+\cdots+x_n\ln x_n),$$

i.e.

$$x_1\ln x_1+x_2\ln x_2+\cdots+x_n\ln x_n\geq \frac{x_1+x_2+\cdots+x_n}{n}(\ln x_1+\ln x_2+\cdots+\ln x_n).$$

■

194 Let $a, b, c > 0$ be real numbers such that $a+b+c=1$. Prove the inequality

$$\frac{a^2+b}{b+c}+\frac{b^2+c}{c+a}+\frac{c^2+a}{a+b}\geq 2.$$

Solution 1 Applying the *Cauchy–Schwarz inequality* for the sequences

$$a_1=\sqrt{\frac{a^2+b}{b+c}},\qquad a_2=\sqrt{\frac{b^2+c}{c+a}},\qquad a_3=\sqrt{\frac{c^2+a}{a+b}}$$

and

$$b_1=\sqrt{(a^2+b)(b+c)},\qquad b_2=\sqrt{(b^2+c)(c+a)},\qquad b_3=\sqrt{(c^2+a)(a+b)}$$

we obtain

$$\frac{a^2+b}{b+c}+\frac{b^2+c}{c+a}+\frac{c^2+a}{a+b}\geq \frac{(a^2+b^2+c^2+1)^2}{(a^2+b)(b+c)+(b^2+c)(c+a)+(c^2+a)(a+b)}.$$

So it suffices to show that

$$\frac{(a^2+b^2+c^2+1)^2}{(a^2+b)(b+c)+(b^2+c)(c+a)+(c^2+a)(a+b)}\geq 2.$$

We have

$$\begin{aligned}
&\frac{(a^2+b^2+c^2+1)^2}{(a^2+b)(b+c)+(b^2+c)(c+a)+(c^2+a)(a+b)}\geq 2\\
\Leftrightarrow\quad &(a^2+b^2+c^2+1)^2\geq 2((a^2+b)(b+c)+(b^2+c)(c+a)\\
&\qquad\qquad +(c^2+a)(a+b))\\
\Leftrightarrow\quad &1+(a^2+b^2+c^2)^2\geq 2(a^2(b+c)+b^2(c+a)+c^2(a+b))\\
&\qquad\qquad +2(ab+bc+ca)\\
\Leftrightarrow\quad &1+(a^2+b^2+c^2)^2\geq 2(a^2(1-a)+b^2(1-b)+c^2(1-c))\\
&\qquad\qquad +2(ab+bc+ca)
\end{aligned}$$

$$\begin{aligned}
\Leftrightarrow \quad & 1+(a^2+b^2+c^2)^2 \geq 2(a^2+b^2+c^2-a^3-b^3-c^3) \\
& \qquad + 2(ab+bc+ca) \\
\Leftrightarrow \quad & (a^2+b^2+c^2)^2+2(a^3+b^3+c^3) \geq 2(a^2+b^2+c^2+ab+bc+ca)-1 \\
\Leftrightarrow \quad & (a^2+b^2+c^2)^2+2(a^3+b^3+c^3) \geq 2(a(1-c)+b(1-a) \\
& \qquad + c(1-b))-1 \\
\Leftrightarrow \quad & (a^2+b^2+c^2)^2+2(a^3+b^3+c^3) \geq 1-2(ab+bc+ca) \\
\Leftrightarrow \quad & (a^2+b^2+c^2)^2+2(a^3+b^3+c^3) \geq (a+b+c)^2-2(ab+bc+ca) \\
& \qquad = a^2+b^2+c^2.
\end{aligned}$$

So we need to show that

$$(a^2+b^2+c^2)^2+2(a^3+b^3+c^3) \geq a^2+b^2+c^2. \tag{1}$$

By *Chebishev's inequality* we deduce

$$(a+b+c)(a^2+b^2+c^2) \leq 3(a^3+b^3+c^3), \quad \text{i.e.} \quad a^3+b^3+c^3 \geq \frac{a^2+b^2+c^2}{3},$$

and clearly $(a^2+b^2+c^2)^2 \geq \frac{a^2+b^2+c^2}{3}$.

Adding these inequalities gives us inequality (1). ■

Solution 2 Take $a+b+c=p=1, ab+bc+ca=q, abc=r$ and use the method from Chap. 14. ■

195 Let $a,b,c>1$ be positive real numbers such that $\frac{1}{a^2-1}+\frac{1}{b^2-1}+\frac{1}{c^2-1}=1$. Prove the inequality

$$\frac{1}{a+1}+\frac{1}{b+1}+\frac{1}{c+1} \leq 1.$$

Solution Without loss of generality we may assume that $a \geq b \geq c$. Then we have

$$\frac{a-2}{a+1} \geq \frac{b-2}{b+1} \geq \frac{c-2}{c+1} \quad \text{and} \quad \frac{a+2}{a-1} \leq \frac{b+2}{b-1} \leq \frac{c+2}{c-1}.$$

Now by *Chebishev's inequality* we get

$$\begin{aligned}
3\left(\frac{a^2-4}{a^2-1}+\frac{b^2-4}{b^2-1}+\frac{c^2-4}{c^2-1}\right) &\leq \left(\frac{a-2}{a+1}+\frac{b-2}{b+1}+\frac{c-2}{c+1}\right) \\
&\quad \times \left(\frac{a+2}{a-1}+\frac{b+2}{b-1}+\frac{c+2}{c-1}\right).
\end{aligned}$$

Since

$$\frac{a^2-4}{a^2-1}+\frac{b^2-4}{b^2-1}+\frac{c^2-4}{c^2-1}=3-3\left(\frac{1}{a^2-1}+\frac{1}{b^2-1}+\frac{1}{c^2-1}\right)=0$$

and

$$\frac{a+2}{a-1}+\frac{b+2}{b-1}+\frac{c+2}{c-1}>0$$

we must have

$$\frac{a-2}{a+1}+\frac{b-2}{b+1}+\frac{c-2}{c+1}\geq 0,$$

which is equivalent to $\frac{1}{a+1}+\frac{1}{b+1}+\frac{1}{c+1}\leq 1$, as required.

Equality holds iff $a=b=c=2$. ∎

196 Let a,b,c,d be positive real numbers such that $a^2+b^2+c^2+d^2=4$. Prove the inequality

$$\frac{1}{5-a}+\frac{1}{5-b}+\frac{1}{5-c}+\frac{1}{5-d}\leq 1.$$

Solution The given inequality is equivalent to

$$\frac{1}{5-a}-\frac{1}{4}+\frac{1}{5-b}-\frac{1}{4}+\frac{1}{5-c}-\frac{1}{4}+\frac{1}{5-d}-\frac{1}{4}\leq 0,$$

i.e.

$$\frac{a-1}{5-a}+\frac{b-1}{5-b}+\frac{c-1}{5-c}+\frac{d-1}{5-d}\leq 0.$$

Without loss of generality we may assume that $a\geq b\geq c\geq d$.

Then we have $a^2-1\geq b^2-1\geq c^2-1\geq d^2-1$.

We'll show that $\frac{1}{4a-a^2+5}\leq\frac{1}{4b-b^2+5}$.

We have

$$4a-a^2+5\geq 4b-b^2+5\quad\Leftrightarrow\quad a+b\leq 4,$$

which is obviously true since $a^2+b^2\leq 4$.

So we have

$$\frac{1}{4a-a^2+5}\leq\frac{1}{4b-b^2+5}\leq\frac{1}{4c-c^2+5}\leq\frac{1}{4d-d^2+5}.$$

Now by *Chebishev's inequality* we obtain

$$3\left(\frac{a^2-1}{4a-a^2+5}+\frac{b^2-1}{4b-b^2+5}+\frac{a^2-1}{4c-c^2+5}+\frac{a^2-1}{4d-d^2+5}\right)$$
$$\leq\sum_{cyc}(a^2-1)\sum_{cyc}\left(\frac{1}{4a-a^2+5}\right)=0.$$

Thus

$$0 \geq \frac{a^2-1}{4a-a^2+5}+\frac{b^2-1}{4b-b^2+5}+\frac{a^2-1}{4c-c^2+5}+\frac{a^2-1}{4d-d^2+5}$$
$$=\frac{a-1}{5-a}+\frac{b-1}{5-b}+\frac{c-1}{5-c}+\frac{d-1}{5-d},$$

as required.

Equality holds iff $a=b=c=d=1$. ∎

197 Let $a,b,c,d \in \mathbb{R}$ such that $\frac{1}{4+a}+\frac{1}{4+b}+\frac{1}{4+c}+\frac{1}{4+d}+\frac{1}{4+e}=1$. Prove the inequality

$$\frac{a}{4+a^2}+\frac{b}{4+b^2}+\frac{c}{4+c^2}+\frac{d}{4+d^2}+\frac{e}{4+e^2}\leq 1.$$

Solution We have

$$\frac{1-a}{4+a}+\frac{1-b}{4+b}+\frac{1-c}{4+c}+\frac{1-d}{4+d}+\frac{1-e}{4+e}$$
$$=\frac{5-(4+a)}{4+a}+\frac{5-(4+b)}{4+b}+\frac{5-(4+c)}{4+c}+\frac{5-(4+d)}{4+d}+\frac{5-(4+e)}{4+e}$$
$$=5-5=0.$$

We'll prove that

$$\frac{a}{4+a^2}+\frac{b}{4+b^2}+\frac{c}{4+c^2}+\frac{d}{4+d^2}+\frac{e}{4+e^2}\leq \frac{1}{4+a}+\frac{1}{4+b}+\frac{1}{4+c}+\frac{1}{4+d}+\frac{1}{4+e}. \quad (1)$$

Inequality (1) is equivalent to

$$\frac{1-a}{(4+a)(4+a^2)}+\frac{1-b}{(4+b)(4+b^2)}+\frac{1-c}{(4+c)(4+c^2)}+\frac{1-d}{(4+d)(4+d^2)}+\frac{1-e}{(4+e)(4+e^2)}\geq 0. \quad (2)$$

Without loss of generality we may assume that $a\geq b\geq c\geq d\geq e$, and then we easily deduce that

$$\frac{1-a}{4+a}\leq\frac{1-b}{4+b}\leq\frac{1-c}{4+c}\leq\frac{1-d}{4+d}\leq\frac{1-e}{4+e}\quad\text{and}$$
$$\frac{1}{4+a^2}\leq\frac{1}{4+b^2}\leq\frac{1}{4+c^2}\leq\frac{1}{4+d^2}\leq\frac{1}{4+e^2}.$$

So by *Chebishev's inequality* we get

$$5\sum_{\text{sym}} \frac{1-a}{(4+a)(4+a^2)} \geq \sum_{\text{sym}} \frac{1-a}{4+a} \cdot \sum_{\text{sym}} \frac{1}{4+a^2} = 0,$$

which means that inequality (2) holds, i.e. inequality (1) is true and since $\frac{1}{4+a} + \frac{1}{4+b} + \frac{1}{4+c} + \frac{1}{4+d} + \frac{1}{4+e} = 1$ we obtain the required result.

Equality occurs iff $a = b = c = d = e = 1$. ■

198 Let a, b, c be real numbers different from 1, such that $a + b + c = 1$. Prove the inequality

$$\frac{1+a^2}{1-a^2} + \frac{1+b^2}{1-b^2} + \frac{1+c^2}{1-c^2} \geq \frac{15}{4}.$$

Solution Since $a, b, c > 0, a \neq 1, b \neq 1, c \neq 1$ and $a + b + c = 1$ it follows that $0 < a, b, c < 1$.

The given inequality is symmetric, so without loss of generality we may assume that $a \leq b \leq c$.

Then we have

$$1 + a^2 \leq 1 + b^2 \leq 1 + c^2 \quad \text{and} \quad 1 - c^2 \leq 1 - b^2 \leq 1 - a^2.$$

Hence

$$\frac{1}{1-a^2} \leq \frac{1}{1-b^2} \leq \frac{1}{1-c^2}.$$

Now by *Chebishev's inequality* we have

$$\begin{aligned} A &= \frac{1+a^2}{1-a^2} + \frac{1+b^2}{1-b^2} + \frac{1+c^2}{1-c^2} \\ &\geq \frac{1}{3}(1+a^2+1+b^2+1+c^2)\left(\frac{1}{1-a^2} + \frac{1}{1-b^2} + \frac{1}{1-c^2}\right), \end{aligned}$$

i.e.

$$A \geq \frac{(a^2+b^2+c^2+3)}{3}\left(\frac{1}{1-a^2} + \frac{1}{1-b^2} + \frac{1}{1-c^2}\right). \tag{1}$$

Also we have the well-known inequality

$$a^2 + b^2 + c^2 \geq \frac{(a+b+c)^2}{3} = \frac{1}{3}.$$

Therefore by (1) we obtain

$$A \geq \frac{(1/3+3)}{3}\left(\frac{1}{1-a^2} + \frac{1}{1-b^2} + \frac{1}{1-c^2}\right) = \frac{10}{9}\left(\frac{1}{1-a^2} + \frac{1}{1-b^2} + \frac{1}{1-c^2}\right). \tag{2}$$

Since $1-a^2, 1-b^2, 1-c^2 > 0$, by using $AM \geq HM$ we deduce

$$\frac{1}{1-a^2}+\frac{1}{1-b^2}+\frac{1}{1-c^2} \geq \frac{9}{3-(a^2+b^2+c^2)} \geq \frac{9}{3-1/3}=\frac{27}{8}. \qquad (3)$$

Finally from (2) and (3) we get

$$A \geq \frac{10}{9}\left(\frac{1}{1-a^2}+\frac{1}{1-b^2}+\frac{1}{1-c^2}\right) \geq \frac{10}{9}\cdot\frac{27}{8}=\frac{15}{4},$$

with equality iff $a=b=c=1/3$. ■

199 Let $x, y, z > 0$, such that $xyz = 1$. Prove the inequality

$$\frac{x^3}{(1+y)(1+z)}+\frac{y^3}{(1+z)(1+x)}+\frac{z^3}{(1+x)(1+y)} \geq \frac{3}{4}.$$

Solution Let $x \geq y \geq z$. Then

$$x^3 \geq y^3 \geq z^3 \quad \text{and} \quad \frac{1}{(1+y)(1+z)} \geq \frac{1}{(1+z)(1+x)} \geq \frac{1}{(1+x)(1+y)}.$$

Applying *Chebishev's inequality* we get

$$\begin{aligned}
3S &= 3\left(\frac{x^3}{(1+y)(1+z)}+\frac{y^3}{(1+z)(1+x)}+\frac{z^3}{(1+x)(1+y)}\right) \\
&\geq (x^3+y^3+z^3)\left(\frac{1}{(1+y)(1+z)}+\frac{1}{(1+z)(1+x)}+\frac{1}{(1+x)(1+y)}\right) \\
&= (x^3+y^3+z^3)\left(\frac{(1+x)+(1+y)+(1+z)}{(1+x)(1+y)(1+z)}\right) \\
&= (x^3+y^3+z^3)\left(\frac{3+x+y+z}{(1+x)(1+y)(1+z)}\right),
\end{aligned}$$

i.e.

$$S \geq \left(\frac{x^3+y^3+z^3}{3}\right)\left(\frac{3+x+y+z}{(1+x)(1+y)(1+z)}\right). \qquad (1)$$

Let $\frac{x+y+z}{3}=a$. Then we have

$$\frac{x^3+y^3+z^3}{3} \geq \left(\frac{x+y+z}{3}\right)^3 = a^3 \quad \text{and} \quad 3a \geq 3\sqrt[3]{xyz}=3, \quad \text{i.e.} \quad a \geq 1.$$

From $AM \geq GM$ we get

$$(1+x)(1+y)(1+z) \leq \left(\frac{3+x+y+z}{3}\right)^3 = (1+a)^3.$$

So by (1) we obtain

$$S \geq \left(\frac{x^3+y^3+z^3}{3}\right)\left(\frac{3+x+y+z}{(1+x)(1+y)(1+z)}\right) \geq a^3\left(\frac{6}{(1+a)^3}\right).$$

Hence it suffices to show that

$$\frac{6a^3}{(1+a)^3} \geq \frac{3}{4},$$

i.e.

$$6\left(1-\frac{1}{1+a}\right)^3 = \frac{6a^3}{(1+a)^3} \geq \frac{3}{4}.$$

Since $a \geq 1$, and the function $f(x) = 6(1-\frac{1}{1+x})^3$ increases on $[1,\infty]$ (why?), it follows that $f(a) \geq f(1) = \frac{3}{4}$, as required. ■

200 Let $a,b,c,d > 0$ be real numbers. Prove the inequality

$$\frac{a}{b+2c+3d}+\frac{b}{c+2d+3a}+\frac{c}{d+2a+3b}+\frac{d}{a+2b+3c} \geq \frac{2}{3}.$$

Solution Let

$$A = b+2c+3d, \qquad B = c+2d+3a, \qquad C = d+2a+3b, \qquad D = a+2b+3c.$$

By the *Cauchy–Schwarz inequality* we have

$$\left(\frac{a}{A}+\frac{b}{B}+\frac{c}{C}+\frac{d}{D}\right)(aA+bB+cC+dD) \geq (a+b+c+d)^2$$

$$\Leftrightarrow \quad \frac{a}{b+2c+3d}+\frac{b}{c+2d+3a}+\frac{c}{d+2a+3b}+\frac{d}{a+2b+3c} \geq \frac{(a+b+c+d)^2}{aA+bB+cC+dD}. \tag{1}$$

Furthermore

$$aA+bB+cC+dD = 4(ab+ac+ad+bc+bd+cd),$$

and (1) becomes

$$\frac{a}{b+2c+3d}+\frac{b}{c+2d+3a}+\frac{c}{d+2a+3b}+\frac{d}{a+2b+3c} \geq \frac{(a+b+c+d)^2}{4(ab+ac+ad+bc+bd+cd)}.$$

So it suffices to prove that

$$\frac{(a+b+c+d)^2}{4(ab+ac+ad+bc+bd+cd)} \geq \frac{2}{3},$$

i.e.

$$3(a+b+c+d)^2 \geq 8(ab+ac+ad+bc+bd+cd). \qquad (2)$$

We'll use *Maclaurin's theorem.*

We have

$$p_2 = \frac{c_2}{6} = \frac{ab+ac+ad+bc+bd+cd}{6}, \quad \text{i.e.}$$

$$ab+ac+ad+bc+bd+cd = 6p_2$$

and

$$p_1 = \frac{c_1}{4} = \frac{a+b+c+d}{4}, \quad \text{i.e.} \quad a+b+c+d = 4p_1.$$

Now inequality (2) is equivalent to $48p_1^2 \geq 48p_2$, i.e. $p_1 \geq p_2^{1/2}$, which is true due to *Maclaurin's theorem.* ∎

201 Let a, b, c be positive real numbers. Prove the inequality

$$\frac{a^2+bc}{b+c} + \frac{b^2+ca}{c+a} + \frac{c^2+ab}{a+b} \geq a+b+c.$$

Solution Assume $a \geq b \geq c$. Then clearly $a^2 \geq b^2 \geq c^2$ and $\frac{1}{b+c} \geq \frac{1}{c+a} \geq \frac{1}{a+b}$.

According to the *rearrangement inequality* we have

$$\frac{a^2}{b+c} + \frac{b^2}{c+a} + \frac{c^2}{a+b} \geq \frac{b^2}{b+c} + \frac{c^2}{c+a} + \frac{a^2}{a+b},$$

i.e.

$$\frac{a^2+bc}{b+c} + \frac{b^2+ca}{c+a} + \frac{c^2+ab}{a+b} \geq \frac{b^2+bc}{b+c} + \frac{c^2+ca}{c+a} + \frac{a^2+ab}{a+b} = a+b+c.$$

Equality occurs iff $a = b = c$. ∎

202 Let $a, b > 0$, $n \in \mathbb{N}$. Prove the inequality

$$\left(1+\frac{a}{b}\right)^n + \left(1+\frac{b}{a}\right)^n \geq 2^{n+1}.$$

Solution We'll use the fact that the function $f(x) = x^n$ is concave on $(0, \infty)$.

So according to *Jensen's inequality* we have

$$\frac{x^n + y^n}{2} \geq \left(\frac{x+y}{2}\right)^n.$$

Remark Note that this is a *power mean inequality*.

Now we have

$$\frac{1}{2}\left(\left(1+\frac{a}{b}\right)^n + \left(1+\frac{b}{a}\right)^n\right) \geq \left(\frac{1+a/b+1+b/a}{2}\right)^n = \left(\frac{2+a/b+b/a}{2}\right)^n. \tag{1}$$

Using $\frac{a}{b} + \frac{b}{a} \geq 2$ and (1) we deduce

$$\left(1+\frac{a}{b}\right)^n + \left(1+\frac{b}{a}\right)^n \geq 2\left(\frac{2+2}{2}\right)^n = 2^{n+1}.$$

■

203 Let $a, b, c > 0$ be real numbers such that $a + b + c = 1$. Prove the inequality

$$\left(a+\frac{1}{a}\right)^2 + \left(b+\frac{1}{b}\right)^2 + \left(c+\frac{1}{c}\right)^2 \geq \frac{100}{3}.$$

Solution The function $f(x) = x^2$ is convex on $(0, \infty)$.

So according to *Jensen's inequality* we have

$$\frac{1}{3}\left(\left(a+\frac{1}{a}\right)^2 + \left(b+\frac{1}{b}\right)^2 + \left(c+\frac{1}{c}\right)^2\right) \geq \left(\frac{1}{3}\left(a+\frac{1}{a}+b+\frac{1}{b}+c+\frac{1}{c}\right)\right)^2,$$

i.e.

$$\begin{aligned}\left(a+\frac{1}{a}\right)^2 + \left(b+\frac{1}{b}\right)^2 + \left(c+\frac{1}{c}\right)^2 &\geq \frac{1}{3}\left(a+b+c+\frac{1}{a}+\frac{1}{b}+\frac{1}{c}\right)^2 \\ &\geq \frac{1}{3}(1+9)^2 = \frac{100}{3}.\end{aligned}$$

■

204 Let $x, y, z > 0$ be real numbers. Prove the inequality

$$\frac{x}{2x+y+z} + \frac{y}{x+2y+z} + \frac{z}{x+y+2z} \leq \frac{3}{4}.$$

Solution Let $s = x + y + z$.

The given inequality becomes

$$\frac{x}{s+x} + \frac{y}{s+y} + \frac{z}{s+z} \leq \frac{3}{4}.$$

Consider the function $f : (0, +\infty) \to (0, +\infty)$, defined by $f(a) = \frac{a}{s+a}$.

We can easily show that $f''(a) \leq 0$, for every $a \in \mathbb{R}^+$, i.e. f is concave on $\mathbb{R}^+$. By *Jensen's inequality* we have

$$\frac{f(x)+f(y)+f(z)}{3} \leq f\left(\frac{x+y+z}{3}\right),$$

i.e.

$$\begin{aligned}\frac{x}{s+x}+\frac{y}{s+y}+\frac{z}{s+z} &= f(x)+f(y)+f(z)\\ &\leq 3f\left(\frac{x+y+z}{3}\right) = 3f\left(\frac{s}{3}\right) = \frac{s/3}{s+s/3} = \frac{3}{4},\end{aligned}$$

as required. ■

205 Let $a, b, c, d > 0$ be real numbers such that $a \leq 1, a+b \leq 5, a+b+c \leq 14, a+b+c+d \leq 30$. Prove that

$$\sqrt{a}+\sqrt{b}+\sqrt{c}+\sqrt{d} \leq 10.$$

Solution The function $f:(0,+\infty) \to (0,+\infty)$ defined by $f(x)=\sqrt{x}$ is concave on $(0,+\infty)$, so by *Jensen's inequality*, for

$$n=4, \qquad \alpha_1=\frac{1}{10}, \qquad \alpha_2=\frac{2}{10}, \qquad \alpha_3=\frac{3}{10}, \qquad \alpha_4=\frac{4}{10}$$

we get

$$\frac{1}{10}\sqrt{a}+\frac{2}{10}\sqrt{\frac{b}{4}}+\frac{3}{10}\sqrt{\frac{c}{9}}+\frac{4}{10}\sqrt{\frac{d}{16}} \leq \sqrt{\frac{a}{10}+\frac{b}{20}+\frac{c}{30}+\frac{d}{40}},$$

i.e.

$$\sqrt{a}+\sqrt{b}+\sqrt{c}+\sqrt{d} \leq 10\sqrt{\frac{12a+6b+4c+3d}{120}}. \tag{1}$$

On the other hand, we have

$$\begin{aligned}&12a+6b+4c+3d\\ &\quad= 3(a+b+c+d)+(a+b+c)+2(a+b)+6a\\ &\quad\leq 3\cdot 30+14+2\cdot 5+6\cdot 1 = 120.\end{aligned}$$

By (1) and the last inequality we obtain the required result. ■

206 Let a, b, c, d be positive real numbers such that $a+b+c+d=4$. Prove the inequality

$$\frac{a}{b^2+b}+\frac{b}{c^2+c}+\frac{c}{d^2+d}+\frac{d}{a^2+a} \geq \frac{8}{(a+c)(b+d)}.$$

Solution Denote $A=\frac{a}{b^2+b}+\frac{b}{c^2+c}+\frac{c}{d^2+d}+\frac{d}{a^2+a}$.

Consider the function $f(x)=\frac{1}{x(x+1)}$. Then f is convex for $x>0$.

According to *Jensen's inequality*, we have

$$\frac{a}{4}\cdot f(b)+\frac{b}{4}\cdot f(c)+\frac{c}{4}\cdot f(d)+\frac{d}{4}\cdot f(a)\geq f\left(\frac{ab+bc+cd+da}{4}\right),$$

i.e.

$$A\geq\frac{64}{(ab+bc+cd+da)^2+4(ab+bc+cd+da)}.$$

So it remains to prove that

$$\frac{64}{(ab+bc+cd+da)^2+4(ab+bc+cd+da)}\geq\frac{8}{(a+c)(b+d)},$$

i.e.

$$ab+bc+cd+da\leq 4,$$

i.e.

$$(a-b+c-d)^2\geq 0,$$

which is obviously true. Equality holds iff $a=b=c=d=1$. ∎

207 Let $x_1,x_2,\ldots,x_n>0$ and $n\in\mathbb{N}, n>1$, such that $x_1+x_2+\cdots+x_n=1$. Prove the inequality

$$\frac{x_1}{\sqrt{1-x_1}}+\frac{x_2}{\sqrt{1-x_2}}+\cdots+\frac{x_n}{\sqrt{1-x_n}}\geq\frac{\sqrt{x_1}+\sqrt{x_2}+\cdots+\sqrt{x_n}}{\sqrt{n-1}}.$$

Solution The function $f(x)=\frac{x}{\sqrt{1-x}}$ is convex on $(0,\infty)$. (Why?)

Hence by *Jensen's inequality* we have

$$\frac{1}{n}\left(\frac{x_1}{\sqrt{1-x_1}}+\frac{x_2}{\sqrt{1-x_2}}+\cdots+\frac{x_n}{\sqrt{1-x_n}}\right)\geq\left(\frac{\frac{x_1+x_2+\cdots+x_n}{n}}{\sqrt{1-\frac{x_1+x_2+\cdots+x_n}{n}}}\right)$$

$$=\frac{\frac{1}{n}}{\sqrt{1-\frac{1}{n}}}=\frac{1}{\sqrt{n(n-1)}}.$$

It follows that

$$\frac{x_1}{\sqrt{1-x_1}}+\frac{x_2}{\sqrt{1-x_2}}+\cdots+\frac{x_n}{\sqrt{1-x_n}}\geq\sqrt{\frac{n}{n-1}}. \tag{1}$$

By $QM \geq AM$ we have

$$\frac{\sqrt{x_1}+\sqrt{x_2}+\cdots+\sqrt{x_n}}{n} \leq \sqrt{\frac{x_1+x_2+\cdots+x_n}{n}} = \frac{1}{\sqrt{n}},$$

i.e.

$$\sqrt{x_1}+\sqrt{x_2}+\cdots+\sqrt{x_n} \leq \sqrt{n}. \tag{2}$$

By (1) and (2) we deduce

$$\frac{x_1}{\sqrt{1-x_1}}+\frac{x_2}{\sqrt{1-x_2}}+\cdots+\frac{x_n}{\sqrt{1-x_n}} \geq \sqrt{\frac{n}{n-1}} \geq \frac{\sqrt{x_1}+\sqrt{x_2}+\cdots+\sqrt{x_n}}{\sqrt{n-1}},$$

as required. ■

208 Let $n \in \mathbb{N}, n \geq 2$. Determine the minimal value of

$$\frac{x_1^5}{x_2+x_3+\cdots+x_n}+\frac{x_2^5}{x_1+x_3+\cdots+x_n}+\cdots+\frac{x_n^5}{x_1+x_2+\cdots+x_{n-1}},$$

where $x_1, x_2, \ldots, x_n \in \mathbb{R}^+$ such that $x_1^2+x_2^2+\cdots+x_n^2=1$.

Solution Let $S = x_1+x_2+\cdots+x_n$. We may assume that $x_1 \geq x_2 \geq \cdots \geq x_n$.

Let $A = \frac{x_1^5}{S-x_1}+\frac{x_2^5}{S-x_2}+\cdots+\frac{x_n^5}{S-x_n} = \sum_{i=1}^n \frac{x_i^5}{S-x_i}$.

Since

$$x_1^4 \geq x_2^4 \geq \cdots \geq x_n^4 \quad \text{and} \quad \frac{x_1}{S-x_1} \geq \frac{x_2}{S-x_2} \geq \cdots \geq \frac{x_n}{S-x_n}$$

we can use *Chebishev's inequality*.

We have $A = \sum_{i=1}^n x_i^4 \frac{x_i}{S-x_i}$.

So

$$A = \sum_{i=1}^n x_i^4 \frac{x_i}{S-x_i} \geq n \sum_{i=1}^n x_i^4 \cdot \sum_{i=1}^n \frac{x_i}{S-x_i}. \tag{1}$$

By $QM \geq AM$ we have

$$\sqrt{\frac{\sum_{i=1}^n x_i^4}{n}} \geq \frac{\sum_{i=1}^n x_i^2}{n}, \quad \text{i.e.} \quad \sum_{i=1}^n x_i^4 \geq \frac{n}{n^2}\sum_{i=1}^n x_i^2 = \frac{1}{n}. \tag{2}$$

The function $f(x) = \frac{x}{S-x}$ is convex.

So by *Jensen's inequality* we have

$$f\left(\frac{x_1+x_2+\cdots+x_n}{n}\right) \leq \frac{1}{n}\sum_{i=1}^n f(x_i),$$

i.e.

$$\frac{1}{n}\sum_{i=1}^{n}\frac{x_i}{S-x_i}\geq\frac{\frac{x_1+x_2+\cdots+x_n}{n}}{x_1+x_2+\cdots+x_n-\frac{x_1+x_2+\cdots+x_n}{n}}=\frac{\frac{1}{n}}{1-\frac{1}{n}}=\frac{1}{n-1},$$

from which it follows that

$$\sum_{i=1}^{n}\frac{x_i}{S-x_i}\geq\frac{n}{n-1}. \tag{3}$$

Finally using (2), (3) and (1) we obtain

$$A\geq n\cdot\frac{1}{n}\cdot\frac{n}{n-1}=\frac{n}{n-1}.$$

Equality occurs if and only if $x_1=x_2=\cdots=x_n=1/\sqrt{n}$. ■

209 Let P, L, R denote the area, perimeter and circumradius of $\triangle ABC$, respectively. Determine the maximum value of the expression $\frac{LP}{R^3}$.

Solution We have

$$\frac{LP}{R^3}=\frac{(a+b+c)abc}{R^3 4R}=\frac{2R(\sin\alpha+\sin\beta+\sin\gamma)8R^3\sin\alpha\sin\beta\sin\gamma}{4R^4},$$

i.e.

$$\frac{LP}{R^3}=4(\sin\alpha+\sin\beta+\sin\gamma)\sin\alpha\sin\beta\sin\gamma. \tag{1}$$

By $AM\geq GM$ we have

$$\sin\alpha\sin\beta\sin\gamma\leq\left(\frac{\sin\alpha+\sin\beta+\sin\gamma}{3}\right)^3.$$

So by (1) we get

$$\frac{LP}{R^3}\leq\frac{4(\sin\alpha+\sin\beta+\sin\gamma)^4}{27}. \tag{2}$$

The function $f(x)=-\sin x$ is convex on $[0,\pi]$, so by *Jensen's inequality* we have

$$\frac{\sin\alpha+\sin\beta+\sin\gamma}{3}\leq\sin\left(\frac{\alpha+\beta+\gamma}{3}\right)=\frac{\sqrt{3}}{2}.$$

Finally from (2) we obtain

$$\frac{LP}{R^3}\leq\frac{4}{27}\left(\frac{3\sqrt{3}}{2}\right)^4=\frac{27}{4}.$$

Equality occurs iff $a=b=c$. ■

210 Let $a, b, c \in \mathbb{R}^+$ such that $a + b + c = abc$. Prove the inequality

$$\frac{1}{\sqrt{1+a^2}} + \frac{1}{\sqrt{1+b^2}} + \frac{1}{\sqrt{1+c^2}} \leq \frac{3}{2}.$$

Solution 1 After taking $a = \tan\alpha, b = \tan\beta, c = \tan\gamma$ where $\alpha, \beta, \gamma \in (0, \pi/2)$, the given inequality becomes

$$\frac{1}{\sqrt{1+\frac{\sin^2\alpha}{\cos^2\alpha}}} + \frac{1}{\sqrt{1+\frac{\sin^2\beta}{\cos^2\beta}}} + \frac{1}{\sqrt{1+\frac{\sin^2\gamma}{\cos^2\gamma}}} \leq \frac{3}{2},$$

i.e.

$$\cos\alpha + \cos\beta + \cos\gamma \leq \frac{3}{2}.$$

Also

$$\begin{aligned}\tan(\alpha+\beta+\gamma) &= \frac{\tan\alpha + \tan\beta + \tan\gamma - \tan\alpha\tan\beta\tan\gamma}{1 - \tan\alpha\tan\beta - \tan\beta\tan\gamma - \tan\gamma\tan\alpha} \\ &= \frac{a+b+c-abc}{1-ab-bc-ca} = 0,\end{aligned}$$

which means $\alpha + \beta + \gamma = \pi$.

The function $f(x) = -\cos x$ is convex on $[0, \pi/2]$.

So by *Jensen's inequality* we have

$$\frac{\cos\alpha + \cos\beta + \cos\gamma}{3} \leq \cos\frac{\alpha+\beta+\gamma}{3} = \frac{1}{2},$$

i.e. we get

$$\cos\alpha + \cos\beta + \cos\gamma \leq \frac{3}{2},$$

as required. ∎

Solution 2 Let $a = \frac{1}{x}, b = \frac{1}{y}, c = \frac{1}{z}$.

The constraint $a + b + c = abc$ becomes $xy + yz + zx = 1$, and the given inequality becomes equivalent to

$$\frac{x}{\sqrt{x^2+1}} + \frac{y}{\sqrt{y^2+1}} + \frac{z}{\sqrt{z^2+1}} \leq \frac{3}{2},$$

i.e.

$$\frac{x}{\sqrt{x^2+xy+yz+zx}} + \frac{y}{\sqrt{y^2+xy+yz+zx}} + \frac{z}{\sqrt{z^2+xy+yz+zx}} \leq \frac{3}{2},$$

i.e.

$$\frac{x}{\sqrt{(x+y)(x+z)}}+\frac{y}{\sqrt{(y+z)(y+x)}}+\frac{z}{\sqrt{(z+x)(z+y)}}\leq\frac{3}{2}. \tag{1}$$

By $AM \geq GM$ we have

$$\begin{aligned}\frac{x}{\sqrt{(x+y)(x+z)}}&=\frac{x\sqrt{(x+y)(x+z)}}{(x+y)(x+z)}\leq\frac{x((x+y)+(x+z))}{2(x+y)(x+z)}\\&=\frac{1}{2}\left(\frac{x}{x+y}+\frac{x}{x+z}\right).\end{aligned}$$

Analogously we get

$$\frac{y}{\sqrt{(y+z)(y+x)}}\leq\frac{1}{2}\left(\frac{y}{y+z}+\frac{y}{y+x}\right) \quad \text{and}$$

$$\frac{z}{\sqrt{(z+x)(z+y)}}\leq\frac{1}{2}\left(\frac{z}{z+x}+\frac{z}{z+y}\right).$$

Adding these three inequalities we get inequality (1). ■

211 Let $a, b, c \in \mathbb{R}$ such that $abc + a + c = b$. Prove the inequality

$$\frac{2}{a^2+1}-\frac{2}{b^2+1}+\frac{3}{c^2+1}\leq\frac{10}{3}.$$

Solution The given condition is equivalent to $b = \frac{a+c}{1-ac}$.
This suggest the substitutions:

$$a=\tan\alpha, \qquad b=\tan\beta, \qquad c=\tan\gamma,$$

where $\tan\beta = \tan(\alpha+\gamma)$ and $\alpha, \beta, \gamma \in (-\pi/2, \pi/2)$, so we have

$$\begin{aligned}A&=\frac{2}{a^2+1}-\frac{2}{b^2+1}+\frac{3}{c^2+1}=\frac{2}{\tan^2\alpha+1}-\frac{2}{\tan^2(\alpha+\gamma)+1}+\frac{3}{\tan^2\gamma+1}\\&=2\cos^2\alpha-2\cos^2(\alpha+\gamma)+3\cos^2\gamma\\&=(2\cos^2\alpha-1)-(2\cos^2(\alpha+\gamma)-1)+3\cos^2\gamma\\&=\cos 2\alpha-\cos(2\alpha+2\gamma)+3\cos^2\gamma\\&=2\sin(2\alpha+\gamma)\sin\gamma+3\cos^2\gamma.\end{aligned}$$

Let $x = |\sin\gamma|$. Then we have

$$A\leq 2x+3(1-x^2)=-3x^2+2x+3=-3\left(x-\frac{1}{3}\right)^2+\frac{10}{3}\leq\frac{10}{3}.$$

Equality holds if and only if $\sin(2\alpha+\gamma)=1$ and $\sin\gamma=\frac{1}{3}$, from which we deduce $(a,b,c)=(\sqrt{2}/2,\sqrt{2},\sqrt{2}/4)$. ■

212 Let $x,y,z>1$ be real numbers such that $\frac{1}{x}+\frac{1}{y}+\frac{1}{z}=2$. Prove the inequality

$$\sqrt{x-1}+\sqrt{y-1}+\sqrt{z-1}\le\sqrt{x+y+z}.$$

Solution 1 Let $x=a+1$, $y=b+1$, $z=c+1$, and clearly a,b and c are positive real numbers.

The initial condition $\frac{1}{x}+\frac{1}{y}+\frac{1}{z}=2$ becomes $\frac{1}{a+1}+\frac{1}{b+1}+\frac{1}{c+1}=2$, i.e.

$$ab+bc+ca+2abc=1. \tag{1}$$

We need to show that

$$\sqrt{a}+\sqrt{b}+\sqrt{c}\le\sqrt{a+b+c+3}. \tag{2}$$

After squaring inequality (2) we get

$$a+b+c+2\sqrt{ab}+2\sqrt{bc}+2\sqrt{ca}\le a+b+c+3$$

or

$$2\sqrt{ab}+2\sqrt{bc}+2\sqrt{ca}\le 3,$$

i.e.

$$\sqrt{ab}+\sqrt{bc}+\sqrt{ca}\le\frac{3}{2}. \tag{3}$$

Identity (1) is equivalent to

$$(\sqrt{ab})^2+(\sqrt{bc})^2+(\sqrt{ca})^2+2(\sqrt{ab}\cdot\sqrt{bc}\cdot\sqrt{ca})=1,$$

so due to Case 7 (Chap. 8) we may take

$$\sqrt{ab}=\sin\frac{\alpha}{2},\qquad \sqrt{bc}=\sin\frac{\beta}{2},\qquad \sqrt{ca}=\sin\frac{\gamma}{2},$$

where $\alpha,\beta,\gamma\in(0,\pi)$ and $\alpha+\beta+\gamma=\pi$.

Now inequality (3) is equivalent to

$$\sin\frac{\alpha}{2}+\sin\frac{\beta}{2}+\sin\frac{\gamma}{2}\le\frac{3}{2},$$

where $\alpha,\beta,\gamma\in(0,\pi)$, $\alpha+\beta+\gamma=\pi$, which is true by N_3 (Chap. 8). ■

Solution 2 Applying the *Cauchy–Schwarz inequality* we have

$$\left(\frac{x-1}{x}+\frac{y-1}{y}+\frac{z-1}{z}\right)(x+y+z)\ge(\sqrt{x-1}+\sqrt{y-1}+\sqrt{z-1})^2.$$

Also

$$\frac{x-1}{x}+\frac{y-1}{y}+\frac{z-1}{z}=3-\left(\frac{1}{x}+\frac{1}{y}+\frac{1}{z}\right)=1.$$

So

$$x+y+z\geq(\sqrt{x-1}+\sqrt{y-1}+\sqrt{z-1})^2,$$

i.e.

$$\sqrt{x+y+z}\geq\sqrt{x-1}+\sqrt{y-1}+\sqrt{z-1}.$$

Equality occurs iff $\frac{x-1}{x^2}=\frac{y-1}{y^2}=\frac{z-1}{z^2}$ and $\frac{1}{x}+\frac{1}{y}+\frac{1}{z}=2$, i.e. $x=y=z=3/2$. ■

213 Let a, b, c be positive real numbers such that $a+b+c=1$. Prove the inequality

$$\sqrt{\frac{1}{a}-1}\sqrt{\frac{1}{b}-1}+\sqrt{\frac{1}{b}-1}\sqrt{\frac{1}{c}-1}+\sqrt{\frac{1}{c}-1}\sqrt{\frac{1}{a}-1}\geq 6.$$

Solution Let $a=xy, b=yz, c=zx$. Then $xy+yz+zx=1$ and due to *Case 3* (Chap. 8) we may take

$$x=\tan\frac{\alpha}{2},\qquad y=\tan\frac{\beta}{2},\qquad z=\tan\frac{\gamma}{2},$$

where $\alpha,\beta,\gamma\in(0,\pi)$ and $\alpha+\beta+\gamma=\pi$.

We have

$$\begin{aligned}\sqrt{\frac{1}{a}-1}\sqrt{\frac{1}{b}-1}&=\sqrt{\frac{(1-a)(1-b)}{ab}}=\sqrt{\frac{(1-xy)(1-yz)}{xy^2z}}\\&=\sqrt{\frac{(yz+zx)(zx+xy)}{xy^2z}}=\sqrt{\frac{(y+x)(z+y)}{y^2}}=\frac{\sqrt{1+y^2}}{y}\\&=\frac{\sqrt{1+\tan^2\frac{\beta}{2}}}{\tan\frac{\beta}{2}}=\frac{1}{\sin\frac{\beta}{2}}.\end{aligned}$$

Similarly we obtain

$$\sqrt{\frac{1}{b}-1}\sqrt{\frac{1}{c}-1}=\frac{1}{\sin\frac{\gamma}{2}}\quad\text{and}\quad\sqrt{\frac{1}{c}-1}\sqrt{\frac{1}{a}-1}=\frac{1}{\sin\frac{\alpha}{2}}.$$

Now the given inequality becomes

$$\frac{1}{\sin\frac{\alpha}{2}}+\frac{1}{\sin\frac{\beta}{2}}+\frac{1}{\sin\frac{\gamma}{2}}\geq 6.$$

By $AM \geq HM$ we have

$$\frac{1}{\sin\frac{\alpha}{2}} + \frac{1}{\sin\frac{\beta}{2}} + \frac{1}{\sin\frac{\gamma}{2}} \geq \frac{9}{\sin\frac{\alpha}{2} + \sin\frac{\beta}{2} + \sin\frac{\gamma}{2}}.$$

So we need to prove that $\sin\frac{\alpha}{2} + \sin\frac{\beta}{2} + \sin\frac{\gamma}{2} \leq \frac{3}{2}$ which is true according to N_3 (Chap. 8).

Equality occurs if and only if $\alpha = \beta = \gamma = \pi/3$, i.e. $a = b = c = \frac{1}{3}$. ∎

214 Let a, b, c be positive real numbers such that $a + b + c + 1 = 4abc$. Prove the inequalities

$$\frac{1}{a} + \frac{1}{b} + \frac{1}{c} \geq 3 \geq \frac{1}{\sqrt{ab}} + \frac{1}{\sqrt{bc}} + \frac{1}{\sqrt{ca}}.$$

Solution We have

$$\begin{aligned} & a + b + c + 1 = 4abc \\ \Leftrightarrow \quad & \frac{1}{bc} + \frac{1}{ca} + \frac{1}{ab} + \frac{1}{abc} = 4 \\ \Leftrightarrow \quad & \frac{1}{(2\sqrt{ab})^2} + \frac{1}{(2\sqrt{bc})^2} + \frac{1}{(2\sqrt{ca})^2} + \frac{2}{(2\sqrt{ab})(2\sqrt{bc})(2\sqrt{ca})} = 1. \end{aligned}$$

Due to Case 7 (Chap. 8) we can make the substitutions

$$\frac{1}{2\sqrt{bc}} = \sin\frac{\alpha}{2}, \qquad \frac{1}{2\sqrt{ca}} = \sin\frac{\beta}{2}, \qquad \frac{1}{2\sqrt{ab}} = \sin\frac{\gamma}{2}, \tag{1}$$

where $\alpha, \beta, \gamma \in (0, \pi)$ and $\alpha + \beta + \gamma = \pi$.

From (1) we easily obtain

$$\frac{1}{a} = \frac{2\sin\frac{\beta}{2}\sin\frac{\gamma}{2}}{\sin\frac{\alpha}{2}}, \qquad \frac{1}{b} = \frac{2\sin\frac{\gamma}{2}\sin\frac{\alpha}{2}}{\sin\frac{\beta}{2}} \quad \text{and} \quad \frac{1}{c} = \frac{2\sin\frac{\alpha}{2}\sin\frac{\beta}{2}}{\sin\frac{\gamma}{2}}. \tag{2}$$

Now the given inequality becomes

$$2\sin\frac{\alpha}{2} + 2\sin\frac{\beta}{2} + 2\sin\frac{\gamma}{2} \leq 3,$$

i.e.

$$\sin\frac{\alpha}{2} + \sin\frac{\beta}{2} + \sin\frac{\gamma}{2} \leq \frac{3}{2},$$

where $\alpha, \beta, \gamma \in (0, \pi)$ and $\alpha + \beta + \gamma = \pi$, which clearly holds due to N_3.

We need to show the left inequality which, due to (2) is equivalent to

$$\frac{2\sin\frac{\beta}{2}\sin\frac{\gamma}{2}}{\sin\frac{\alpha}{2}} + \frac{2\sin\frac{\gamma}{2}\sin\frac{\alpha}{2}}{\sin\frac{\beta}{2}} + \frac{2\sin\frac{\alpha}{2}\sin\frac{\beta}{2}}{\sin\frac{\gamma}{2}} \geq 3. \tag{3}$$

Let a, b, c be the lengths of the sides of the triangle with angles α, β and γ, let s be its semi-perimeter, and let $x = s - a$, $y = s - b$, $z = s - c$.

Then due to Case 9 (Chap. 8) inequality (3) is equivalent to

$$\frac{x}{y+z} + \frac{y}{z+x} + \frac{z}{x+y} \geq \frac{3}{2},$$

i.e. we obtain the famous *Nesbitt's inequality*, which clearly holds. And we are done. ∎

215 Let a, b, c be non-negative real numbers such that $ab + bc + ca = 1$. Prove the inequality

$$\frac{a}{1+a^2} + \frac{b}{1+b^2} + \frac{c}{1+c^2} \leq \frac{3\sqrt{3}}{4}.$$

Solution Since $ab + bc + ca = 1$ (Case 3, Chap. 8) we take:

$$a = \tan\frac{\alpha}{2}, \qquad b = \tan\frac{\beta}{2}, \qquad c = \tan\frac{\gamma}{2},$$

where $\alpha, \beta, \gamma \in (0, \pi)$ and $\alpha + \beta + \gamma = \pi$.

So we have

$$\frac{a}{1+a^2} + \frac{b}{1+b^2} + \frac{c}{1+c^2} = \frac{1}{2}(\sin\alpha + \sin\beta + \sin\gamma),$$

and the given inequality becomes

$$\frac{1}{2}(\sin\alpha + \sin\beta + \sin\gamma) \leq \frac{3\sqrt{3}}{4},$$

i.e.

$$\sin\alpha + \sin\beta + \sin\gamma \leq \frac{3\sqrt{3}}{2},$$

which is true according to N_1 (Chap. 8).

Equality occurs if and only if $a = b = c = 1/\sqrt{3}$. ∎

Remark This is the same problem as Problem 92.

216 Let a, b, c be positive real numbers such that $a+b+c = 1$. Prove the inequality

$$\sqrt{\frac{ab}{c+ab}} + \sqrt{\frac{bc}{a+bc}} + \sqrt{\frac{ca}{b+ca}} \leq \frac{3}{2}.$$

Solution We have

$$\begin{aligned}(c+a)(c+b) &= c^2 + ca + cb + ab = c^2 + c(a+b) + ab = c^2 + c(1-c) + ab \\ &= c + ab.\end{aligned}$$

Analogously we get

$$(a+b)(a+c)=a+bc \quad \text{and} \quad (b+c)(b+a)=b+ca.$$

Now the given inequality becomes

$$\sqrt{\frac{ab}{(c+a)(c+b)}}+\sqrt{\frac{bc}{(a+b)(a+c)}}+\sqrt{\frac{ca}{(b+c)(b+a)}}\leq\frac{3}{2}.$$

According to Case 9 (Chap. 8) it suffices to show that

$$\sin\frac{\alpha}{2}+\sin\frac{\beta}{2}+\sin\frac{\gamma}{2}\leq\frac{3}{2},$$

where $\alpha,\beta,\gamma\in(0,\pi)$ and $\alpha+\beta+\gamma=\pi$, which is true due to N_3 (Chap. 8). ■

217 Let $a,b,c>0$ be real numbers such that $(a+b)(b+c)(c+a)=1$. Prove the inequality

$$ab+bc+ca\leq\frac{3}{4}.$$

Solution We homogenize as follows

$$(ab+bc+ca)^3\leq\frac{27}{64}(a+b)^2(b+c)^2(c+a)^2. \tag{1}$$

Since inequality (1) is homogenous, we may assume that $ab+bc+ca=1$.

Now, by Case 3 (Chap. 8) we can use the substitutions

$$a=\tan\frac{\alpha}{2},\qquad b=\tan\frac{\beta}{2},\qquad c=\tan\frac{\gamma}{2},$$

where $\alpha,\beta,\gamma\in(0,\pi)$ and $\alpha+\beta+\gamma=\pi$.

Then

$$a+b=\tan\frac{\alpha}{2}+\tan\frac{\beta}{2}=\frac{\sin\frac{\alpha}{2}\cos\frac{\beta}{2}+\cos\frac{\alpha}{2}\sin\frac{\beta}{2}}{\cos\frac{\alpha}{2}\cos\frac{\beta}{2}}=\frac{\sin\frac{\alpha+\beta}{2}}{\cos\frac{\alpha}{2}\cos\frac{\beta}{2}}=\frac{\cos\frac{\gamma}{2}}{\cos\frac{\alpha}{2}\cos\frac{\beta}{2}}.$$

Similarly

$$b+c=\frac{\cos\frac{\alpha}{2}}{\cos\frac{\beta}{2}\cos\frac{\gamma}{2}} \quad \text{and} \quad c+a=\frac{\cos\frac{\beta}{2}}{\cos\frac{\gamma}{2}\cos\frac{\alpha}{2}},$$

i.e. we obtain

$$(a+b)(b+c)(c+a)=\frac{1}{\cos\frac{\alpha}{2}\cos\frac{\beta}{2}\cos\frac{\gamma}{2}}.$$

Therefore inequality (1) becomes

$$\frac{1}{\cos^2\frac{\alpha}{2}\cos^2\frac{\beta}{2}\cos^2\frac{\gamma}{2}} \geq \frac{64}{27}, \quad \text{i.e.} \quad \cos\frac{\alpha}{2}\cos\frac{\beta}{2}\cos\frac{\gamma}{2} \leq \frac{3\sqrt{3}}{8},$$

which is true due to N_8 (Chap. 8). So we are done. ■

218 Let $a, b, c \geq 0$ be real numbers such that $a^2+b^2+c^2+abc=4$. Prove the inequality

$$0 \leq ab+bc+ca-abc \leq 2.$$

Solution Observe that if $a, b, c > 1$ then $a^2+b^2+c^2+abc > 4$.

Therefore at least one number from a, b and c must be less than or equal to 1.

Without loss of generality assume that $a \leq 1$.

Then we have

$$ab+bc+ca-abc \geq bc-abc = bc(1-a) \geq 0.$$

So we have proved the left inequality.

Let $a=2x, b=2y, c=2z$.

Then the condition $a^2+b^2+c^2+abc=4$ becomes

$$x^2+y^2+z^2+2xyz=1 \tag{1}$$

and the given inequality becomes

$$2xy+2yz+2zx-4xyz \leq 1. \tag{2}$$

By (1) and Case 8 (Chap. 8) we can take

$$x=\cos\alpha, \qquad y=\cos\beta, \qquad z=\cos\gamma,$$

where $\alpha, \beta, \gamma \in [0, \pi/2]$ and $\alpha+\beta+\gamma=\pi$.

Therefore inequality (2) becomes

$$2\cos\alpha\cos\beta+2\cos\beta\cos\gamma+2\cos\gamma\cos\alpha-4\cos\alpha\cos\beta\cos\gamma \leq 1,$$

i.e.

$$\cos\alpha\cos\beta+\cos\beta\cos\gamma+\cos\gamma\cos\alpha-2\cos\alpha\cos\beta\cos\gamma \leq \frac{1}{2}. \tag{3}$$

Clearly at least one of the angles α, β and γ is less than or equal to $\pi/3$.

Without loss of generality, we may assume $\alpha \geq \pi/3$ and it follows that $\cos\alpha \leq \frac{1}{2}$.

We have

$$\begin{aligned}&\cos\alpha\cos\beta+\cos\beta\cos\gamma+\cos\gamma\cos\alpha-2\cos\alpha\cos\beta\cos\gamma\\&\quad=\cos\alpha(\cos\beta+\cos\gamma)+\cos\beta\cos\gamma(1-2\cos\alpha).\end{aligned} \tag{4}$$

By N_5 (Chap. 8) we have that

$$\cos\alpha + \cos\beta + \cos\gamma \leq \frac{3}{2}, \quad \text{i.e.} \quad \cos\beta + \cos\gamma \leq \frac{3}{2} - \cos\alpha. \tag{5}$$

Also

$$2\cos\beta\cos\gamma = \cos(\beta - \gamma) + \cos(\beta + \gamma) \leq 1 + \cos(\beta + \gamma) = 1 - \cos\alpha. \tag{6}$$

By (4), (5) and (6) we obtain

$$\begin{aligned}
&\cos\alpha\cos\beta + \cos\beta\cos\gamma + \cos\gamma\cos\alpha - 2\cos\alpha\cos\beta\cos\gamma \\
&\quad = \cos\alpha(\cos\beta + \cos\gamma) + \cos\beta\cos\gamma(1 - 2\cos\alpha) \\
&\quad \leq \cos\alpha\left(\frac{3}{2} - \cos\alpha\right) + \frac{1 - \cos\alpha}{2}(1 - 2\cos\alpha) = 2,
\end{aligned}$$

as required. ■

219 Let a, b, c be positive real numbers. Prove the inequality

$$a^2 + b^2 + c^2 + 2abc + 3 \geq (1 + a)(1 + b)(1 + c).$$

Solution The given inequality is equivalent to

$$a^2 + b^2 + c^2 + abc + 2 \geq a + b + c + ab + bc + ac.$$

Recall the *Turkevicius inequality*:

For any positive real numbers x, y, z, t we have

$$x^4 + y^4 + z^4 + t^4 + 2xyzt \geq x^2y^2 + y^2z^2 + z^2t^2 + t^2x^2 + x^2z^2 + y^2t^2.$$

If we set $a = x^2, b = y^2, c = z^2, t = 1$ we deduce

$$a^2 + b^2 + c^2 + 2\sqrt{abc} + 1 \geq a + b + c + ab + bc + ac. \tag{1}$$

Since $AM \geq GM$ we get

$$2\sqrt{abc} \leq abc + 1. \tag{2}$$

From (1) and (2) we obtain

$$a^2 + b^2 + c^2 + abc + 2 \geq a^2 + b^2 + c^2 + 2\sqrt{abc} + 1 \geq a + b + c + ab + bc + ac.$$

■

220 Let a, b, c be real numbers. Prove the inequality

$$\sqrt{a^2 + (1 - b)^2} + \sqrt{b^2 + (1 - c)^2} + \sqrt{c^2 + (1 - a)^2} \geq \frac{3\sqrt{2}}{2}.$$

Solution By *Minkowski's inequality* we have

$$\sqrt{a^2+(1-b)^2}+\sqrt{b^2+(1-c)^2}+\sqrt{c^2+(1-a)^2}$$
$$\geq \sqrt{(a+b+c)^2+(3-a-b-c)^2}=\sqrt{2\left(a+b+c-\frac{3}{2}\right)^2+\frac{9}{2}}\geq \frac{3\sqrt{2}}{2}.$$ ■

221 Let $a_1, a_2, \ldots, a_n \in \mathbb{R}^+$ such that $\sum_{i=1}^n a_i^3 = 3$ and $\sum_{i=1}^n a_i^5 = 5$. Prove the inequality

$$\sum_{i=1}^n a_i > \frac{3}{2}.$$

Solution We'll use *Hölder's inequality*:

If $a_1, a_2, \ldots, a_n; b_1, b_2, \ldots, b_n \in \mathbb{R}^+$ and $p, q \in (0, 1)$, $1/p + 1/q = 1$ then we have

$$\sum_{i=1}^n a_i b_i \leq \left(\sum_{i=1}^n a_i^p\right)^{\frac{1}{p}} \left(\sum_{i=1}^n b_i^q\right)^{\frac{1}{q}}.$$

We have

$$\sum_{i=1}^n a_i^3 = \sum_{i=1}^n a_i a_i^2 \leq \left(\sum_{i=1}^n a_i^{5/3}\right)^{3/5} \left(\sum_{i=1}^n (a_i^2)^{5/2}\right)^{2/5},$$

i.e.

$$3 \leq \left(\sum_{i=1}^n a_i^{5/3}\right)^{3/5} \cdot 5^{2/5} \quad \text{i.e.} \quad \frac{3}{5^{2/5}} \leq \left(\sum_{i=1}^n a_i^{5/3}\right)^{3/5}. \tag{1}$$

We'll show that

$$\sum_{i=1}^n a_i^{5/3} \leq \left(\sum_{i=1}^n a_i\right)^{5/3}.$$

Let $S = \sum_{i=1}^n a_i$.

Since $0 < \frac{a_i}{S} \leq 1$ and $\frac{5}{3} > 1$ we have that $(\frac{a_i}{S})^{5/3} \leq \frac{a_i}{S} = 1$ from which we deduce

$$\sum_{i=1}^n \left(\frac{a_i}{S}\right)^{5/3} \leq \sum_{i=1}^n \frac{a_i}{S} = 1.$$

So

$$\sum_{i=1}^n a_i^{5/3} \leq S^{5/3} = \left(\sum_{i=1}^n a_i\right)^{5/3},$$

since $2^5 > 5^2, 2 > 5^{2/5}$ and by (1) we obtain the required inequality. ■

222 Let a, b, c be positive real numbers such that $ab + bc + ca = 3$. Prove the inequality

$$(1 + a^2)(1 + b^2)(1 + c^2) \geq 8.$$

Solution By *Hölder's inequality* we have

$$(a^2b^2 + a^2 + b^2 + 1)(b^2 + c^2 + b^2c^2 + 1)(a^2 + a^2c^2 + c^2 + 1) \geq (1 + ab + bc + ca)^3$$

i.e.

$$(1 + a^2)^2(1 + b^2)^2(1 + c^2)^2 \geq 2^6,$$

as required. ■

223 Let a, b, c be positive real numbers such that $ab + bc + ca = 1$. Prove the inequality

$$(a^2 + ab + b^2)(b^2 + bc + c^2)(c^2 + ca + a^2) \geq 1.$$

Solution We'll show stronger inequality, i.e.

$$(a^2 + ab + b^2)(b^2 + bc + c^2)(c^2 + ca + a^2) \geq (ab + bc + ca)^3.$$

By *Hölder's inequality* we have

$$\begin{aligned}(a^2 + ab + b^2)&(b^2 + bc + c^2)(c^2 + ca + a^2) \\ &= (ab + a^2 + b^2)(b^2 + c^2 + bc)(a^2 + ca + c^2) \geq (ab + bc + ca)^3,\end{aligned}$$

as required. ■

224 Let a, b, c be positive real numbers such that $abc = 1$. Prove the inequality

$$\frac{a}{\sqrt{7 + b^2 + c^2}} + \frac{b}{\sqrt{7 + c^2 + a^2}} + \frac{c}{\sqrt{7 + a^2 + b^2}} \geq 1.$$

Solution Denote

$$A = \frac{a}{\sqrt{7 + b^2 + c^2}} + \frac{b}{\sqrt{7 + c^2 + a^2}} + \frac{c}{\sqrt{7 + a^2 + b^2}}$$

and

$$B = a(7 + b^2 + c^2) + b(7 + c^2 + a^2) + c(7 + a^2 + b^2).$$

By *Hölder's inequality* we have

$$A^2B \geq (a + b + c)^3. \tag{1}$$

Furthermore

$$B = 7(a+b+c) + (a+b+c)(ab+bc+ca) - 3$$
$$\leq 7(a+b+c) + \frac{(a+b+c)^3}{3} - 3 \leq (a+b+c)^3 \tag{2}$$

and by (1) and (2) we obtain

$$A^2 \geq \frac{(a+b+c)^3}{B} \geq 1, \quad \text{i.e.} \quad A \geq 1,$$

as required. ■

225 Let $a_1, a_2, \ldots, a_n$ be positive real numbers such that $a_1 + a_2 + \cdots + a_n = 1$. Prove the inequality

$$\frac{a_1}{\sqrt{1-a_1}} + \frac{a_2}{\sqrt{1-a_2}} + \cdots + \frac{a_n}{\sqrt{1-a_n}} \geq \sqrt{\frac{n}{n-1}}.$$

Solution Let us denote

$$A = \frac{a_1}{\sqrt{1-a_1}} + \frac{a_2}{\sqrt{1-a_2}} + \cdots + \frac{a_n}{\sqrt{1-a_n}},$$
$$B = a_1(1-a_1) + a_2(1-a_2) + \cdots + a_n(1-a_n).$$

By *Hölder's inequality* we have

$$A^2 B \geq (a_1 + a_2 + \cdots + a_n)^3 = 1. \tag{1}$$

Applying $QM \geq AM$ we deduce

$$B = 1 - (a_1^2 + a_2^2 + \cdots + a_n^2) \leq 1 - \frac{(a_1 + a_2 + \cdots + a_n)^2}{n} = \frac{n-1}{n}. \tag{2}$$

By (1) and (2) we obtain

$$\frac{n-1}{n} \cdot A^2 \geq A^2 B \geq 1, \quad \text{i.e.} \quad A \geq \sqrt{\frac{n}{n-1}}.$$

Equality holds iff $a_i = \frac{1}{n}$, for every $i = 1, 2, \ldots, n$. ■

226 Let a, b, c be positive real numbers. Prove the inequality

$$\frac{a}{\sqrt{2b^2+2c^2-a^2}} + \frac{b}{\sqrt{2c^2+2a^2-b^2}} + \frac{c}{\sqrt{2a^2+2b^2-c^2}} \geq \sqrt{3}.$$

Solution Denote

$$A = \frac{a}{\sqrt{2b^2+2c^2-a^2}} + \frac{b}{\sqrt{2c^2+2a^2-b^2}} + \frac{c}{\sqrt{2a^2+2b^2-c^2}}$$

and

$$B = a(2b^2 + 2c^2 - a^2) + b(2c^2 + 2a^2 - b^2) + c(2a^2 + 2b^2 - c^2)$$
$$= 2ab(a+b) + 2bc(b+c) + 2ca(c+a) - a^3 - b^3 - c^3.$$

By *Hölder's inequality* we have

$$A^2B \geq (a+b+c)^3. \tag{1}$$

We'll show that

$$(a+b+c)^3 \geq 3B, \tag{2}$$

and then by (1) we'll obtain the required inequality.

Inequality (2) is equivalent to

$$4(a^3+b^3+c^3) + 6abc \geq 4(ab(a+b) + bc(b+c) + ca(c+a)). \tag{3}$$

The following inequalities are true:

$$3((a^3+b^3+c^3) + 3abc) \geq 4(ab(a+b) + bc(b+c) + ca(c+a)) \quad (Schur),$$
$$a^3+b^3+c^3 \geq 3abc \quad (AM \geq GM).$$

Adding the last two inequalities we obtain inequality (3).

Equality occurs iff $a=b=c$. ■

227 Let a,b,c be positive real numbers such that $ab+bc+ca \geq 3$. Prove the inequality

$$\frac{a}{\sqrt{a+b}} + \frac{b}{\sqrt{b+c}} + \frac{c}{\sqrt{c+a}} \geq \frac{3}{\sqrt{2}}.$$

Solution By *Hölder's inequality* we have

$$\left(\frac{a}{\sqrt{a+b}} + \frac{b}{\sqrt{b+c}} + \frac{c}{\sqrt{c+a}}\right)^{2/3} (a(a+b)+b(b+c)+c(c+a))^{1/3} \geq a+b+c,$$

i.e.

$$\left(\frac{a}{\sqrt{a+b}} + \frac{b}{\sqrt{b+c}} + \frac{c}{\sqrt{c+a}}\right)^2 \geq \frac{(a+b+c)^3}{a^2+b^2+c^2+ab+bc+ca}.$$

It is enough to show that

$$\frac{(a+b+c)^3}{a^2+b^2+c^2+ab+bc+ca} \geq \frac{9}{2},$$

i.e.

$$2(a+b+c)^3 \geq 9(a^2+b^2+c^2+ab+bc+ca). \tag{1}$$

Let $p=a+b+c$ and $q=ab+bc+ca$.

Using the initial condition we have $q \geq 3$, and then inequality (1) is equivalent to

$$2p^3 \geq 9(p^2-2q+q) \quad \text{or} \quad 2p^3+9q \geq 9p^2.$$

Applying $AM \geq GM$ we obtain

$$2p^3+9q \geq 2p^3+27 = p^3+p^3+27 \geq 3\sqrt[3]{27p^6} = 9p^2,$$

as required. ■

228 Let $a, b, c \geq 1$ be real numbers such that $a+b+c=2abc$. Prove the inequality

$$\sqrt[3]{(a+b+c)^2} \geq \sqrt[3]{ab-1}+\sqrt[3]{bc-1}+\sqrt[3]{ca-1}.$$

Solution By the initial condition we have

$$a+b+c=2abc \quad \text{or} \quad \frac{1}{ab}+\frac{1}{bc}+\frac{1}{ca}=2 \quad \Leftrightarrow \quad \frac{ab-1}{ab}+\frac{bc-1}{bc}+\frac{ca-1}{ca}=1.$$

By *Hölder's inequality* for triples

$$(a,b,c), (b,c,a), \left(\frac{ab-1}{ab}, \frac{bc-1}{bc}, \frac{ca-1}{ca}\right)$$

we obtain

$$(a+b+c)^{1/3}(b+c+a)^{1/3}\left(\frac{ab-1}{ab}+\frac{bc-1}{bc}+\frac{ca-1}{ca}\right)^{1/3}$$
$$\geq (ab-1)^{1/3}+(bc-1)^{1/3}+(ca-1)^{1/3}.$$

Since

$$\frac{ab-1}{ab}+\frac{bc-1}{bc}+\frac{ca-1}{ca}=1$$

we get

$$\sqrt[3]{(a+b+c)^2} \geq \sqrt[3]{ab-1}+\sqrt[3]{bc-1}+\sqrt[3]{ca-1}.$$ ■

229 Let t_a, t_b, t_c be the lengths of the medians, and a, b, c be the lengths of the sides of a given triangle. Prove the inequality

$$t_a t_b + t_b t_c + t_c t_a < \frac{5}{4}(ab+bc+ca).$$

Solution We can easily show the inequalities

$$t_a < \frac{b+c}{2}, \qquad t_b < \frac{a+c}{2}, \qquad t_c < \frac{b+a}{2}.$$

After adding these we get

$$t_a + t_b + t_c < a + b + c. \tag{1}$$

By squaring (1) we deduce

$$t_a^2 + t_b^2 + t_c^2 + 2(t_a t_b + t_b t_c + t_c t_a) < a^2 + b^2 + c^2 + 2(ab + bc + ca). \tag{2}$$

On the other hand, we have

$$t_a^2 = \frac{2(b^2+c^2)-a^2}{4}, \qquad t_b^2 = \frac{2(a^2+c^2)-b^2}{4}, \qquad t_c^2 = \frac{2(b^2+a^2)-c^2}{4}$$

so

$$t_a^2 + t_b^2 + t_c^2 = \frac{3}{4}(a^2 + b^2 + c^2).$$

Now using the previous result and (2) we get

$$t_a t_b + t_b t_c + t_c t_a < \frac{1}{8}(a^2 + b^2 + c^2) + (ab + bc + ca). \tag{3}$$

Also we have $a^2 + b^2 + c^2 < 2(ab + bc + ca)$, since

$$a^2 + b^2 + c^2 - 2(ab + bc + ca) = a(a - b - c) + b(b - a - c) + c(c - a - b) < 0.$$

Finally by (3) and the previous inequality we obtain

$$t_a t_b + t_b t_c + t_c t_a < \frac{5}{4}(ab + bc + ca).$$

■

230 Let a, b, c and t_a, t_b, t_c be the lengths of the sides and lengths of the medians of an arbitrary triangle, respectively. Prove the inequality

$$at_a + bt_b + ct_c \leq \frac{\sqrt{3}}{2}(a^2 + b^2 + c^2).$$

Solution By the *Cauchy–Schwarz inequality* we have

$$(a^2 + b^2 + c^2)(t_a^2 + t_b^2 + t_c^2) \geq (at_a + bt_b + ct_c)^2. \tag{1}$$

Also

$$t_a^2 + t_b^2 + t_c^2 = \frac{3}{4}(a^2 + b^2 + c^2). \tag{2}$$

From (1) and (2) we get

$$(at_a+bt_b+ct_c)^2 \le \frac{3}{4}(a^2+b^2+c^2)^2 \quad \text{i.e.} \quad at_a+bt_b+ct_c \le \frac{\sqrt{3}}{2}(a^2+b^2+c^2),$$

as required. ■

231 Let a, b, c be the lengths of the sides of a triangle. Prove the inequality

$$\sqrt{a+b-c}+\sqrt{c+a-b}+\sqrt{b+c-a} \le \sqrt{a}+\sqrt{b}+\sqrt{c}.$$

Solution We'll use *Ravi's substitutions*, i.e. let $a=x+y, b=y+z, c=z+x$, where $x, y, z \in \mathbb{R}^+$.

Now the given inequality is equivalent to

$$\sqrt{2x}+\sqrt{2y}+\sqrt{2z} \le \sqrt{x+y}+\sqrt{y+z}+\sqrt{z+x}.$$

By $QM \ge AM$ we have $\sqrt{\frac{x+y}{2}} \ge \frac{\sqrt{x}+\sqrt{y}}{2}$, from which we deduce that

$$\sqrt{x+y} \ge \frac{\sqrt{x}+\sqrt{y}}{\sqrt{2}}.$$

Analogously we get

$$\sqrt{y+z} \ge \frac{\sqrt{y}+\sqrt{z}}{\sqrt{2}} \quad \text{and} \quad \sqrt{z+x} \ge \frac{\sqrt{z}+\sqrt{x}}{\sqrt{2}}.$$

After adding these three inequalities we obtain

$$\sqrt{x+y}+\sqrt{y+z}+\sqrt{z+x} \ge 2\frac{\sqrt{x}}{\sqrt{2}}+2\frac{\sqrt{y}}{\sqrt{2}}+2\frac{\sqrt{z}}{\sqrt{2}},$$

i.e.

$$\sqrt{x+y}+\sqrt{y+z}+\sqrt{z+x} \ge \sqrt{2x}+\sqrt{2y}+\sqrt{2z},$$

as required. ■

232 Let P be the area of the triangle with side lengths a, b and c, and T be the area of the triangle with side lengths $a+b, b+c$ and $c+a$. Prove that $T \ge 4P$.

Solution We have

$$P^2 = s(s-a)(s-b)(s-c), \quad \text{where } s=\frac{a+b+c}{2},$$

i.e.

$$16P^2 = (a+b+c)(a+b-c)(a+c-b)(b+c-a).$$

Let s_1 be the semi-perimeter of the triangle with side lengths $a+b, a+c, b+c$.

Then

$$s_1 = \frac{a+b+a+c+b+c}{2} = a+b+c = 2s.$$

So we get

$$\begin{aligned} T^2 &= s_1(s_1-(a+b))(s_1-(a+c))(s_1-(b+c)) \\ &= 2s(2s-(a+b))(2s-(a+c))(2s-(b+c)) = abc(a+b+c). \end{aligned}$$

It suffices to show that $T^2 \geq 16P^2$ i.e.

$$abc(a+b+c) \geq (a+b+c)(a+b-c)(a+c-b)(b+c-a).$$

We have

$$a^2 \geq a^2-(b-c)^2 = (a-b+c)(a+b-c) = (a+c-b)(a+b-c).$$

Analogously

$$b^2 \geq (a+b-c)(b+c-a) \quad \text{and} \quad c^2 \geq (b+c-a)(a+c-b).$$

If we multiply the last three inequalities (Can we do this?) we obtain

$$a^2b^2c^2 \geq (a+b-c)^2(a+c-b)^2(b+c-a)^2,$$

i.e.

$$abc \geq (a+b-c)(a+c-b)(b+c-a),$$

as required.

Equality occurs iff $a=b=c$. ■

233 Let a, b, c be the lengths of the sides of a triangle, such that $a+b+c=3$. Prove the inequality

$$a^2+b^2+c^2+\frac{4abc}{3} \geq \frac{13}{3}.$$

Solution Let $a=x+y$, $b=y+z$ and $c=z+x$.

So we have $x+y+z=\frac{3}{2}$ and since $AM \geq GM$ we get $xyz \leq (\frac{x+y+z}{3})^3 = \frac{1}{8}$.

Now we obtain

$$\begin{aligned}
a^2+b^2+c^2+\frac{4abc}{3} \\
&= \frac{(a^2+b^2+c^2)(a+b+c)+4abc}{3} \\
&= \frac{2((x+y)^2+(y+z)^2+(z+x)^2)(x+y+z)+4(x+y)(y+z)(z+x)}{3} \\
&= \frac{4}{3}((x+y+z)^3-xyz) \geq \frac{4}{3}\left(\left(\frac{3}{2}\right)^3-\frac{1}{8}\right)=\frac{13}{3}.
\end{aligned}$$

Equality occurs iff $x=y=z$, i.e. $a=b=c=1$. ■

234 Let a, b, c be the lengths of the sides of a triangle. Prove that

$$\sqrt[3]{\frac{a^3+b^3+c^3+3abc}{2}} \geq \max\{a, b, c\}.$$

Solution Without loss of generality we may assume that $a \geq b \geq c$.

We need to show that

$$\sqrt[3]{\frac{a^3+b^3+c^3+3abc}{2}} \geq a,$$

i.e.

$$-a^3+b^3+c^3+3abc \geq 0.$$

Since

$$\begin{aligned}
-a^3+b^3+c^3+3abc &= (-a)^3+b^3+c^3-3(-a)bc \\
&= \frac{1}{2}(-a+b+c)((a+b)^2+(a+c)^2+(b-c)^2),
\end{aligned}$$

and since $b+c>a$ we obtain

$$-a^3+b^3+c^3+3abc \geq 0,$$

as required. ■

235 Let a, b, c be the lengths of the sides of a triangle. Prove the inequality

$$abc < a^2(s-a)+b^2(s-a)+c^2(s-a) \leq \frac{3}{2}abc.$$

Solution Since

$$2(a^2(s-a)+b^2(s-a)+c^2(s-a)) = a^2b+a^2c+b^2a+b^2c+c^2a+c^2b - (a^3+b^3+c^3)$$

and

$$(b+c-a)(c+a-b)(a+b-c) = a^2b+a^2c+b^2a+b^2c+c^2a+c^2b-(a^3+b^3+c^3)-2abc$$

we have

$$2(a^2(s-a)+b^2(s-a)+c^2(s-a)) = (b+c-a)(c+a-b)(a+b-c)+2abc.$$

Hence

$$a^2(s-a)+b^2(s-a)+c^2(s-a) = \frac{(b+c-a)(c+a-b)(a+b-c)}{2}+abc > abc.$$

Recalling the well-known inequality

$$(b+c-a)(c+a-b)(a+b-c) \le abc,$$

we have

$$a^2(s-a)+b^2(s-a)+c^2(s-a) = \frac{(b+c-a)(c+a-b)(a+b-c)}{2}+abc \le \frac{3}{2}abc.$$

Equality holds if and only if the triangle is equilateral. ■

236 Let a, b, c be the lengths of the sides of a triangle. Prove that

$$\frac{1}{\sqrt{a}+\sqrt{b}-\sqrt{c}}+\frac{1}{\sqrt{b}+\sqrt{c}-\sqrt{a}}+\frac{1}{\sqrt{c}+\sqrt{a}-\sqrt{b}} \ge \frac{3(\sqrt{a}+\sqrt{b}+\sqrt{c})}{a+b+c}.$$

Solution Firstly it is easy to show that if there exists a triangle with lengths sides a, b, c then there also exists a triangle with length sides $\sqrt{a}, \sqrt{b}, \sqrt{c}$.

Furthermore

$$(\sqrt{a}+\sqrt{b}+\sqrt{c})^2 = a+b+c+2(\sqrt{ab}+\sqrt{bc}+\sqrt{ca}) \le 3(a+b+c)$$

i.e.

$$\frac{1}{\sqrt{a}+\sqrt{b}+\sqrt{c}} \ge \frac{\sqrt{a}+\sqrt{b}+\sqrt{c}}{3(a+b+c)}. \tag{1}$$

Applying $AM \geq HM$ we deduce

$$\frac{3}{\frac{1}{\sqrt{a}+\sqrt{b}-\sqrt{c}}+\frac{1}{\sqrt{b}+\sqrt{c}-\sqrt{a}}+\frac{1}{\sqrt{c}+\sqrt{a}-\sqrt{b}}} \leq \frac{\sqrt{a}+\sqrt{b}+\sqrt{c}}{3},$$

i.e.

$$\frac{1}{\sqrt{a}+\sqrt{b}-\sqrt{c}}+\frac{1}{\sqrt{b}+\sqrt{c}-\sqrt{a}}+\frac{1}{\sqrt{c}+\sqrt{a}-\sqrt{b}} \geq \frac{9}{\sqrt{a}+\sqrt{b}+\sqrt{c}}. \quad (2)$$

By (1) and (2) we get the required inequality. ∎

237 Let a, b, c be the lengths of the sides of a triangle with area P. Prove that

$$a^2+b^2+c^2 \geq 4\sqrt{3}P.$$

Solution After setting $a = x+y, b = y+z, c = z+x$ where $x, y, z > 0$, the given inequality becomes

$$((x+y)^2+(y+z)^2+(z+x)^2)^2 \geq 48xyz(x+y+z).$$

From $AM \geq GM$ we have

$$((x+y)^2+(y+z)^2+(z+x)^2)^2 \geq (4xy+4yz+4zx)^2 = 16(xy+yz+zx)^2. \quad (1)$$

Since for every $p, q, r \in \mathbb{R}$ we have $(p+q+r)^2 \geq 3(pq+qr+rp)$, by (1) we get

$$\begin{aligned}
&((x+y)^2+(y+z)^2+(z+x)^2)^2 \\
&\quad \geq 16(xy+yz+zx)^2 \\
&\quad \geq 16 \cdot 3((xy)(yz)+(yz)(zx)+(zx)(xy)) = 48xyz(x+y+z),
\end{aligned}$$

as required.

Equality holds iff $x = y = z$, i.e. iff $a = b = c$. ∎

238 (*Hadwinger–Finsler*) Let a, b, c be the lengths of the sides of a triangle. Prove the inequality

$$a^2+b^2+c^2 \geq 4\sqrt{3}P+(a-b)^2+(b-c)^2+(c-a)^2.$$

Solution 1 The given inequality is equivalent to

$$2(ab+bc+ca)-(a^2+b^2+c^2) \geq 4\sqrt{3}P.$$

We'll use *Ravi's substitutions*, i.e. $a = x+y, b = y+z, c = z+x$, where $x, y, z > 0$.

Then the previous inequality becomes

$$xy+yz+zx \geq \sqrt{3xyz(x+y+z)},$$

which is true due to

$$(xy+yz+zx)^2 - 3xyz(x+y+z) = \frac{(xy-yz)^2+(yz-zx)^2+(zx-xy)^2}{2}.$$

Clearly equality holds iff $x=y=z$, i.e. iff $a=b=c$. ∎

Solution 2 The given inequality can be rewritten as

$$2(ab+bc+ca) \geq 4\sqrt{3}P + a^2+b^2+c^2. \tag{1}$$

Using $\frac{ab\sin\gamma}{2} = \frac{ac\sin\beta}{2} = \frac{bc\sin\alpha}{2} = P$ it follows that

$$ab = \frac{2P}{\sin\gamma}, \qquad ac = \frac{2P}{\sin\beta}, \qquad bc = \frac{2P}{\sin\alpha}.$$

From

$$\cot\alpha = \frac{\cos\alpha}{\sin\alpha} = \frac{\frac{b^2+c^2-a^2}{2bc}}{\frac{a}{2R}} = \frac{R}{abc}(b^2+c^2-a^2)$$

we get

$$\cot\alpha+\cot\beta+\cot\gamma = \frac{R}{abc}(a^2+b^2+c^2),$$

i.e.

$$a^2+b^2+c^2 = 4P(\cot\alpha+\cot\beta+\cot\gamma),$$

and inequality (1) becomes

$$4P\left(\frac{1}{\sin\alpha}+\frac{1}{\sin\gamma}+\frac{1}{\sin\beta}\right) \geq 4\sqrt{3}P + 4P(\cot\alpha+\cot\beta+\cot\gamma),$$

i.e.

$$\left(\frac{1}{\sin\alpha}-\cot\alpha\right)+\left(\frac{1}{\sin\beta}-\cot\beta\right)+\left(\frac{1}{\sin\gamma}-\cot\gamma\right) \geq \sqrt{3}$$

$$\Leftrightarrow \quad \frac{1-\cos\alpha}{\sin\alpha}+\frac{1-\cos\beta}{\sin\beta}+\frac{1-\cos\gamma}{\sin\gamma} \geq \sqrt{3}. \tag{2}$$

But $1-\cos\alpha = 2\sin^2\frac{\alpha}{2}$ and $\sin\alpha = 2\sin\frac{\alpha}{2}\cos\frac{\alpha}{2}$, so we have

$$\frac{1-\cos\alpha}{\sin\alpha} = \frac{2\sin^2\frac{\alpha}{2}}{2\sin\frac{\alpha}{2}\cos\frac{\alpha}{2}} = \tan\frac{\alpha}{2}.$$

Now inequality (2) is equivalent to

$$\tan\frac{\alpha}{2}+\tan\frac{\beta}{2}+\tan\frac{\gamma}{2} \geq \sqrt{3},$$

which is true, since $\tan x$ is convex on $(0,\pi/2)$ (*Jensen's inequality*). ∎

239 Let a, b, c be the lengths of the sides of a triangle. Prove that

$$\frac{1}{8abc+(a+b-c)^3}+\frac{1}{8abc+(b+c-a)^3}+\frac{1}{8abc+(c+a-b)^3}\leq\frac{1}{3abc}.$$

Solution The given inequality is equivalent to

$$\frac{1}{8abc}-\frac{1}{8abc+(a+b-c)^3}+\frac{1}{8abc}-\frac{1}{8abc+(b+c-a)^3}+\frac{1}{8abc}-\frac{1}{8abc+(c+a-b)^3}\geq\frac{3}{8abc}-\frac{1}{3abc},$$

i.e.

$$\frac{(a+b-c)^3}{8abc+(a+b-c)^3}+\frac{(b+c-a)^3}{8abc+(b+c-a)^3}+\frac{(c+a-b)^3}{8abc+(c+a-b)^3}\geq\frac{1}{3}. \quad (1)$$

■

Lemma 21.3 *Let $a, b, c, x, y, z \in \mathbb{R}^+$. Then*

$$\frac{a^3}{x}+\frac{b^3}{y}+\frac{c^3}{z}\geq\frac{(a+b+c)^3}{3(x+y+z)}.$$

Proof We'll use the *generalized Hölder inequality*, i.e.

If $(a_i), (b_i), (c_i), i = 1, 2, \ldots, n$, are positive real numbers and p, q, r are such that $p+q+r=1$, then

$$\left(\sum_{i=1}^n a_i\right)^p\cdot\left(\sum_{i=1}^n b_i\right)^q\cdot\left(\sum_{i=1}^n c_i\right)^r\geq\sum_{i=1}^n a_i^p b_i^q c_i^r.$$

For $n=3$, $p=q=r=1/3$ and

$$a_1=a_2=a_3=1;\qquad b_1=x,\qquad b_2=y,\qquad b_3=z;$$

$$c_1=\frac{a^3}{x},\qquad c_2=\frac{b^3}{y},\qquad c_3=\frac{c^3}{z}$$

we get

$$\sqrt[3]{(1+1+1)(x+y+z)\left(\frac{a^3}{x}+\frac{b^3}{y}+\frac{c^3}{z}\right)}\geq\sqrt[3]{1\cdot x\cdot\frac{a^3}{x}}+\sqrt[3]{1\cdot y\cdot\frac{b^3}{y}}+\sqrt[3]{1\cdot z\cdot\frac{c^3}{z}},$$

i.e.

$$3(x+y+z)\left(\frac{a^3}{x}+\frac{b^3}{y}+\frac{c^3}{z}\right)\geq(a+b+c)^3 \quad\Leftrightarrow\quad \frac{a^3}{x}+\frac{b^3}{y}+\frac{c^3}{z}\geq\frac{(a+b+c)^3}{3(x+y+z)}.$$

According to (1) and Lemma 21.3, we have

$$\begin{aligned}&\frac{(a+b-c)^3}{8abc+(a+b-c)^3}+\frac{(b+c-a)^3}{8abc+(b+c-a)^3}+\frac{(c+a-b)^3}{8abc+(c+a-b)^3}\\&\quad\geq\frac{(a+b-c+b+c-a+c+a-b)^3}{3(24abc+(a+b-c)^3+(b+c-a)^3+(c+a-b)^3)}=\frac{1}{3}.\end{aligned}$$

□

240 In the triangle ABC, $\overline{AC}^2$ is the arithmetic mean of $\overline{BC}^2$ and $\overline{AB}^2$. Prove that

$$\cot^2\beta\geq\cot\alpha\cdot\cot\gamma.$$

Solution Let $\overline{BC}=a$, $\overline{AC}=b$, $\overline{AB}=c$. Then we have $2b^2=a^2+c^2$. By the *law of sines and cosines* we have

$$\cot\beta=\frac{\cos\beta}{\sin\beta}=\frac{\frac{a^2+c^2-b^2}{2ac}}{\frac{b}{2R}}=\frac{(a^2+c^2-b^2)R}{abc},$$

$$\cot\alpha=\frac{(b^2+c^2-a^2)R}{abc}\quad\text{and}\quad\cot\gamma=\frac{(b^2+a^2-c^2)R}{abc}.$$

So we need to prove that

$$\frac{(b^2+c^2-a^2)R}{abc}\cdot\frac{(b^2+a^2-c^2)R}{abc}\leq\frac{(a^2+c^2-b^2)^2R^2}{(abc)^2},$$

i.e.

$$(b^2+c^2-a^2)\cdot(b^2+a^2-c^2)\leq(a^2+c^2-b^2)^2.$$

Applying $AM\geq GM$ we have

$$(b^2+c^2-a^2)\cdot(b^2+a^2-c^2)\leq\left(\frac{b^2+c^2-a^2+b^2+a^2-c^2}{2}\right)^2,$$

i.e.

$$(b^2+c^2-a^2)\cdot(b^2+a^2-c^2)\leq(b^2)^2=(2b^2-b^2)^2=(a^2+c^2-b^2)^2,$$

as required.

Equality occurs iff $a=b=c$. ■

241 Let d_1, d_2 and d_3 be the distances from an arbitrary point to the sides BC, CA, AB, respectively, of the triangle ABC. Prove the inequality

$$\frac{9}{4}(d_1^2+d_2^2+d_3^2) \geq \left(\frac{P}{R}\right)^2.$$

Solution We have $P = \frac{ad_1+bd_2+cd_3}{2}$, i.e.

$$P^2 = \frac{1}{4}(ad_1+bd_2+cd_3)^2. \qquad (1)$$

By the *Cauchy–Schwarz inequality* we have

$$(ad_1+bd_2+cd_3)^2 \leq (a^2+b^2+c^2)(d_1^2+d_2^2+d_3^2). \qquad (2)$$

Also

$$a^2+b^2+c^2 \leq 9R^2. \qquad (3)$$

Finally by (1), (2) and (3) we obtain the required inequality.

Equality holds iff the triangle is equilateral and the given point is the center of the triangle. ■

242 Let a, b, c be the side lengths, and h_a, h_b, h_c be the lengths of the altitudes (respectively) of a given triangle. Prove the inequality

$$\frac{h_a+h_b+h_c}{a+b+c} \leq \frac{\sqrt{3}}{2}.$$

Solution We have

$$(a+b+c)^2 \geq 3(ab+bc+ca) = \frac{3abc}{2P}(h_a+h_b+h_c) = 6R \cdot (h_a+h_b+h_c). \quad (1)$$

Recall the well-known inequality $a^2+b^2+c^2 \leq 9R^2$.

Then we have

$$(a+b+c)^2 \leq 3(a^2+b^2+c^2) \leq 27R^2, \quad \text{i.e.} \quad a+b+c \leq 3\sqrt{3}R. \qquad (2)$$

Now by (1) and (2) we get

$$(a+b+c)^2 \geq 6\frac{(a+b+c)}{3\sqrt{3}} \cdot (h_a+h_b+h_c), \quad \text{i.e.} \quad \frac{h_a+h_b+h_c}{a+b+c} \leq \frac{\sqrt{3}}{2}.$$

Equality occurs iff $a = b = c$. ■

243 Let O be an arbitrary point in the interior of $\triangle ABC$. Let x, y and z be the distances from O to the sides BC, CA, AB, respectively, and let R be the circumradius of the triangle $\triangle ABC$. Prove the inequality

$$\sqrt{x}+\sqrt{y}+\sqrt{z} \leq 3\sqrt{\frac{R}{2}}.$$

Solution Let $\overline{BC} = a, \overline{CA} = b, \overline{AB} = c$.

By the *Cauchy–Schwarz inequality* we have

$$(\sqrt{x} + \sqrt{y} + \sqrt{z})^2 \le (ax + by + cz)\left(\frac{1}{a} + \frac{1}{b} + \frac{1}{c}\right).$$

Since $ax + by + cz = 2P$ and $P = \frac{abc}{4R}$ we have

$$(\sqrt{x} + \sqrt{y} + \sqrt{z})^2 \le 2P \cdot \frac{ab + bc + ca}{abc} = \frac{ab + bc + ca}{2R}. \quad (1)$$

Also we have

$$ab + bc + ca \le a^2 + b^2 + c^2 \le 9R^2. \quad (2)$$

By (1) and (2) it follows that

$$(\sqrt{x} + \sqrt{y} + \sqrt{z})^2 \le \frac{9}{2}R, \quad \text{i.e.} \quad \sqrt{x} + \sqrt{y} + \sqrt{z} \le 3\sqrt{\frac{R}{2}}.$$

Equality holds iff the triangle is equilateral. ■

244 Let D, E and F be the feet of the altitudes of the triangle ABC dropped from the vertices A, B and C, respectively. Prove the inequality

$$\left(\frac{\overline{EF}}{a}\right)^2 + \left(\frac{\overline{FD}}{b}\right)^2 + \left(\frac{\overline{DE}}{c}\right)^2 \ge \frac{3}{4}.$$

Solution Clearly $\overline{EF} = a\cos\alpha, \overline{FD} = b\cos\beta, \overline{DE} = c\cos\gamma$, and the given inequality becomes

$$\cos^2\alpha + \cos^2\beta + \cos^2\gamma \ge \frac{3}{4},$$

which is true according to N_{11} (Chap. 8). ■

245 Let a, b, c be the side-lengths and h_a, h_b, h_c be the lengths of the respective altitudes, and s be the semi-perimeter of a given triangle. Prove the inequality

$$\frac{h_a}{a} + \frac{h_b}{b} + \frac{h_c}{c} \le \frac{s}{2r}.$$

Solution From $\sqrt{(s-b)(s-c)} \le \frac{s-b+s-c}{2} = \frac{a}{2}$ (equality holds iff $b = c$), we have

$$\frac{1}{a} \le \frac{1}{2\sqrt{(s-b)(s-c)}}.$$

Hence

$$\frac{h_a}{a} = \frac{2P}{a^2} \le \frac{P}{2(s-b)(s-c)}.$$

Analogously we get

$$\frac{h_b}{b} \leq \frac{P}{2(s-c)(s-a)} \quad \text{and} \quad \frac{h_c}{c} \leq \frac{P}{2(s-a)(s-b)}.$$

Hence

$$\frac{h_a}{a} + \frac{h_b}{b} + \frac{h_c}{c} \leq \frac{P}{2}\left(\frac{1}{(s-b)(s-c)} + \frac{1}{(s-c)(s-a)} + \frac{1}{(s-a)(s-b)}\right)$$
$$= \frac{sP}{2(s-a)(s-b)(s-c)} = \frac{s^2 P}{2P^2} = \frac{s^2}{2P} = \frac{s^2}{2sr} = \frac{s}{2r}.$$

Equality occurs iff the triangle is equilateral. ■

246 Let a, b, c be the side lengths, and h_a, h_b, h_c be the altitudes, respectively, of a triangle. Prove the inequality

$$\frac{a^2}{h_b^2 + h_c^2} + \frac{b^2}{h_a^2 + h_c^2} + \frac{c^2}{h_a^2 + h_b^2} \geq 2.$$

Solution We have

$$\frac{a^2}{h_b^2 + h_c^2} + \frac{b^2}{h_a^2 + h_c^2} + \frac{c^2}{h_a^2 + h_b^2} = \frac{a^2b^2c^2}{4P^2(b^2+c^2)} + \frac{a^2b^2c^2}{4P^2(a^2+c^2)} + \frac{a^2b^2c^2}{4P^2(a^2+b^2)}$$
$$= \frac{a^2b^2c^2}{4P^2}\left(\frac{1}{b^2+c^2} + \frac{1}{a^2+c^2} + \frac{1}{a^2+b^2}\right).$$

Also

$$a^2b^2c^2 = 16P^2R^2$$

and

$$\frac{1}{b^2+c^2} + \frac{1}{a^2+c^2} + \frac{1}{a^2+b^2} \geq \frac{9}{2(a^2+b^2+c^2)} \quad (\text{since } AM \geq HM).$$

Therefore

$$\frac{a^2}{h_b^2 + h_c^2} + \frac{b^2}{h_a^2 + h_c^2} + \frac{c^2}{h_a^2 + h_b^2} \geq \frac{16P^2R^2}{4P^2} \cdot \frac{9}{2(a^2+b^2+c^2)} = \frac{18R^2}{a^2+b^2+c^2} \geq 2,$$

where the last inequality is true since $a^2 + b^2 + c^2 \leq 9R^2$.

Equality holds iff the triangle is equilateral. ■

247 Let a, b, c be the side lengths, h_a, h_b, h_c be the altitudes, respectively and r be the inradius of a triangle. Prove the inequality

$$\frac{1}{h_a - 2r} + \frac{1}{h_b - 2r} + \frac{1}{h_c - 2r} \geq \frac{3}{r}.$$

Solution By $\frac{1}{h_a}+\frac{1}{h_b}+\frac{1}{h_c}=\frac{1}{r}$ we obtain

$$\frac{h_a-2r}{h_a}+\frac{h_b-2r}{h_b}+\frac{h_c-2r}{h_c}=1.$$

Applying $AM \geq HM$ we get

$$\left(\frac{h_a-2r}{h_a}+\frac{h_b-2r}{h_b}+\frac{h_c-2r}{h_c}\right)\left(\frac{h_a}{h_a-2r}+\frac{h_b}{h_b-2r}+\frac{h_c}{h_c-2r}\right)\geq 9,$$

i.e.

$$\frac{h_a}{h_a-2r}+\frac{h_b}{h_b-2r}+\frac{h_c}{h_c-2r}\geq 9.$$

Therefore

$$\begin{aligned}&\frac{2r}{h_a-2r}+\frac{2r}{h_b-2r}+\frac{2r}{h_c-2r}\\&\quad=\frac{h_a-(h_a-2r)}{h_a-2r}+\frac{h_b-(h_b-2r)}{h_b-2r}+\frac{h_c-(h_c-2r)}{h_c-2r}\\&\quad=\frac{h_a}{h_a-2r}+\frac{h_b}{h_b-2r}+\frac{h_c}{h_c-2r}-3\geq 9-3=6,\end{aligned}$$

i.e.

$$\frac{1}{h_a-2r}+\frac{1}{h_b-2r}+\frac{1}{h_c-2r}\geq\frac{3}{r}.$$

■

248 Let $a,b,c;l_\alpha,l_\beta,l_\gamma$ be the lengths of the sides and the bisectors of the respective angles. Let s be the semi-perimeter and r denote the inradius of a given triangle. Prove the inequality

$$\frac{l_\alpha}{a}+\frac{l_\beta}{b}+\frac{l_\gamma}{c}\leq\frac{s}{2r}.$$

Solution The following identities hold:

$$l_\alpha=\frac{2\sqrt{bc}}{b+c}\sqrt{s(s-a)},\qquad l_\beta=\frac{2\sqrt{ca}}{c+a}\sqrt{s(s-b)}\quad\text{and}\quad l_\gamma=\frac{2\sqrt{ab}}{a+b}\sqrt{s(s-c)}.$$

From the obvious inequality $\frac{2\sqrt{xy}}{x+y}\leq 1$ and the previous identities we obtain that

$$l_\alpha\leq\sqrt{s(s-a)},\qquad l_\beta\leq\sqrt{s(s-b)}\quad\text{and}\quad l_\gamma\leq\sqrt{s(s-c)}.\tag{1}$$

Also

$$h_a\leq l_\alpha,\qquad h_b\leq l_\beta\quad\text{and}\quad h_c\leq l_\gamma.\tag{2}$$

So we have

$$\frac{l_\alpha}{a}+\frac{l_\beta}{b}+\frac{l_\gamma}{c}=\frac{l_\alpha h_a}{2P}+\frac{l_\beta h_b}{2P}+\frac{l_\gamma h_c}{2P}\overset{(2)}{\leq}\frac{l_\alpha^2+l_\beta^2+l_\gamma^2}{2P}$$
$$\overset{(1)}{\leq}\frac{s(s-a)+s(s-b)+s(s-c)}{2P}$$
$$=\frac{3s^2-s(a+b+c)}{2rs}=\frac{3s^2-2s^2}{2rs}=\frac{s^2}{2rs}=\frac{s}{2r}.$$

Equality occurs iff the triangle is equilateral. ■

249 Let $a,b,c;l_\alpha,l_\beta,l_\gamma$ be the lengths of the sides and of the bisectors of respective angles. Let R and r be the circumradius and inradius, respectively, of a given triangle. Prove the inequality

$$18r^2\sqrt{3}\leq al_\alpha+bl_\beta+cl_\gamma<9R^2.$$

Solution We have

$$a^2\geq a^2-(b-c)^2=(a+b-c)(a+c-b)=4(s-c)(s-b).$$

Hence

$$a\geq 2\sqrt{(s-c)(s-b)},$$

with equality if and only if $b=c$.

Since $l_\alpha=2\sqrt{bc}\frac{\sqrt{s(s-a)}}{b+c}$ and by the previous inequality we get

$$al_\alpha\geq\frac{4\sqrt{bc}}{b+c}\sqrt{s(s-a)(s-c)(s-b)}=\frac{4\sqrt{bc}}{b+c}P.$$

Analogously we obtain

$$bl_\beta\geq\frac{4\sqrt{ac}}{a+c}P\quad\text{and}\quad cl_\gamma\geq\frac{4\sqrt{ab}}{a+b}P.$$

Therefore

$$al_\alpha+bl_\beta+cl_\gamma\geq 4P\left(\frac{4\sqrt{bc}}{b+c}+\frac{4\sqrt{ac}}{a+c}+\frac{4\sqrt{ab}}{a+b}\right).\tag{1}$$

By $AM\geq GM$ we have

$$\frac{4\sqrt{bc}}{b+c}+\frac{4\sqrt{ac}}{a+c}+\frac{4\sqrt{ab}}{a+b}\geq 3\sqrt[3]{\frac{abc}{(a+b)(b+c)(c+a)}}.\tag{2}$$

Also we have

$$4s=(a+b)+(b+c)+(c+a)\geq 3\sqrt[3]{(a+b)(b+c)(c+a)}.$$

Hence

$$\sqrt[3]{\frac{1}{(a+b)(b+c)(c+a)}} \geq \frac{3}{4s}. \tag{3}$$

By (1), (2) and (3) we obtain

$$al_\alpha + bl_\beta + cl_\gamma \geq \frac{9P}{s}\sqrt[3]{abc} = \frac{9sr}{s}\sqrt[3]{4PR} = 9r\sqrt[3]{4srR}. \tag{4}$$

According to Exercise 13.2 (Chap. 3) we have that $s \geq 3r\sqrt{3}$, and clearly $R \geq 2r$.
Now by (4) we get

$$al_\alpha + bl_\beta + cl_\gamma \geq 9r\sqrt[3]{4srR} \geq 9r\sqrt[3]{24r^3\sqrt{3}} = 18r^2\sqrt{3}.$$

Equality occurs iff $a = b = c$.
We need to show the right-hand side inequality.
We have

$$\sqrt{s(s-a)} \leq \frac{s+s-a}{2} = \frac{b+c}{2}.$$

Note that we have a strict inequality since $s \neq s-a$.
Now we have

$$l_\alpha = 2\sqrt{bc}\frac{\sqrt{s(s-a)}}{b+c} < \sqrt{bc} \leq \frac{b+c}{2}, \quad \text{i.e.} \quad al_\alpha < a\frac{b+c}{2}.$$

Analogously we obtain

$$bl_\beta < b\frac{a+c}{2} \quad \text{and} \quad cl_\gamma < c\frac{a+b}{2}.$$

So

$$al_\alpha + bl_\beta + cl_\gamma < ab + bc + ca. \tag{5}$$

If we consider the well-known inequalities

$$ab + bc + ca \leq a^2 + b^2 + c^2 \quad \text{and} \quad a^2 + b^2 + c^2 \leq 9R^2,$$

from (5) we obtain the required inequality. ■

250 Let a, b, c be the lengths of the sides of triangle, with circumradius $r = 1/2$. Prove the inequality

$$\frac{a^4}{b+c-a} + \frac{b^4}{a+c-b} + \frac{c^4}{a+b-c} \geq 9\sqrt{3}.$$

Solution Let s be the semi-perimeter of the given triangle. The given inequality becomes

$$A=\frac{a^4}{2(s-a)}+\frac{b^4}{2(s-b)}+\frac{c^4}{2(s-c)}\geq 9\sqrt{3}.$$

By the *Cauchy–Schwarz inequality* we obtain

$$A\cdot(2(s-a)+2(s-b)+2(s-c))\geq(a^2+b^2+c^2)^2$$
$$\Leftrightarrow\quad 2s\cdot A\geq(a^2+b^2+c^2)^2,$$

i.e.

$$A\geq\frac{(a^2+b^2+c^2)^2}{a+b+c}. \tag{1}$$

Applying $QM\geq AM$ we deduce

$$\frac{a^2+b^2+c^2}{3}\geq\left(\frac{a+b+c}{3}\right)^2,\quad\text{i.e.}\quad a^2+b^2+c^2\geq\frac{(a+b+c)^2}{3}.$$

Then by (1) we get

$$A\geq\frac{(a^2+b^2+c^2)^2}{a+b+c}\geq\frac{(a+b+c)^4}{9(a+b+c)}=\frac{(a+b+c)^3}{9}. \tag{2}$$

Let's introduce *Ravi's substitutions*, i.e. let us take $a=x+y, b=y+z, c=z+x$. Then clearly $s=\frac{a+b+c}{2}=x+y+z$.

By *Heron's formula* we obtain

$$P^2=s(s-a)(s-b)(s-c)=xyz(x+y+z). \tag{3}$$

Also

$$P^2=s^2r^2=\frac{(x+y+z)^2}{4}. \tag{4}$$

By (3) and (4) we get

$$x+y+z=4xyz. \tag{5}$$

Since $AM\geq GM$ and using (5) we obtain

$$\left(\frac{x+y+z}{3}\right)^3\geq xyz=\frac{x+y+z}{4},$$

i.e.

$$x+y+z\geq\frac{3\sqrt{3}}{2}.$$

Thus

$$a+b+c=2(x+y+z)\geq 3\sqrt{3}. \tag{6}$$

Finally according to (2) and (6) it follows that

$$A\geq \frac{(a+b+c)^3}{9}=\frac{(3\sqrt{3})^3}{9}\geq 9\sqrt{3}.$$

Equality occurs if and only if the triangle is equilateral with side equal to $\sqrt{3}$. ■

251 Let a, b, c be the side-lengths of a triangle. Prove the inequality

$$\frac{a}{3a-b+c}+\frac{b}{3b-c+a}+\frac{c}{3c-a+b}\geq 1.$$

Solution We have

$$\begin{aligned}&\frac{4a}{3a-b+c}+\frac{4b}{3b-c+a}+\frac{4c}{3c-a+b}\\&\quad=3+\frac{a+b-c}{3a-b+c}+\frac{b+c-a}{3b-c+a}+\frac{c+a-b}{3c-a+b}.\end{aligned}$$

So it remains to show that

$$\frac{a+b-c}{3a-b+c}+\frac{b+c-a}{3b-c+a}+\frac{c+a-b}{3c-a+b}\geq 1.$$

By the *Cauchy–Schwarz inequality* (Corollary 4.3) we have

$$\begin{aligned}&\frac{a+b-c}{3a-b+c}+\frac{b+c-a}{3b-c+a}+\frac{c+a-b}{3c-a+b}\\&\quad=\frac{(a+b-c)^2}{(3a-b+c)(a+b-c)}+\frac{(b+c-a)^2}{(3b-c+a)(b+c-a)}\\&\qquad+\frac{(c+a-b)^2}{(3c-a+b)(c+a-b)}\\&\quad\geq\frac{(a+b+c)^2}{(3a-b+c)(a+b-c)+(3b-c+a)(b+c-a)+(3c-a+b)(c+a-b)}\\&\quad=1,\end{aligned}$$

as required.

Equality holds iff $a=b=c=1$. ■

252 Let h_a, h_b and h_c be the lengths of the altitudes, and R and r be the circumradius and inradius, respectively, of a given triangle. Prove the inequality

$$h_a+h_b+h_c\leq 2R+5r.$$

Solution

Lemma 21.4 *In an arbitrary triangle we have*

$$ab + bc + ca = r^2 + s^2 + 4rR \quad \text{and} \quad a^2 + b^2 + c^2 = 2(s^2 - 4Rr - r^2).$$

Proof We have

$$\begin{aligned} r^2 + s^2 + 4rR &= \frac{P^2}{s^2} + s^2 + \frac{abc}{P} \cdot \frac{P}{s} = \frac{(s-a)(s-b)(s-c)}{s} + s^2 + \frac{abc}{s} \\ &= \frac{s^3 - as^2 - bs^2 - cs^2 + abs + bcs + cas - abc + s^3 + abc}{s} \\ &= 2s^2 - s(a+b+c) + ab + bc + ca \\ &= 2s^2 - 2s^2 + ab + bc + ca = ab + bc + ca. \end{aligned}$$

Hence

$$ab + bc + ca = r^2 + s^2 + 4rR. \tag{1}$$

Now by (1) we have

$$\begin{aligned} ab + bc + ca = r^2 + s^2 + 4rR &= \frac{1}{2}\left(2r^2 + 8rR + \frac{(a+b+c)^2}{2}\right) \\ &= \frac{1}{2}\left(2r^2 + 8rR + \frac{a^2+b^2+c^2}{2}\right) + \frac{ab+bc+ca}{2}, \end{aligned}$$

from which it follows that

$$ab + bc + ca = 2r^2 + 8rR + \frac{a^2+b^2+c^2}{2}. \tag{2}$$

Now (1) and (2) yields

$$a^2 + b^2 + c^2 = 2(s^2 - 4Rr - r^2). \tag{3}$$

□

Without proof we will give the following lemma (the proof can be found in [6]).

Lemma 21.5 *In an arbitrary triangle we have*

$$s^2 \le 4R^2 + 4Rr + 3r^2. \tag{4}$$

Lemma 21.6 *In an arbitrary triangle we have* $a^2 + b^2 + c^2 \le 8R^2 + 4r^2$.

Proof From (3) and (4) we have

$$a^2 + b^2 + c^2 = 2(s^2 - 4Rr - r^2) \le 2(4R^2 + 4Rr + 3r^2 - 4Rr - r^2) = 8R^2 + 4r^2.$$

Hence

$$a^2 + b^2 + c^2 \leq 8R^2 + 4r^2. \tag{5}$$

□

Now let us consider our problem.
We have

$$\begin{aligned}
2R(h_a + h_b + h_c) &= 2R\left(\frac{2P}{a} + \frac{2P}{b} + \frac{2P}{c}\right) = 4PR\frac{ab+bc+ca}{abc} \\
&= ab + bc + ca \\
&\overset{(2)}{=} 2r^2 + 8rR + \frac{a^2+b^2+c^2}{2} \\
&\overset{(4)}{\leq} 2r^2 + 8rR + 4R^2 + 2r^2
\end{aligned}$$

$$\Leftrightarrow \quad R(h_a + h_b + h_c) \leq 2R^2 + 4Rr + 2r^2 \leq 2R^2 + 4Rr + Rr \leq R(2R + 5r).$$

Hence

$$h_a + h_b + h_c \leq 2R + 5r.$$

Equality occurs iff $a = b = c$. ■

253 Let a, b, c be the side-lengths, and α, β and γ be the angles of a given triangle, respectively. Prove the inequality

$$a\left(\frac{1}{\beta} + \frac{1}{\gamma}\right) + b\left(\frac{1}{\gamma} + \frac{1}{\alpha}\right) + c\left(\frac{1}{\alpha} + \frac{1}{\beta}\right) \geq 2\left(\frac{a}{\alpha} + \frac{b}{\beta} + \frac{c}{\gamma}\right).$$

Solution If $a \geq b$ then $\alpha \geq \beta$ and analogously if $a \leq b$ then we have $\alpha \leq \beta$.
So we have $(a - b)(\alpha - \beta) \geq 0$, i.e. we have

$$a\alpha + b\beta \geq a\beta + b\alpha$$

i.e.

$$\frac{a}{\beta} + \frac{b}{\alpha} \geq \frac{a}{\alpha} + \frac{b}{\beta}. \tag{1}$$

Analogously we have

$$\frac{a}{\gamma} + \frac{c}{\alpha} \geq \frac{a}{\alpha} + \frac{c}{\gamma} \tag{2}$$

and

$$\frac{c}{\beta} + \frac{b}{\gamma} \geq \frac{c}{\beta} + \frac{b}{\gamma}. \tag{3}$$

Adding (1), (2) and (3) we obtain the required inequality.
Equality occurs iff $a = b = c$, i.e. if the triangle is equilateral. ■

254 Let a, b, c be the lengths of the sides of a given triangle, and α, β, γ be the respective angles (in radians). Prove the inequalities

1° $\frac{1}{\alpha}+\frac{1}{\beta}+\frac{1}{\gamma}\geq\frac{9}{\pi}$.
2° $\frac{b+c-a}{\alpha}+\frac{c+a-b}{\beta}+\frac{a+b-c}{\gamma}\geq\frac{6s}{\pi}$, where $s=\frac{a+b+c}{2}$.
3° $\frac{b+c-a}{a\alpha}+\frac{c+a-b}{b\beta}+\frac{a+b-c}{c\gamma}\geq\frac{9}{\pi}$.

Solution 1° Since $AM \geq HM$ we have

$$\frac{1}{\alpha}+\frac{1}{\beta}+\frac{1}{\gamma}\geq\frac{9}{\alpha+\beta+\gamma}=\frac{9}{\pi}.$$

2° Let $x=b+c-a$, $y=c+a-b$ and $z=a+b-c$.
Without loss the generality we may assume that $a\leq b\leq c$. Then clearly $\alpha\leq\beta\leq\gamma$.
Also $x\geq y\geq z$ and $\frac{1}{\alpha}\geq\frac{1}{\beta}\geq\frac{1}{\gamma}$.
Chebishev's inequality gives us

$$(x+y+z)\left(\frac{1}{\alpha}+\frac{1}{\beta}+\frac{1}{\gamma}\right)\leq 3\left(\frac{x}{\alpha}+\frac{y}{\beta}+\frac{z}{\gamma}\right)$$

i.e.

$$\begin{aligned}\left(\frac{x}{\alpha}+\frac{y}{\beta}+\frac{z}{\gamma}\right)&\geq\frac{1}{3}(x+y+z)\left(\frac{1}{\alpha}+\frac{1}{\beta}+\frac{1}{\gamma}\right)\\&\geq\frac{1}{3}\cdot\frac{9(x+y+z)}{\alpha+\beta+\gamma}=\frac{6s}{\pi}.\end{aligned}$$

3° Let $x=\frac{b+c-a}{a}$, $y=\frac{c+a-b}{b}$ and $z=\frac{a+b-c}{c}$.
Without loss the generality we may assume $a\leq b\leq c$. Then $\alpha\leq\beta\leq\gamma$.
Also $x\geq y\geq z$ and $\frac{1}{\alpha}\geq\frac{1}{\beta}\geq\frac{1}{\gamma}$.
Chebishev's inequality gives us

$$\begin{aligned}\left(\frac{x}{\alpha}+\frac{y}{\beta}+\frac{z}{\gamma}\right)&\geq\frac{1}{3}(x+y+z)\left(\frac{1}{\alpha}+\frac{1}{\beta}+\frac{1}{\gamma}\right)\\&\geq\frac{1}{3}\left(\frac{b+c-a}{a}+\frac{c+a-b}{b}+\frac{a+b-c}{c}\right)\cdot\frac{9}{\pi}\\&=\frac{3}{\pi}\left(\frac{a}{b}+\frac{b}{a}+\frac{a}{c}+\frac{c}{a}+\frac{b}{c}+\frac{c}{b}-3\right)\geq\frac{3}{\pi}(2+2+2-3)=\frac{9}{\pi}.\end{aligned}$$ ■

255 Let X be an arbitrary interior point of a given regular n-gon with side-length a. Let $h_1, h_2, \ldots, h_n$ be the distances from X to the sides of the n-gon. Prove that

$$\frac{1}{h_1}+\frac{1}{h_2}+\cdots+\frac{1}{h_n}>\frac{2\pi}{a}.$$

Solution Let S be the area of the given n-gon, and let r be the inradius of its inscribed circle.

Then $S = \frac{nar}{2}$.

On the other hand, we have

$$S = \frac{1}{2}a(h_1 + h_2 + \cdots + h_n).$$

Applying $AM \geq HM$ we have

$$\frac{n}{\frac{1}{h_1} + \frac{1}{h_2} + \cdots + \frac{1}{h_n}} \leq \frac{h_1 + h_2 + \cdots + h_n}{n} = \frac{2S}{na} = r,$$

i.e.

$$\frac{1}{h_1} + \frac{1}{h_2} + \cdots + \frac{1}{h_n} \geq \frac{n}{r}. \tag{1}$$

The perimeter of the n-gon is larger than the perimeter of its inscribed circle, so we have

$$na > 2\pi r, \quad \text{i.e.} \quad \frac{n}{r} > \frac{2\pi}{a}.$$

Now by (1) we obtain

$$\frac{1}{h_1} + \frac{1}{h_2} + \cdots + \frac{1}{h_n} \geq \frac{n}{r} > \frac{2\pi}{a}.$$

■

256 Prove that among the lengths of the sides of an arbitrary n-gon ($n \geq 3$), there always exist two of them (let's denote them by b and c), such that $1 \leq \frac{b}{c} < 2$.

Solution Let $a_1, a_2, \ldots, a_n$ be the lengths of the sides of the given n-gon.

Without loss of generality we may assume that $a_1 \geq a_2 \geq \cdots \geq a_n$.

Suppose that such a side does not exist, i.e. let us suppose that for any two sides b and c we have $\frac{b}{c} \geq 2$ $(b > c)$, i.e. let us suppose that for every $i \in \{1, 2, \ldots, n-1\}$ we have $\frac{a_i}{a_{i+1}} \geq 2$.

So it follows that

$$a_2 \leq \frac{a_1}{2}, \qquad a_3 \leq \frac{a_2}{2} \leq \frac{a_1}{4}, \ldots, \qquad a_n \leq \frac{a_{n-1}}{2} \leq \frac{a_1}{2^{n-1}}.$$

If we add these inequalities we obtain

$$a_2 + \cdots + a_n \leq a_1\left(\frac{1}{2} + \frac{1}{2^2} + \cdots + \frac{1}{2^{n-1}}\right) = a_1\left(1 - \frac{1}{2^{n-1}}\right) < a_1,$$

which is impossible (why?).

■

257 Let a_1, a_2, a_3, a_4 be the lengths of the sides, and s be the semi-perimeter of an arbitrary quadrilateral. Prove that

$$\sum_{i=1}^{4} \frac{1}{s+a_i} \leq \frac{2}{9} \sum_{1\leq i<j\leq 4} \frac{1}{\sqrt{(s-a_i)(s-a_j)}}.$$

Solution From $AM \geq GM$ we have

$$\begin{aligned}\frac{2}{9} \sum_{1\leq i<j\leq 4} \frac{1}{\sqrt{(s-a_i)(s-a_j)}} &\geq \frac{2}{9} \cdot 2 \sum_{1\leq i<j\leq 4} \frac{1}{(s-a_i)+(s-a_j)} \\ &= \frac{4}{9} \sum_{1\leq i<j\leq 4} \frac{1}{a_i+a_j}. \qquad (1)\end{aligned}$$

Let $a_1 = a, a_2 = b, a_3 = c, a_4 = d$.

We'll show that

$$\begin{aligned}&\frac{2}{9}\left(\frac{1}{a+b}+\frac{1}{a+c}+\frac{1}{a+d}+\frac{1}{b+c}+\frac{1}{b+d}+\frac{1}{c+d}\right) \\ &\quad \geq \frac{1}{3a+b+c+d}+\frac{1}{a+3b+c+d}+\frac{1}{a+b+3c+d}+\frac{1}{a+b+c+3d}.\end{aligned}$$

From $AM \geq HM$ we deduce

$$\left(\frac{1}{a+b}+\frac{1}{a+c}+\frac{1}{a+d}\right)((a+b)+(a+c)+(a+d)) \geq 9,$$

i.e.

$$\frac{1}{9}\left(\frac{1}{a+b}+\frac{1}{a+c}+\frac{1}{a+d}\right) \geq \frac{1}{3a+b+c+d}.$$

Similarly we obtain

$$\frac{1}{9}\left(\frac{1}{a+b}+\frac{1}{b+c}+\frac{1}{b+d}\right) \geq \frac{1}{a+3b+c+d},$$

$$\frac{1}{9}\left(\frac{1}{a+c}+\frac{1}{b+c}+\frac{1}{c+d}\right) \geq \frac{1}{a+b+3c+d},$$

$$\frac{1}{9}\left(\frac{1}{a+d}+\frac{1}{b+d}+\frac{1}{c+d}\right) \geq \frac{1}{a+b+c+3d}.$$

Adding these inequalities we get

$$\begin{aligned}&\frac{2}{9}\left(\frac{1}{a+b}+\frac{1}{a+c}+\frac{1}{a+d}+\frac{1}{b+c}+\frac{1}{b+d}+\frac{1}{c+d}\right)\\&\quad\geq\frac{1}{3a+b+c+d}+\frac{1}{a+3b+c+d}+\frac{1}{a+b+3c+d}+\frac{1}{a+b+c+3d}\\&\quad=\frac{1}{2}\left(\frac{1}{s+a}+\frac{1}{s+b}+\frac{1}{s+c}+\frac{1}{s+d}\right),\end{aligned}$$

i.e.

$$\frac{4}{9}\sum_{1\leq i<j\leq 4}\frac{1}{a_i+a_j}\geq\sum_{i=1}^{4}\frac{1}{s+a_i}. \tag{2}$$

From (1) and (2) we obtain the given inequality.

Equality holds iff $a=b=c=d$. ■

258 Let $n\in\mathbb{N}$, and α,β,γ be the angles of a given triangle. Prove the inequality

$$\cot^n\frac{\alpha}{2}+\cot^n\frac{\beta}{2}+\cot^n\frac{\gamma}{2}\geq 3^{\frac{n+2}{2}}.$$

Solution We use the identity

$$\cot\frac{\alpha}{2}+\cot\frac{\beta}{2}+\cot\frac{\gamma}{2}=\cot\frac{\alpha}{2}\cdot\cot\frac{\beta}{2}\cdot\cot\frac{\gamma}{2}.$$

Since $\frac{\alpha}{2},\frac{\beta}{2},\frac{\gamma}{2}\in(0,\pi/2)$ it follows that $\cot\frac{\alpha}{2},\cot\frac{\beta}{2},\cot\frac{\gamma}{2}\geq 0$.

Applying $AM\geq GM$ we have

$$\cot\frac{\alpha}{2}+\cot\frac{\beta}{2}+\cot\frac{\gamma}{2}\geq 3\sqrt[3]{\cot\frac{\alpha}{2}\cdot\cot\frac{\beta}{2}\cdot\cot\frac{\gamma}{2}}$$

or

$$\cot\frac{\alpha}{2}\cdot\cot\frac{\beta}{2}\cdot\cot\frac{\gamma}{2}\geq 3\sqrt[3]{\cot\frac{\alpha}{2}\cdot\cot\frac{\beta}{2}\cdot\cot\frac{\gamma}{2}},$$

i.e.

$$\cot\frac{\alpha}{2}\cdot\cot\frac{\beta}{2}\cdot\cot\frac{\gamma}{2}\geq 3^{3/2}. \tag{1}$$

Furthermore, using the *power mean inequality* we get

$$\cot^n\frac{\alpha}{2}+\cot^n\frac{\beta}{2}+\cot^n\frac{\gamma}{2}\geq 3\left(\cot\frac{\alpha}{2}\cdot\cot\frac{\beta}{2}\cdot\cot\frac{\gamma}{2}\right)^{n/3}.$$

Now from the previous inequality and (1) we obtain

$$\cot^n \frac{\alpha}{2} + \cot^n \frac{\beta}{2} + \cot^n \frac{\gamma}{2} \geq 3^{\frac{n+2}{2}}.$$

Equality occurs iff $\alpha = \beta = \gamma = \pi/3$. ■

259 Let α, β, γ be the angles of an arbitrary acute triangle. Prove that

$$2(\sin\alpha + \sin\beta + \sin\gamma) > 3(\cos\alpha + \cos\beta + \cos\gamma).$$

Solution Clearly $\alpha + \beta > \frac{\pi}{2}$.

Since $\sin x$ is an increasing function on $[0, \pi/2]$ we have

$$\sin\alpha > \sin\left(\frac{\pi}{2} - \beta\right) = \cos\beta. \tag{1}$$

Analogously

$$\sin\beta > \sin\left(\frac{\pi}{2} - \alpha\right) = \cos\alpha. \tag{2}$$

Now (1) and (2) give us

$$1 - \cos\beta > 1 - \sin\alpha \quad \text{and} \quad 1 - \cos\alpha > 1 - \sin\beta.$$

If we multiply these inequalities we get

$$(1 - \cos\beta)(1 - \cos\alpha) > (1 - \sin\alpha)(1 - \sin\beta)$$

or

$$1 - \cos\beta - \cos\alpha + \cos\alpha\cos\beta > 1 - \sin\beta - \sin\alpha + \sin\alpha\sin\beta$$

or

$$\begin{aligned}\sin\alpha + \sin\beta &> \cos\alpha + \cos\beta - \cos\alpha\cos\beta + \sin\alpha\sin\beta \\ &= \cos\alpha + \cos\beta - \cos(\alpha + \beta) = \cos\alpha + \cos\beta + \cos\gamma.\end{aligned}$$

Analogously we obtain

$$\sin\beta + \sin\gamma > \cos\alpha + \cos\beta + \cos\gamma \quad \text{and} \quad \sin\gamma + \sin\alpha > \cos\alpha + \cos\beta + \cos\gamma.$$

After adding these inequalities we get

$$2(\sin\alpha + \sin\beta + \sin\gamma) > 3(\cos\alpha + \cos\beta + \cos\gamma),$$

as required. ■

260 Let α, β, γ be the angles of a triangle. Prove the inequality

$$\sin\alpha + \sin\beta + \sin\gamma \geq \sin 2\alpha + \sin 2\beta + \sin 2\gamma.$$

Solution Applying the sine law we obtain

$$\sin\alpha + \sin\beta + \sin\gamma = \frac{a+b+c}{2R} = \frac{P}{rR}.$$

Also

$$\begin{aligned}\sin 2\alpha + \sin 2\beta + \sin 2\gamma &= 2(\sin\alpha\cos\alpha + \sin\beta\cos\beta + \sin\gamma\cos\gamma)\\ &= \frac{1}{R}(a\cos\alpha + b\cos\beta + c\cos\gamma).\end{aligned}$$

Since

$$a\cos\alpha + b\cos\beta + c\cos\gamma = \frac{2P}{R}$$

we have

$$\sin 2\alpha + \sin 2\beta + \sin 2\gamma = \frac{2P}{R^2}.$$

Therefore

$$\frac{\sin\alpha + \sin\beta + \sin\gamma}{\sin 2\alpha + \sin 2\beta + \sin 2\gamma} = \frac{R}{2r} \geq 1.$$

Equality holds if and only if the triangle is equilateral. ∎

261 Let α, β, γ be the angles of a triangle. Prove the inequality

$$\cos\alpha + \sqrt{2}(\cos\beta + \cos\gamma) \leq 2.$$

Solution Since $\alpha + \beta + \gamma = \pi$, we have

$$\begin{aligned}\cos\alpha + \sqrt{2}(\cos\beta + \cos\gamma) &= \cos\alpha + 2\sqrt{2}\cos\frac{\beta+\gamma}{2}\cos\frac{\beta-\gamma}{2}\\ &= \cos\alpha + 2\sqrt{2}\sin\frac{\alpha}{2}\cos\frac{\beta-\gamma}{2}\\ &\leq \cos\alpha + 2\sqrt{2}\sin\frac{\alpha}{2} = 2 - 2\left(\sin\frac{\alpha}{2} - \frac{\sqrt{2}}{2}\right)^2 \leq 2.\end{aligned}$$

Equality holds if and only if $\alpha = \pi/2, \beta = \gamma$. ∎

262 Let α, β, γ be the angles of a triangle and let t be a real number. Prove the inequality

$$\cos\alpha + t(\cos\beta + \cos\gamma) \leq 1 + \frac{t^2}{2}.$$

Solution For any three real numbers β, γ, t, the following inequality holds:

$$(\cos\beta + \cos\gamma - t)^2 + (\sin\beta - \sin\gamma)^2 \geq 0,$$

i.e.

$$-\cos(\beta+\gamma) + t(\cos\beta + \cos\gamma) \leq 1 + \frac{t^2}{2}.$$

Since $\alpha + \beta + \gamma = \pi$ we have

$$\cos\alpha + t(\cos\beta + \cos\gamma) \leq 1 + \frac{t^2}{2}.$$

Equality occurs iff $0 < t < 2$, $\cos\alpha = 1 - \frac{t^2}{2}$, $\cos\beta = \cos\gamma$. ■

263 Let $0 \leq \alpha, \beta, \gamma \leq 90°$ such that $\sin\alpha + \sin\beta + \sin\gamma = 1$. Prove the inequality

$$\tan^2\alpha + \tan^2\beta + \tan^2\gamma \geq \frac{3}{8}.$$

Solution We have

$$\tan^2 x = \frac{\sin^2 x}{\cos^2 x} = \frac{1-\cos^2 x}{\cos^2 x} = \frac{1}{\cos^2 x} - 1.$$

The given inequality becomes

$$\frac{1}{\cos^2\alpha} + \frac{1}{\cos^2\beta} + \frac{1}{\cos^2\gamma} \geq \frac{3}{8} + 3 = \frac{27}{8}.$$

Applying $AM \geq HM$ we get

$$\frac{3}{\frac{1}{\cos^2\alpha} + \frac{1}{\cos^2\beta} + \frac{1}{\cos^2\gamma}} \leq \frac{\cos^2\alpha + \cos^2\beta + \cos^2\gamma}{3} = 1 - \frac{\sin^2\alpha + \sin^2\beta + \sin^2\gamma}{3}, \tag{1}$$

and since $\sin x \geq 0$ for $x \in [0, \pi]$ we have

$$\sqrt{\frac{\sin^2\alpha + \sin^2\beta + \sin^2\gamma}{3}} \geq \frac{\sin\alpha + \sin\beta + \sin\gamma}{3} = \frac{1}{3},$$

i.e.

$$\sin^2\alpha + \sin^2\beta + \sin^2\gamma \geq \frac{1}{3}.$$

So in (1) we obtain

$$\frac{3}{\frac{1}{\cos^2\alpha} + \frac{1}{\cos^2\beta} + \frac{1}{\cos^2\gamma}} = 1 - \frac{\sin^2\alpha + \sin^2\beta + \sin^2\gamma}{3} \leq 1 - \frac{1}{9} = \frac{8}{9},$$

i.e.

$$\frac{1}{\cos^2\alpha}+\frac{1}{\cos^2\beta}+\frac{1}{\cos^2\gamma}\geq\frac{27}{8},$$

as required. ∎

264 Let a, b, c be positive real numbers such that $a+b+c=3$. Prove the inequality

$$(1+a+a^2)(1+b+b^2)(1+c+c^2)\geq 9(ab+bc+ca).$$

Solution Let us denote $x=a+b+c=3$, $y=ab+bc+ca$, $z=abc$.
Now the given inequality can be rewritten as

$$z^2-2z-2xz+z(x+y)+x^2+x+y^2-y+3xy+1\geq 9y,$$

i.e.

$$(z-1)^2-(z-1)(x-y)+(x-y)^2\geq 0,$$

which is obviously true. Equality holds iff $a=b=c=1$. ∎

265 Let $a, b, c>0$ such that $a+b+c=1$. Prove the inequality

$$6(a^3+b^3+c^3)+1\geq 5(a^2+b^2+c^2).$$

Solution Let $a+b+c=p=1$, $ab+bc+ca=q$, $abc=r$.
By I_1 and I_2 (Chap. 14) we have

$$a^3+b^3+c^3=p(p^2-3q)+3r=1-3q+3r$$

and

$$a^2+b^2+c^2=p^2-2q=1-2q.$$

Now the given inequality becomes

$$18r+1-2q-6q+1\geq 0,$$

i.e.

$$9r+1\geq 4q$$

which is true due to N_1 (Chap. 14). ∎

266 Let $x, y, z\in\mathbb{R}^+$ such that $x+y+z=1$. Prove the inequality

$$(1-x^2)^2+(1-z^2)^2+(1-z^2)^2\leq(1+x)(1+y)(1+z).$$

Solution Let $p = x + y + z = 1, q = xy + yz + zx, r = xyz$.
The given inequality is equivalent to

$$3 - 2(x^2 + y^2 + z^2) + x^4 + y^4 + z^4 \leq (1 + x)(1 + y)(1 + z).$$

By I_1, I_4 and I_9 (Chap. 14) we have

$$x^2 + y^2 + z^2 = p^2 - 2q = 1 - 2q,$$
$$x^4 + y^4 + z^4 = (p^2 - 2q)^2 - 2(q^2 - 2pr) = (1 - 2q)^2 - 2(q^2 - 2r),$$
$$(1 + x)(1 + y)(1 + z) = 1 + p + q + r = 2 + q + r.$$

So we need to show that

$$3 - 2(1 - 2q) + (1 - 2q)^2 - 2(q^2 - 2r) \leq 2 + q + r,$$

i.e.

$$3 - 2 + 4q + 1 - 4q + 4q^2 - 2q^2 + 4r \leq 2 + q + r$$
$$\Leftrightarrow \quad 2q^2 - q + 3r \leq 0.$$

By N_1 and N_3 (Chap. 14) we have

$$3q \leq p^2 = 1, \quad \text{i.e.} \quad q \leq \frac{1}{3}, \tag{1}$$

and

$$pq \geq 9r, \quad \text{i.e.} \quad q \geq 9r, \quad \text{i.e.} \quad r \leq \frac{q}{9}. \tag{2}$$

By (2) we have

$$2q^2 - q + 3r \leq 2q^2 - q + 3\frac{q}{9} = 2q\left(q - \frac{1}{3}\right) \leq 0.$$

The last inequality is true due to (1) and the fact that $q \geq 0$, so we are done. ■

267 Let x, y, z be non-negative real numbers such that $x^2 + y^2 + z^2 = 1$. Prove the inequality

$$(1 - xy)(1 - yz)(1 - zx) \geq \frac{8}{27}.$$

Solution Let $p = x + y + z, q = xy + yz + zx, r = xyz$. Clearly $p, q, r \geq 0$.
Then $x^2 + y^2 + z^2 = p^2 - 2q$, and the constraint becomes

$$p^2 - 2q = 1. \tag{1}$$

We can easily show that

$$(1-xy)(1-yz)(1-zx) = 1-q+pr-r^2.$$

Now the given inequality becomes

$$1-q+pr-r^2 \geq \frac{8}{27}. \tag{2}$$

By $N_1 : p^3 - 4pq + 9r \geq 0$ and (1), we have

$$\begin{aligned} & p(p^2-4q)+9r \geq 0 \\ \Leftrightarrow \quad & p(1-2q)+9r \geq 0 \\ \Leftrightarrow \quad & 9r \geq p(2q-1). \end{aligned} \tag{3}$$

By $N_4 : p^2 \geq 3q$ and $p^2 - 2q = 1$ we obtain

$$2q+1 \geq 3q, \quad \text{i.e.} \quad q \leq 1. \tag{4}$$

From (4) and $N_3 : pq - 9r \geq 0$ we obtain

$$p \geq pq \geq 9r \quad \Leftrightarrow \quad 9p - 9r \geq 8p \quad \Leftrightarrow \quad p - r \geq \frac{8}{9}p,$$

from which we deduce

$$r(p-r) \geq \frac{8}{9}pr \overset{(3)}{\geq} \frac{8}{9}p\frac{p(2q-1)}{9} = \frac{8p^2(2q-1)}{81} \overset{(1)}{=} \frac{8(2q+1)(2q-1)}{81}. \tag{5}$$

Now we have

$$1-q+pr-r^2 = 1-q+r(p-r) \geq 1-q+\frac{8(2q+1)(2q-1)}{81}. \tag{6}$$

By (2) and (6), we have that it suffices to show that

$$1-q+\frac{8(2q+1)(2q-1)}{81} \geq \frac{8}{27},$$

which is equivalent to

$$(1-q)(49-32q) \geq 0,$$

which clearly holds, due to (4). ■

268 Let $a, b, c \in \mathbb{R}^+$ such that $\frac{1}{a+1} + \frac{1}{b+1} + \frac{1}{c+1} = 2$. Prove the inequalities:

1° $\frac{1}{8a^2+1} + \frac{1}{8b^2+1} + \frac{1}{8c^2+1} \geq 1$
2° $\frac{1}{4ab+1} + \frac{1}{4bc+1} + \frac{1}{4ca+1} \geq \frac{3}{2}$

Solution Let $p=a+b+c, q=ab+bc+ca, r=abc$.

From $\frac{1}{a+1}+\frac{1}{b+1}+\frac{1}{c+1}=2$ we deduce

$$(a+1)(b+1)+(b+1)(c+1)+(c+1)(a+1)=2(a+1)(b+1)(c+1). \quad (1)$$

According to I_9 and I_{10} (Chap. 14), (1) is equivalent to

$$3+2p+q=2(1+p+q+r)$$

i.e.

$$q+2r=1. \quad (2)$$

1° We easily get that

$$(8a^2+1)(8b^2+1)+(8b^2+1)(8c^2+1)+(8c^2+1)(8a^2+1)$$
$$=64(q^2-2pr)+16(p^2-2q)+3$$

and

$$(8a^2+1)(8b^2+1)(8c^2+1)=512r^2+64(q^2-2pr)+8(p^2-2q)+1.$$

So inequality 1° becomes

$$64(q^2-2pr)+16(p^2-2q)+3\geq 512r^2+64(q^2-2pr)+8(p^2-2q)+1,$$

i.e.

$$8(p^2-2q)+2\geq 512r^2. \quad (3)$$

Using that $q^3\geq 27r^2$ and $q=1-2r$ we get

$$(1-2r)^3\geq 27r^2 \quad \Leftrightarrow \quad 8r^3+15r^2+6r-1\leq 0$$
$$\Leftrightarrow \quad (8r-1)(r^2+2r+1)\leq 0,$$

from where we deduce that

$$8r-1\leq 0, \quad \text{i.e.} \quad r\leq \frac{1}{8}. \quad (4)$$

Since $AM\geq HM$ we have

$$((a+1)+(b+1)+(c+1))\left(\frac{1}{a+1}+\frac{1}{b+1}+\frac{1}{c+1}\right)\geq 9$$

or

$$2(a+b+c+3)\geq 9,$$

i.e.

$$p = a + b + c \geq \frac{3}{2}. \tag{5}$$

From $N_1 : p^2 \geq 3q$ (Chap. 14) it follows that

$$\frac{p^2}{3} \geq q. \tag{6}$$

By (5) and (6) we have

$$8(p^2 - 2q) + 2 \geq 8\left(p^2 - 2\frac{p^2}{3}\right) + 2 = \frac{8}{3}p^2 + 2 \geq \frac{8}{3}\frac{9}{4} + 2 = 8. \tag{7}$$

From (3) and (7) we have that it suffices to show that

$$8 \geq 512r^2$$

or

$$r \leq \frac{1}{8},$$

which is true according to (4). And we are done.

2° We have

$$(4ab + 1)(4bc + 1) + (4bc + 1)(4ca + 1) + (4ca + 1)(4ab + 1) = 64pr + 8q + 3$$

and

$$(4ab + 1)(4bc + 1)(4ca + 1) = 64r^2 + 16pr + 4q + 1.$$

We need to show that

$$64pr + 8q + 3 \geq \frac{3}{2}(64r^2 + 16pr + 4q + 1)$$

or

$$32pr + 16q + 6 \geq 192r^2 + 48pr + 12q + 3,$$

i.e.

$$192r^2 + 16pr - 4q - 3 \leq 0. \tag{8}$$

By $N_7 : q^2 \geq 3pr$ (Chap. 14), it follows that $pr \leq \frac{q^2}{3}$.

Now since $q = 1 - 2r$ we get

$$\begin{aligned}192r^2 + 16pr - 4q - 3 &\leq 192r^2 + 16\frac{q^2}{3} - 4q - 3\\ &= 192r^2 + 16\frac{(1-2r)^2}{3} - 4(1-2r) - 3\end{aligned}$$

$$= \frac{5}{3}(128r^2 - 8r - 1)$$

$$= \frac{5}{3} \cdot 128\left(r - \frac{1}{8}\right)\left(r + \frac{1}{16}\right). \tag{9}$$

From (9) and $r \leq \frac{1}{8}$ it follows that $192r^2 + 16pr - 4q - 3 \leq 0$, which means that inequality (8), i.e. inequality 2°, holds. ■

269 Let $a, b, c > 0$ be real numbers such that $ab + bc + ca = 1$. Prove the inequality

$$\frac{1}{a+b} + \frac{1}{b+c} + \frac{1}{c+a} - \frac{1}{a+b+c} \geq 2.$$

Solution Let $p = a + b + c, q = ab + bc + ca = 1, r = abc$.

The given inequality is equivalent to

$$\frac{(a+b)(b+c) + (b+c)(c+a) + (c+a)(a+b)}{(a+b)(b+c)(c+a)} - \frac{1}{a+b+c} \geq 2. \tag{1}$$

By I_5, I_6 (Chap. 14) and (1) we have that it is enough to prove that

$$\frac{p^2+q}{pq-r} - \frac{1}{p} \geq 2,$$

i.e.

$$\frac{p^2+1}{p-r} - \frac{1}{p} \geq 2,$$

which is equivalent as follows

$$\begin{aligned} & p^3 + p - p + r \geq 2p^2 - 2pr \\ \Leftrightarrow \quad & p^3 - 2p^2 + 2pr + r \geq 0 \\ \Leftrightarrow \quad & p^3 - 2p^2 + r(2p+1) \geq 0. \end{aligned} \tag{2}$$

Let

$$f(p) = p^3 - 2p^2 + r(2p+1). \tag{3}$$

From $N_4 : p^2 \geq 3q = 3$ (Chap. 14) it follows that $p \geq \sqrt{3}$.

If $p \geq 2$ then clearly $f(p) \geq 0$.

Let $\sqrt{3} \leq p < 2$.

By $N_1 : p^3 - 4pq + 9r \geq 0$ we have

$$p^3 - 4p + 9r \geq 0, \quad \text{i.e.} \quad r \geq \frac{4p - p^3}{9}. \tag{4}$$

By (3) and (4) we obtain

$$f(p) = p^3 - 2p^2 + r(2p+1) \geq p^3 - 2p^2 + \left(\frac{4p - p^3}{9}\right)(2p+1)$$
$$= -2p(p-2)(p-1)^2 \geq 0.$$

The last inequality holds, since $p < 2$. So we have proved (2), and we are done. ■

270 Let $a, b, c \geq 0$ be real numbers. Prove the inequality

$$\frac{ab+4bc+ca}{a^2+bc} + \frac{bc+4ca+ab}{b^2+ca} + \frac{ca+4ab+bc}{c^2+ab} \geq 6.$$

Solution Let $p = a+b+c, q = ab+bc+ca, r = abc$.

Since the given inequality is homogenous we may assume that $p = 1$.

After elementary algebraic operations we can easily rewrite the given inequality in the form

$$7pq - 12r^2 \geq 4q^3 - q^2. \tag{1}$$

By $N_1 : p^3 - 4pq + 9r \geq 0$ (Chap. 14) we have $9r \geq 4q - 1$ and clearly $0 \leq q \leq \frac{1}{3}$.

So

$$9rq^2 \geq q^2(4q-1) \quad \Leftrightarrow \quad \frac{9rq}{3} \geq q^2(4q-1) \quad \Leftrightarrow \quad 3rq \geq q^2(4q-1). \tag{2}$$

From $N_3 : pq - 9r \geq 0$ (Chap. 14) it follows that $q \geq 9r$, i.e. we have

$$4rq \geq 36r^2 \geq 12r^2. \tag{3}$$

By (2) and (3) we obtain

$$7pq - 12r^2 = 3rq + 4rq - 12r^2 \geq 3rq \geq q^2(4q-1),$$

i.e. inequality (1) holds, as required. ■

271 Let a, b, c be positive real numbers such that $a+b+c+1 = 4abc$. Prove the inequality

$$\frac{1}{a^4+b+c} + \frac{1}{b^4+c+a} + \frac{1}{c^4+a+b} \leq \frac{3}{a+b+c}.$$

Solution By the *Cauchy-Schwarz inequality* we have

$$\frac{1}{a^4+b+c} = \frac{1+b^3+c^3}{(a^4+b+c)(1+b^3+c^3)} \leq \frac{1+b^3+c^3}{(a^2+b^2+c^2)^2}.$$

Similarly we get

$$\frac{1}{b^4+c+a} \leq \frac{1+c^3+b^3}{(a^2+b^2+c^2)^2} \quad \text{and} \quad \frac{1}{c^4+a+b} \leq \frac{1+a^3+b^3}{(a^2+b^2+c^2)^2}.$$

After adding the last three inequalities we obtain

$$\frac{1}{a^4+b+c} + \frac{1}{b^4+c+a} + \frac{1}{c^4+a+b} \leq \frac{3+2(a^3+b^3+c^3)}{(a^2+b^2+c^2)^2},$$

so it suffices to prove that

$$\frac{3+2(a^3+b^3+c^3)}{(a^2+b^2+c^2)^2} \leq \frac{3}{a+b+c},$$

i.e.

$$3(a^2+b^2+c^2)^2 \geq (a+b+c)(3+2(a^3+b^3+c^3)).$$

Let $a+b+c=p, ab+bc+ca=q$ and $abc=r$.

Then since $a+b+c+1=4abc$, by $AM \geq GM$ it follows that

$$4r = a+b+c+1 \geq 4\sqrt[4]{r}, \quad \text{i.e.} \quad r \geq 1.$$

Now we have

$$\begin{aligned}
A &= 3(a^2+b^2+c^2)^2 - (a+b+c)(3+2(a^3+b^3+c^3)) \\
&= 3(p^2-2q)^2 - p(3+2p(p^2-3q)+6r) \\
&= 3(p^2-2q)^2 - 3p - 2p^2(p^2-3q) - 6pr \\
&= 3p^4 - 12p^2q + 12q^2 - 3p - 2p^4 + 6p^2q - 6pr \\
&= p^4 - 6p^2q + 12q^2 - 3p - 6pr \\
&= (p^2-3q)^2 + q^2 - 3p + 2(q^2-3pr).
\end{aligned}$$

Since $r \geq 1$ we have $q^2 - 3p \geq q^2 - 3pr$ and it follows that

$$A = (p^2-3q)^2 + q^2 - 3p + 2(q^2-3pr) \geq (p^2-3q)^2 + 3(q^2-3pr).$$

According to $N_7 : q^2 - 3pr \geq 0$ we deduce that

$$A \geq (p^2-3q)^2 + 3(q^2-3pr) \geq 0,$$

as required. ■

272 Let $x, y, z > 0$ be real numbers such that $x+y+z=1$. Prove the inequality

$$(x^2+y^2)(y^2+z^2)(z^2+x^2) \leq \frac{1}{32}.$$

Solution Let $p = x + y + z = 1, q = xy + yz + zx, r = xyz$.

Then we have

$$x^2 + y^2 = (x+y)^2 - 2xy = (1-z)^2 - 2xy = 1 - 2z + z^2 - 2xy$$
$$= 1 - z - z(1-z) - 2xy = 1 - z - z(x+y) - 2xy = 1 - z - q - xy.$$

Analogously we deduce

$$y^2 + z^2 = 1 - x - q - yz \quad \text{and} \quad z^2 + x^2 = 1 - y - q - zx.$$

So the given inequality becomes

$$(1 - z - q - xy)(1 - x - q - yz)(1 - y - q - zx) \le \frac{1}{32}. \tag{1}$$

After algebraic transformations we find that inequality (1) is equivalent to

$$q^2 - 2q^3 - r(2 + r - 4q) \le \frac{1}{32}. \tag{2}$$

Assume that $q \le \frac{1}{4}$.

Using $N_1 : p^3 - 4pq + 9r \ge 0$ (Chap. 14), it follows that

$$9r \ge 4q - 1, \quad \text{i.e.} \quad r \ge \frac{4q-1}{9},$$

and clearly $q \le \frac{1}{3}$.

It follows that

$$2 + r - 4q \ge 2 + \frac{4q-1}{9} - 4q = \frac{17 - 32q}{9} \ge \frac{17 - \frac{32}{3}}{9} > 0.$$

So we have

$$q^2 - 2q^3 - r(2 + r - 4q) \le q^2 - 2q^3 = q^2(1 - 2q)$$
$$= \frac{q}{2} \cdot 2q(1-2q) \le \frac{q}{2}\left(\frac{2q + (1-2q)}{2}\right)^2 = \frac{q}{8} \le \frac{1}{32},$$

i.e. inequality (2) holds for $q \le \frac{1}{4}$.

We need just to consider the case when $q > \frac{1}{4}$.

Let

$$f(r) = q^2 - 2q^3 - r(2 + r - 4q). \tag{3}$$

Clearly $r \ge \frac{4q-1}{9}$.

Using $N_3 : pq - 9r \ge 0$ (Chap. 14) it follows that $9r \le q$, i.e. $r \le \frac{q}{9}$.

We have

$$f'(r) = 4q - 2 - 2r \le \frac{4}{3} - 2 - 2r \le 0.$$

This means that f is a strictly decreasing function on $(\frac{4q-1}{9}, \frac{q}{9})$, from which it follows that

$$f(r) \leq f\left(\frac{4q-1}{9}\right) = q^2 - 2q^3 - \frac{1}{81}(4q-1)(17-32q),$$

i.e.

$$f(r) \leq \frac{81(q^2-2q^3)-(4q-1)(17-32q)}{81}. \tag{4}$$

Let

$$g(q) = 81(q^2-2q^3)-(4q-1)(17-32q). \tag{5}$$

Then

$$g'(q) = -486q^2 + 418q - 100.$$

Since $\frac{1}{4} < q \leq \frac{1}{3}$, we get

$$g'(q) = -486q^2 + 418q - 100 < \frac{-486}{16} + \frac{418}{3} - 100 < 0.$$

So g decreases on $(1/4, 1/3)$, i.e. we have

$$g(q) < g\left(\frac{1}{4}\right) = \frac{81}{32}. \tag{6}$$

Finally by (3), (4), (5) and (6) we obtain

$$\begin{aligned} q^2 - 2q^3 - r(2+r-4q) = f(r) &\leq f\left(\frac{4q-1}{9}\right) \\ &= \frac{81(q^2-2q^3)-(4q-1)(17-32q)}{81} \\ &= \frac{g(q)}{81} < \frac{g(\frac{1}{4})}{81} = \frac{\frac{81}{32}}{81} = \frac{1}{32}, \end{aligned}$$

as required. ∎

273 Let $x, y, z \in \mathbb{R}^+$ such that $x + y + z = 1$. Prove the inequalities:

$$1 \leq \frac{x}{1-yz} + \frac{y}{1-zx} + \frac{z}{1-xy} \leq \frac{9}{8}.$$

Solution Let $p = x + y + z = 1, q = xy + yz + zx, r = xyz$.

We have

$$\begin{aligned}
&x(1-zx)(1-xy)+y(1-yz)(1-xy)+z(1-zx)(1-yz)\\
&\quad = x(1-xy-zx+x^2yz)+y(1-xy-yz+y^2xz)+z(1-zx-zy+z^2xy)\\
&\quad = x+y+z-x^2(y+z)-y^2(z+x)-z^2(x+y)+x^3yz+y^3zx+z^3xy\\
&\quad = p-x^2(p-x)-y^2(p-y)-z^2(p-z)+xyz(x^2+y^2+z^2)\\
&\quad = p-(p-xyz)(x^2+y^2+z^2)+x^3+y^3+z^3\\
&\quad = p-(p-r)(p^2-2q)+p(p^2-3q)+3r\\
&\quad = 1-(1-r)(1-2q)+1-3q+3r; \qquad (1)
\end{aligned}$$

also we have

$$\begin{aligned}
(1-xy)(1-yz)(1-zx) &= (1-xy-yz+y^2xz)(1-zx)\\
&= 1-zx-xy-yz+x^2yz+y^2zx+z^2xy-x^2y^2z^2\\
&= 1-q+pr-r^2 = 1-q+r-r^2. \qquad (2)
\end{aligned}$$

By (1) and (2) we have that the left inequality is equivalent to

$$1-q+r-r^2 \le 1-(1-r)(1-2q)+1-3q+3r \quad \Leftrightarrow \quad r-2q+3 \ge 0. \quad (3)$$

Using $N_5 : p^3 \ge 27r$ (Chap. 14), it follows that $r \le \frac{1}{27}$.

Also by $N_1 : p^3 - 4pq + 9r \ge 0$ (Chap. 14), we have$q \le \frac{9r+1}{4}$.

Now we deduce

$$\begin{aligned}
r-2q+3 &\ge r-2\frac{9r+1}{4}+3 = \frac{4r-18r-2+12}{4} = \frac{10-14r}{4} = \frac{5-7r}{2}\\
&\ge \frac{5-\frac{7}{27}}{2} > 0,
\end{aligned}$$

i.e. inequality (3) holds.

We need to show the right side inequality from (1), which, using identities (1) and (2) is

$$9r^2+23r+q-16qr \le 1. \qquad (4)$$

Let us denote $f(r) = 9r^2 + r(23-16q) + q$.

By $N_7 : q^2 \ge 3pr = 3r$ (Chap. 14), it follows that

$$r \le \frac{q^2}{3}, \quad \text{i.e.} \quad 0 \le r \le \frac{q^2}{3}.$$

We have

$$f'(r) = 18r + 23 - 16q. \qquad (5)$$

Using $N_4 : p^2 \ge 3q$ (Chap. 14), it follows that $q \le \frac{1}{3}$.

By (5) we have

$$f'(r) = 18r + 23 - 16q \geq 18r + 23 - \frac{16}{3} > 0,$$

i.e. f increases on $(0, \frac{q^2}{3})$, where $q \leq \frac{1}{3}$.

So we obtain $f(r) \leq f(\frac{q^2}{3})$.

It suffices to show that $f(\frac{q^2}{3}) \leq 1$.

We have $f(\frac{q^2}{3}) = q^4 - \frac{16}{3}q^3 + \frac{23}{3}q^2 + q$.

Now we get

$$\begin{aligned}
& f\left(\frac{q^2}{3}\right) \leq 1 \\
\Leftrightarrow \quad & q^4 - \frac{16}{3}q^3 + \frac{23}{3}q^2 + q - 1 \leq 0 \\
\Leftrightarrow \quad & (3q-1)(q^3 - 5q^2 + 6q + 3) \leq 0 \\
\Leftrightarrow \quad & (3q-1)(q(q-2)(q-3) + 3) \leq 0,
\end{aligned}$$

which clearly holds since $0 \leq q \leq \frac{1}{3}$. This complete the proof. ■

274 Let $x, y, z \in \mathbb{R}^+$, such that $xyz = 1$. Prove the inequality

$$\frac{1}{(1+x)^2} + \frac{1}{(1+y)^2} + \frac{1}{(1+z)^2} + \frac{2}{(1+x)(1+y)(1+z)} \geq 1.$$

Solution Let $x + y + z = p, xy + yz + zx = q$ and $xyz = r = 1$.

The given inequality becomes

$$\begin{aligned}
& (1+x)^2(1+y)^2 + (1+y)^2(1+z)^2 + (1+z)^2(1+x)^2 + 2(1+x)(1+y)(1+z) \\
& \quad \geq (1+x)^2(1+y)^2(1+z)^2. \qquad (1)
\end{aligned}$$

By I_9 and I_{11} (Chap. 14), we have

$$(1+x)(1+y)(1+z) = 1 + p + q + r = 2 + p + q$$

and

$$\begin{aligned}
& (1+x)^2(1+y)^2 + (1+y)^2(1+z)^2 + (1+z)^2(1+x)^2 \\
& \quad = (3 + 2p + q)^2 - 2(3+p)(1+p+q+r) \\
& \quad = (3 + 2p + q)^2 - 2(3+p)(2+p+q).
\end{aligned}$$

So inequality (1) becomes

$$(3+2p+q)^2-2(3+p)(2+p+q)+2(2+p+q)\geq(2+p+q)^2$$
$$\Leftrightarrow \quad p^2\geq 2q+3.$$

According to $N_6: q^3\geq 27r^2=27$ (Chap. 14), it follows that

$$q\geq 3. \tag{2}$$

By $N_4: p^2\geq 3q$ (Chap. 14), we obtain

$$p^2\geq 3q=2q+q\overset{(2)}{\geq}2q+3,$$

as required. ■

275 Let $a,b,c\geq 0$ such that $a+b+c=1$. Prove the inequalities:

1° $ab+bc+ca\leq a^3+b^3+c^3+6abc$
2° $a^3+b^3+c^3+6abc\leq a^2+b^2+c^2$
3° $a^2+b^2+c^2\leq 2(a^3+b^3+c^3)+3abc$.

Solution Let $p=a+b+c=1, q=ab+bc+ca, r=abc$.
1° Using $I_2: a^3+b^3+c^3=p(p^2-3q)+3r=1-3q+3r$ we have that inequality 1° is equivalent to

$$q\leq 1-3q+3r+6r \quad\Leftrightarrow\quad 9r+1\geq 4q,$$

which is true since N_1 (Chap. 14).
2° Using $I_1: a^2+b^2+c^2=p^2-2q=1-2q$ we get the equivalent form

$$1-3q+9r\leq 1-2q \quad\Leftrightarrow\quad 9r\leq q$$

which is true since N_3 (Chap. 14).
3° The given inequality is equivalent to

$$1-2q\leq 2(1-3q+3r)+3r \quad\Leftrightarrow\quad 4q\leq 1+9r,$$

which is true since N_1 (Chap. 14). ■

276 Let $x,y,z\geq 0$ be real numbers such that $xy+yz+zx+xyz=4$. Prove the inequality

$$3(x^2+y^2+z^2)+xyz\geq 10.$$

Solution Let $p=x+y+z=1, q=xy+yz+zx, r=xyz$.
The given inequality becomes

$$3(p^2-2q)+r\geq 10, \quad \text{with constraint } q+r=4.$$

So it is enough to show that

$$3p^2 - 6q + 4 - q \geq 10, \quad \text{i.e.} \quad 3p^2 - 7q - 6 \geq 0. \tag{1}$$

Applying $N_1 : p^3 - 4pq + 9r \geq 0$ (Chap. 14), and since $q + r = 4$ we deuce

$$p^3 - 4pq + 9(4 - q) \geq 0, \quad \text{i.e.} \quad q \leq \frac{p^3 + 36}{4p + 9}.$$

So

$$3p^2 - 7q - 6 \geq 3p^2 - 7\frac{p^3 + 36}{4p + 9} - 6 = \frac{(p - 3)(5p^2 + 42p + 102)}{4p + 9}. \tag{2}$$

Applying $AM \geq GM$ we obtain

$$4 = xy + yz + zx + xyz \geq 4\sqrt[4]{(xyz)^3}$$
$$\Leftrightarrow \quad 1 \geq xyz. \tag{3}$$

Also $(x + y + z)^2 \geq 3(xy + yz + zx)$, so we deduce

$$p = x + y + z \geq \sqrt{3(4 - xyz)} \overset{(3)}{\geq} \sqrt{3(4 - 1)} = 3.$$

Finally by using (2) we obtain that $3p^2 - 7q - 6 \geq 0$, i.e. inequality (1) holds. ■

277 Let $a, b, c \in \mathbb{R}^+$. Prove the inequality

$$x^4(y + z) + y^4(z + x) + z^4(x + y) \leq \frac{1}{12}(x + y + z)^5.$$

Solution Let $p = a + b + c, q = ab + bc + ca, r = abc$.

Since the given inequality is homogenous, without loss of generality we may assume that $p = 1$.

We have

$$\begin{aligned}
x^4(y + z) + y^4(z + x) + z^4(x + y) &= x^3(xy + xz) + y^3(yz + yx) + z^3(zx + zy) \\
&= x^3(q - yz) + y^3(q - zx) + z^3(q - xy) \\
&= q(x^3 + y^3 + z^3) - xyz(x^2 + y^2 + z^2) \\
&= q(p(p^2 - 3q) + 3r) - r(p^2 - 2q) \\
&= q(1 - 3q + 3r) - r(1 - 2q) \\
&= q(1 - 3q) + r(5q - 1).
\end{aligned}$$

Now the given inequality becomes

$$q(1-3q)+r(5q-1)\leq \frac{1}{12}. \tag{1}$$

From $3q \leq p^2$ it follows that

$$q \leq \frac{1}{3}. \tag{2}$$

If $q \leq \frac{1}{5}$ then $r(5q-1) \leq 0$, so we have

$$q(1-3q)+r(5q-1)\leq q(1-3q)=\frac{1}{3}(1-3q)\cdot 3q \overset{G\leq A}{\leq} \frac{1}{3}\left(\frac{(1-3q)+3q}{2}\right)^2=\frac{1}{12},$$

i.e. inequality (1) holds.

Let

$$q > \frac{1}{5}, \tag{3}$$

i.e. let $q \in (1/5, 1/3]$ and denote

$$f(q)=q(1-3q)+5rq-r.$$

Then

$$f'(q)=1-6q+5r. \tag{4}$$

Using $N_3 : pq \geq 9r$ (Chap. 14), we get

$$q \geq 9r. \tag{5}$$

Now according to (3), (4) and (5) we deduce

$$f'(q)=1-6q+5r\leq 1-6q+\frac{5}{9}q=1-\frac{49}{9}q<1-\frac{49}{9}\cdot\frac{1}{5}<0,$$

i.e. f is strictly decreasing on $q \in (1/5, 1/3]$, so it follows that $f(q) < f(\frac{1}{5})$, i.e. we deduce that

$$q(1-3q)+r(5q-1)<\frac{1}{5}\left(1-\frac{3}{5}\right)+r\left(5\frac{1}{5}-1\right)=\frac{2}{25}<\frac{1}{12},$$

as required. ∎

278 Let $a, b, c \in \mathbb{R}^+$ such that $a+b+c=1$. Prove the inequality

$$\frac{1}{a}+\frac{1}{b}+\frac{1}{c}+48(ab+bc+ca)\geq 25.$$

Solution Setting $ab+bc+ca=\frac{1-q^2}{3}\geq 0, q\geq 0$, it follows that $q\in[0,1]$.
We have

$$\frac{1}{a}+\frac{1}{b}+\frac{1}{c}+48(ab+bc+ca)=\frac{ab+bc+ca}{abc}+48(ab+bc+ca)$$
$$=\frac{1-q^2}{3r}+16(1-q^2).$$

So it suffices to show that

$$\frac{1-q^2}{3r}+16(1-q^2)\geq 25.$$

Due to Theorem 15.1 (Chap. 15) we have

$$\frac{1-q^2}{3r}+16(1-q^2)\geq 27\frac{1-q^2}{3(1-q)^2(1+2q)}+16(1-q^2)$$
$$=9\frac{1+q}{(1-q)(1+2q)}+16(1-q^2)$$
$$=\frac{2q^2(4q-1)^2}{(1-q)(1+2q)}+25\geq 25.$$

Equality occurs if and only if $(a,b,c)=(1/3,1/3,1/3)$ or $(a,b,c)=(1/2,1/4,1/4)$ (up to permutation). ■

279 Let a,b,c be non-negative real numbers such that $a+b+c=2$. Prove the inequality

$$a^4+b^4+c^4+abc\geq a^3+b^3+c^3.$$

Solution Applying *Schur's inequality* (fourth degree) we have that

$$a^4+b^4+c^4+abc(a+b+c)\geq a^3(b+c)+b^3(c+a)+c^3(a+b),$$

i.e.

$$2(a^4+b^4+c^4)+abc(a+b+c)\geq(a^3+b^3+c^3)(a+b+c)$$

from which, using the initial condition, we obtain the result as required.
Equality holds iff $a=b=c=2/3$ or $a=b=1, c=0$ (over all permutations). ■

280 Let a,b,c be non-negative real numbers. Prove the inequality

$$2(a^2+b^2+c^2)+abc+8\geq 5(a+b+c).$$

Solution We'll use *Schur's inequality*, i.e.

$$x^3 + y^3 + z^3 + 3xyz \geq xy(x+y) + yz(y+z) + zx(z+x), \quad \text{for all } x, y, z \geq 0.$$

By $AM \geq GM$ and $QM \geq AM$ we have

$$\begin{aligned}
&6(2(a^2+b^2+c^2) + abc + 8 - 5(a+b+c)) \\
&\quad = 12(a^2+b^2+c^2) + 6abc + 48 - 30(a+b+c) \\
&\quad = 12(a^2+b^2+c^2) + 3(2abc+1) + 45 - 30(a+b+c) \\
&\quad \geq 12(a^2+b^2+c^2) + 9\sqrt[3]{(abc)^2} + 45 - 5((a+b+c)^2+9) \\
&\quad = \frac{9abc}{\sqrt[3]{abc}} + 3(a^2+b^2+c^2) - 6(ab+bc+ca) \\
&\qquad + 2((a-b)^2+(b-c)^2+(c-a)^2) \\
&\quad \geq \frac{9abc}{\sqrt[3]{abc}} + 3(a^2+b^2+c^2) - 6(ab+bc+ca) \\
&\quad \geq \frac{27abc}{a+b+c} + 3(a+b+c)^2 - 12(ab+bc+ca) \\
&\quad = \frac{3}{a+b+c}(9abc + (a+b+c)^3 - 4(ab+bc+ca)(a+b+c)) \\
&\quad = \frac{3}{a+b+c}(a^3+b^3+c^3+3abc - ab(a+b) + bc(b+c) + ca(c+a)) \geq 0.
\end{aligned}$$

And we are done. Equality holds iff $a = b = c = 1$. ■

281 Let a, b, c be non-negative real numbers. Prove the inequality

$$a^3 + b^3 + c^3 + 4(a+b+c) + 9abc \geq 8(ab+bc+ca).$$

Solution We'll use *Schur's inequality*, i.e. for all $a, b, c \geq 0$ we have

$$a^3 + b^3 + c^3 + abc(a+b+c) \geq ab(a^2+b^2) + bc(b^2+c^2) + ca(c^2+a^2).$$

By $AM \geq GM$ we have

$$4(a+b+c) + \frac{4(ab+bc+ca)^2}{(a+b+c)} \geq 8(ab+bc+ca).$$

So it suffices to prove that

$$a^3 + b^3 + c^3 + 9abc \geq \frac{4(ab+bc+ca)^2}{(a+b+c)}.$$

The previous inequality is equivalent to

$$\begin{aligned}&a^4+b^4+c^4+abc(a+b+c)+ab(a^2+b^2)+bc(b^2+c^2)+ca(c^2+a^2)\\&\quad\geq 4(a^2b^2+b^2c^2+c^2a^2).\end{aligned}$$

Applying *Schur's inequality* and $AM \geq GM$ we obtain

$$\begin{aligned}&a^4+b^4+c^4+abc(a+b+c)+ab(a^2+b^2)+bc(b^2+c^2)+ca(c^2+a^2)\\&\quad\geq 2(ab(a^2+b^2)+bc(b^2+c^2)+ca(c^2+a^2))\\&\quad\geq 2(ab(2ab)+bc(2bc)+ca(2ca))=4(a^2b^2+b^2c^2+c^2a^2),\end{aligned}$$

as required.

Equality holds iff $a=b=c=1$ or $a=b=2$, $c=0$ (up to permutation). ■

282 Let a,b,c be non-negative real numbers. Prove the inequality

$$\frac{a^3}{b^2-bc+c^2}+\frac{b^3}{c^2-ca+a^2}+\frac{c^3}{a^2-ab+b^2}\geq a+b+c.$$

Solution Applying the *Cauchy–Schwarz inequality* (Corollary 4.3) we deduce

$$\begin{aligned}&\frac{a^3}{b^2-bc+c^2}+\frac{b^3}{c^2-ca+a^2}+\frac{c^3}{a^2-ab+b^2}\\&\quad=\frac{a^4}{a(b^2-bc+c^2)}+\frac{b^4}{b(c^2-ca+a^2)}+\frac{c^4}{c(a^2-ab+b^2)}\\&\quad\geq\frac{(a^2+b^2+c^2)^2}{a(b^2-bc+c^2)+b(c^2-ca+a^2)+c(a^2-ab+b^2)}.\end{aligned}$$

So it suffices to prove that

$$(a^2+b^2+c^2)^2\geq(a(b^2-bc+c^2)+b(c^2-ca+a^2)+c(a^2-ab+b^2))(a+b+c).$$

The previous inequality is equivalent to

$$\begin{aligned}&a^4+b^4+c^4+2(a^2b^2+b^2c^2+c^2a^2)\\&\quad\geq(a+b+c)(a^2(b+c)+b^2(c+a)+c^2(a+b))-3abc(a+b+c)\end{aligned}$$

or

$$a^4+b^4+c^4+abc(a+b+c)\geq a^3(b+c)+b^3(c+a)+c^3(a+b),$$

and it is *Schur's inequality* (fourth degree).

Equality holds iff $a=b=c$ or $a=b$, $c=0$ (up to permutation). ■

283 Let a,b,c be non-negative real numbers such that $a+b+c=2$. Prove the inequality

$$a^3+b^3+c^3+\frac{15abc}{4}\geq 2.$$

Solution Applying *Schur's inequality* we have that the following inequality holds

$$a^3+b^3+c^3+\frac{15abc}{4}\geq\frac{(a+b+c)^3}{4},$$

from which we obtain the required inequality. Equality holds iff $a=b=c=2/3$ or $a=b=1, c=0$ (over all permutations). ■

284 Let a,b,c be positive real numbers such that $abc=1$. Prove the inequality

$$\frac{a^2+bc}{a^2(b+c)}+\frac{b^2+ca}{b^2(c+a)}+\frac{c^2+ab}{c^2(a+b)}\geq ab+bc+ca.$$

Solution We'll show that

$$\frac{a^2+bc}{a^2(b+c)}+\frac{b^2+ca}{b^2(c+a)}+\frac{c^2+ab}{c^2(a+b)}\geq\frac{1}{a}+\frac{1}{b}+\frac{1}{c}. \tag{1}$$

We have

$$\frac{a^2+bc}{a^2(b+c)}-\frac{1}{a}=\frac{(a-b)(a-c)}{a^2(b+c)}.$$

Analogously we deduce

$$\frac{b^2+ca}{b^2(c+a)}-\frac{1}{b}=\frac{(b-c)(b-a)}{b^2(c+a)} \quad \text{and} \quad \frac{c^2+ab}{c^2(a+b)}-\frac{1}{c}=\frac{(c-a)(c-b)}{c^2(a+b)}.$$

Applying the previous identities and Corollary 12.1 from *Schur's inequality* we obtain (1). From (1) and $abc=1$ we obtain the required inequality.

Equality holds iff $a=b=c=1$. ■

285 Let a,b,c be positive real numbers such that $a^2+b^2+c^2=3$. Prove the inequality

$$\frac{a^3+abc}{(b+c)^2}+\frac{b^3+abc}{(c+a)^2}+\frac{c^3+abc}{(a+b)^2}\geq\frac{3}{2}.$$

Solution We'll show that

$$\frac{a^3+abc}{(b+c)^2}+\frac{b^3+abc}{(c+a)^2}+\frac{c^3+abc}{(a+b)^2}\geq\frac{a^2}{b+c}+\frac{b^2}{c+a}+\frac{c^2}{a+b}. \tag{1}$$

We have

$$\frac{a^3+abc}{(b+c)^2}-\frac{a^2}{b+c}=\frac{a}{(b+c)^2}(a-b)(a-c);$$

analogously we get the other two identities.

Now (1) is equivalent to

$$\frac{a}{(b+c)^2}(a-b)(a-c)+\frac{b}{(c+a)^2}(b-c)(b-a)+\frac{c}{(a+b)^2}(c-a)(c-b)\geq 0. \quad (2)$$

Assume that $a \geq b \geq c$.

Then we easily deduce that $\frac{a}{(b+c)^2} \geq \frac{b}{(c+a)^2} \geq \frac{c}{(a+b)^2}$, and the correctness of (2) will follow from Corollary 12.1 of *Schur's inequality*.

Furthermore, we'll show that

$$\frac{a^2}{b+c}+\frac{b^2}{c+a}+\frac{c^2}{a+b} \geq \frac{\sqrt{3(a^2+b^2+c^2)}}{2}. \quad (3)$$

Assume that $a \geq b \geq c$. Then

$$a^2 \geq b^2 \geq c^2 \quad \text{and} \quad \frac{1}{b+c} \geq \frac{1}{c+a} \geq \frac{1}{a+b}.$$

Applying *Chebishev's inequality* and $AM \geq HM$ we get

$$\begin{aligned}\frac{a^2}{b+c}+\frac{b^2}{c+a}+\frac{c^2}{a+b} &\geq \frac{1}{3}(a^2+b^2+c^2)\left(\frac{1}{b+c}+\frac{1}{c+a}+\frac{1}{a+b}\right)\\ &\geq \frac{1}{3}(a^2+b^2+c^2)\frac{9}{2(a+b+c)}\\ &\geq \frac{3(a^2+b^2+c^2)}{2\sqrt{3(a^2+b^2+c^2)}} = \frac{\sqrt{3(a^2+b^2+c^2)}}{2}.\end{aligned}$$

So inequality (3) is proved.

By (1), (3) and the initial condition we obtain

$$\begin{aligned}\frac{a^3+abc}{(b+c)^2}+\frac{b^3+abc}{(c+a)^2}+\frac{c^3+abc}{(a+b)^2} &\geq \frac{a^2}{b+c}+\frac{b^2}{c+a}+\frac{c^2}{a+b} \geq \frac{\sqrt{3(a^2+b^2+c^2)}}{2}\\ &= \frac{3}{2}.\end{aligned}$$

Equality holds iff $a=b=c=1$. ■

286 Let a, b, c be positive real numbers such that $a^4+b^4+c^4=3$. Prove the inequality

$$\frac{1}{4-ab}+\frac{1}{4-bc}+\frac{1}{4-ca} \leq 1.$$

Solution 1 After clearing denominators the given inequality becomes

$$48-8\sum_{sym} ab+abc\sum_{sym} a \leq 64-16\sum_{sym} ab+4abc\sum_{sym} a-a^2b^2c^2,$$

i.e.

$$16 + 3abc(a+b+c) \ge a^2b^2c^2 + 8(ab+bc+ca). \tag{1}$$

Applying *Schur's inequality* we have that

$$(a^3+b^3+c^3+3abc)(a+b+c) \ge (ab(a+b)+bc(b+c)+ca(c+a))(a+b+c),$$

and since $a^4+b^4+c^4=3$ we deduce

$$3 + 3abc(a+b+c) \ge (ab+ac)^2 + (ac+bc)^2 + (bc+ab)^2. \tag{2}$$

Using $AM \ge GM$ we get

$$(ab+ac)^2 + (ac+bc)^2 + (bc+ab)^2 + 12 \ge 8(ab+bc+ca). \tag{3}$$

Now from (2) and (3) we deduce

$$15 + 3abc(a+b+c) \ge 8(ab+bc+ca). \tag{4}$$

Once more we apply $AM \ge GM$, and we get

$$3 = a^4+b^4+c^4 \ge 3\sqrt[3]{(abc)^4}, \quad \text{i.e.} \quad 1 \ge abc$$

or

$$1 \ge a^2b^2c^2. \tag{5}$$

Finally using (4) and (5) we get inequality (1).

Equality holds iff $a=b=c=1$. ■

Solution 2 Let $x=ab$, $y=bc$ and $z=ac$. The given inequality is equivalent to

$$\frac{1-x}{4-x} + \frac{1-y}{4-y} + \frac{1-z}{4-z} \ge 0$$

or

$$\frac{1-x^2}{4+3x-x^2} + \frac{1-y^2}{4+3y-y^2} + \frac{1-z^2}{4+3z+z^2} \ge 0.$$

Notice that

$$x^2+y^2+z^2 = (ab)^2+(bc)^2+(ca)^2 \le a^4+b^4+c^4 = 3.$$

Assume that $x \ge y \ge z$. Then clearly

$$1-x^2 \le 1-y^2 \le 1-z^2 \quad \text{and} \quad \frac{1}{4+3x-x^2} \le \frac{1}{4+3y-y^2} \le \frac{1}{4+3z+z^2}.$$

Therefore by *Chebishev's inequality* we obtain

$$3\left(\frac{1-x^2}{4+3x-x^2}+\frac{1-y^2}{4+3y-y^2}+\frac{1-z^2}{4+3z+z^2}\right)$$
$$\geq (1-x^2+1-y^2+1-z^2)\left(\frac{1}{4+3x-x^2}+\frac{1}{4+3y-y^2}+\frac{1}{4+3z+z^2}\right)$$
$$\geq 0,$$

as required.

Equality occurs iff $a=b=c=1$. ■

287 Let a, b, c be positive real numbers such that $ab+bc+ca=3$. Prove the inequality

$$(a^3-a+5)(b^5-b^3+5)(c^7-c^5+5)\geq 125.$$

Solution For any real number x, the numbers $x-1, x^2-1, x^3-1$ and x^5-1 are of the same sign.

Therefore

$$(x-1)(x^2-1)\geq 0, \qquad (x^2-1)(x^3-1)\geq 0 \quad \text{and} \quad (x^2-1)(x^5-1)\geq 0,$$

i.e.

$$a^3-a^2-a+1\geq 0,$$
$$b^5-b^3-b^2+1\geq 0,$$
$$c^7-c^5-c^2+1\geq 0.$$

So it follows that

$$a^3-a+5\geq a^2+4, \qquad b^5-b^3+5\geq b^2+4 \quad \text{and} \quad c^7-c^5+5\geq c^2+4.$$

Multiplying these inequalities gives us

$$(a^3-a+5)(b^5-b^3+5)(c^7-c^5+5)\geq (a^2+4)(b^2+4)(c^2+4). \tag{1}$$

We'll prove that

$$(a^2+4)(b^2+4)(c^2+4)\geq 25(ab+bc+ca+2). \tag{2}$$

We have

$$(a^2+4)(b^2+4)(c^2+4)$$
$$=a^2b^2c^2+4(a^2b^2+b^2c^2+c^2a^2)+16(a^2+b^2+c^2)+64$$

$$= a^2b^2c^2 + (a^2 + b^2 + c^2) + 2 + 4(a^2b^2 + b^2c^2 + c^2a^2 + 3)$$
$$+ 15(a^2 + b^2 + c^2) + 50. \tag{3}$$

By the obvious inequalities

$$(a-b)^2 + (b-c)^2 + (c-a)^2 \geq 0 \quad \text{and} \quad (ab-1)^2 + (bc-1)^2 + (ca-1)^2 \geq 0$$

we obtain

$$a^2 + b^2 + c^2 \geq ab + bc + ca, \tag{4}$$
$$a^2b^2 + b^2c^2 + c^2a^2 + 3 \geq 2(ab + bc + ca). \tag{5}$$

We'll prove that

$$a^2b^2c^2 + (a^2 + b^2 + c^2) + 2 \geq 2(ab + bc + ca). \tag{6}$$

Lemma 21.7 *Let $x, y, z > 0$. Then*

$$3xyz + x^3 + y^3 + z^3 \geq 2((xy)^{3/2} + (yz)^{3/2} + (zx)^{3/2}).$$

Proof By *Schur's inequality* and $AM \geq GM$ we have

$$x^3 + y^3 + z^3 + 3xyz \geq (x^2y + y^2x) + (z^2y + y^2z) + (x^2z + z^2x)$$
$$\geq 2((xy)^{3/2} + (yz)^{3/2} + (zx)^{3/2}). \qquad \square$$

By Lemma 21.7 for $x = a^{2/3}, y = b^{2/3}, z = c^{2/3}$ we deduce

$$3(abc)^{2/3} + a^2 + b^2 + c^2 \geq 2(ab + bc + ca).$$

Therefore it suffices to prove that

$$a^2b^2c^2 + 2 \geq 3(abc)^{2/3},$$

which follows immediately by $AM \geq GM$.

Thus we have proved inequality (6).

Now by (3), (4), (5) and (6) we obtain inequality (2).

Finally by (1), (2) and since $ab + bc + ca = 3$ we obtain the required inequality.

Equality occurs if and only if $a = b = c = 1$. ■

288 Let x, y, z be positive real numbers. Prove the inequality

$$\frac{1}{x^2 + xy + y^2} + \frac{1}{y^2 + yz + z^2} + \frac{1}{z^2 + zx + x^2} \geq \frac{9}{(x + y + z)^2}.$$

Solution It is true that $x^2+xy+y^2=(x+y+z)^2-(xy+yz+zx)-(x+y+z)z$.

Now we have

$$\frac{(x+y+z)^2}{x^2+xy+y^2}=\frac{1}{1-\frac{xy+yz+zx}{(x+y+z)^2}-\frac{z}{x+y+z}},$$

i.e.

$$\frac{(x+y+z)^2}{x^2+xy+y^2}=\frac{1}{1-(ab+bc+ca)-c}$$

where $a=\frac{x}{x+y+z}$, $b=\frac{y}{x+y+z}$, $c=\frac{z}{x+y+z}$.

The given inequality can be written in the form

$$\frac{1}{1-d-c}+\frac{1}{1-d-b}+\frac{1}{1-d-a}\geq 9 \tag{1}$$

where a, b, c are positive real numbers such that

$$a+b+c=1 \quad \text{and} \quad d=ab+bc+ca.$$

After clearing the denominators, inequality (1) becomes

$$9d^3-6d^2-3d+1+9abc\geq 0 \quad \text{or} \quad d(3d-1)^2+(1-4d+9abc)\geq 0,$$

which is true since $1-4d+9abc\geq 0$ (the last inequality is a direct consequences of *Schur's inequality*). ■

289 Let x, y, z be positive real numbers such that $xyz=x+y+z+2$. Prove the inequalities

1° $xy+yz+zx\geq 2(x+y+z)$
2° $\sqrt{x}+\sqrt{y}+\sqrt{z}\leq \frac{3\sqrt{xyz}}{2}$.

Solution 1° The identity $xyz=x+y+z+2$ can be rewritten as

$$\frac{1}{1+x}+\frac{1}{1+y}+\frac{1}{1+z}=1.$$

Let's denote $\frac{1}{1+x}=a$, $\frac{1}{1+y}=b$, $\frac{1}{1+z}=c$.

Then

$$a+b+c=1 \quad \text{and} \quad x=\frac{b+c}{a}, \qquad y=\frac{c+a}{b}, \qquad z=\frac{a+b}{c}.$$

Now we have

$$\begin{aligned} &xy+yz+zx \geq 2(x+y+z) \\ &\quad\Leftrightarrow\quad \frac{b+c}{a}\cdot\frac{c+a}{b}+\frac{c+a}{b}\cdot\frac{a+b}{c}+\frac{a+b}{c}\cdot\frac{b+c}{a} \\ &\qquad\quad \geq 2\left(\frac{b+c}{a}+\frac{c+a}{b}+\frac{a+b}{c}\right) \\ &\quad\Leftrightarrow\quad a^3+b^3+c^3+3abc \geq ab(a+b)+bc(b+c)+ca(c+a), \end{aligned}$$

which clearly holds (*Schur's inequality*).

2° The given inequality is equivalent to

$$\begin{aligned} &\frac{1}{\sqrt{yz}}+\frac{1}{\sqrt{zx}}+\frac{1}{\sqrt{xy}} \leq \frac{3}{2} \\ &\quad\Leftrightarrow\quad \sqrt{\frac{a}{b+c}\cdot\frac{b}{c+a}}+\sqrt{\frac{b}{c+a}\cdot\frac{c}{a+b}}+\sqrt{\frac{c}{a+b}\cdot\frac{a}{b+c}} \leq \frac{3}{2}. \end{aligned} \tag{1}$$

Using $AM \geq GM$ we have

$$\begin{aligned} &\sqrt{\frac{a}{b+c}\cdot\frac{b}{c+a}} \leq \frac{1}{2}\left(\frac{a}{a+c}+\frac{b}{c+b}\right), \\ &\sqrt{\frac{b}{c+a}\cdot\frac{c}{a+b}} \leq \frac{1}{2}\left(\frac{b}{a+b}+\frac{c}{c+a}\right) \quad \text{and} \\ &\sqrt{\frac{c}{a+b}\cdot\frac{a}{b+c}} \leq \frac{1}{2}\left(\frac{c}{b+c}+\frac{a}{a+b}\right). \end{aligned}$$

Adding the last three inequalities we obtain inequality (1), as required. ■

290 Let x, y, z be positive real numbers. Prove the inequality

$$8(x^3+y^3+z^3) \geq (x+y)^3+(y+z)^3+(z+x)^3.$$

Solution 1 The given inequality is equivalent to

$$\begin{aligned} &2(x^3+y^3+z^3) \geq x^2y+x^2z+y^2x+y^2z+z^2x+z^2y \\ &\quad\Leftrightarrow\quad T[3,0,0] \geq T[2,1,0], \end{aligned} \tag{1}$$

which obviously holds according to *Muirhead's inequality*. ■

Solution 2 Let $p=x+y+z, q=xy+yz+zx, r=xyz$.

Since the given inequality is homogenous we may assume that $p=1$.

Using I_2 we get

$$x^3+y^3+z^3=p(p^2-3q)+3r=1-3q+3r$$

and

$$\begin{aligned}x^2y+x^2z+y^2x+y^2z+z^2x+z^2y &= xy(x+y)+yz(y+z)+zx(z+x)\\&= xy(1-z)+yz(1-x)+zx(1-y)\\&= xy+yz+zx-3xyz=q-3r.\end{aligned}$$

Now inequality (1) becomes

$$2(1-3q+3r)\geq q-3r \quad \Leftrightarrow \quad 2+9r\geq 7q,$$

which is true according to N_8, and we are done. ■

Solution 3 We can easily deduce that

$$4(x^3+y^3)-(x+y)^3=3(x+y)(x-y)^2\geq 0, \quad \text{i.e.} \quad 4(x^3+y^3)\geq (x+y)^3.$$

Analogously we get

$$4(y^3+z^3)\geq (y+z)^3 \quad \text{and} \quad 4(z^3+x^3)\geq (z+x)^3.$$

Adding these three inequalities we obtain the result. ■

Solution 4 According to *Jensen's inequality* for the convex function $f(x)=x^3$, we obtain

$$\begin{aligned}&\frac{1}{2}f(x)+\frac{1}{2}f(y)\geq f\left(\frac{x+y}{2}\right) \quad \text{or} \quad \frac{x^3+y^3}{2}\geq\left(\frac{x+y}{2}\right)^3\\&\Leftrightarrow \quad 4(x^3+y^3)\geq (x+y)^3.\end{aligned}$$

Now the solution follows as in the previous solution. ■

291 Let a,b,c be non-negative real numbers. Prove the inequality

$$a^3+b^3+c^3+abc\geq\frac{1}{7}(a+b+c)^3.$$

Solution We have

$$\begin{aligned}(a+b+c)^3 &= a^3+b^3+c^3+3(a^2(b+c)+b^2(c+a)+c^2(a+b))+6abc\\&= \frac{T[3,0,0]}{2}+3T[2,1,0]+T[1,1,1]\end{aligned}$$

and

$$a^3+b^3+c^3+abc=\frac{T[3,0,0]}{2}+\frac{T[1,1,1]}{6}.$$

So we need to prove that

$$7\left(\frac{T[3,0,0]}{2}+\frac{T[1,1,1]}{6}\right)\geq\frac{T[3,0,0]}{2}+3T[2,1,0]+T[1,1,1],$$

i.e.

$$3T[3,0,0]+\frac{T[1,1,1]}{6}\geq 3T[2,1,0],$$

which is true according to $T[3,0,0]\geq T[2,1,0]$ and $T[1,1,1]\geq 0$ (*Muirhead's theorem*). ■

292 Let a, b, c be positive real numbers such that $a+b+c=1$. Prove the inequality

$$a^2+b^2+c^2+3abc\geq\frac{4}{9}.$$

Solution We will normalize as follows

$$9(a+b+c)(a^2+b^2+c^2)+27abc\geq 4(a+b+c)^3$$

which is equivalent to

$$5(a^3+b^3+c^3)+3abc\geq 3(ab(a+b)+bc(b+c)+ca(c+a)). \qquad (1)$$

According to *Schur's inequality* we have that

$$a^3+b^3+c^3+3abc\geq ab(a+b)+bc(b+c)+ca(c+a) \qquad (2)$$

and by *Muirhead's theorem* we have that

$$2T[3,0,0]\geq 2T[2,1,0],$$

i.e.

$$4(a^3+b^3+c^3)\geq 2(ab(a+b)+bc(b+c)+ca(c+a)). \qquad (3)$$

Adding these two inequalities gives us inequality (1). ■

293 Let $a_1, a_2, \ldots, a_n$ be positive real numbers. Prove the inequality

$$(1+a_1)(1+a_2)\cdots(1+a_n)\leq\left(1+\frac{a_1^2}{a_2}\right)\left(1+\frac{a_2^2}{a_3}\right)\cdots\left(1+\frac{a_n^2}{a_1}\right).$$

Solution Let $x_i=\ln a_i$, then given inequality becomes

$$(1+e^{x_1})(1+e^{x_2})\cdots(1+e^{x_n})\leq(1+e^{2x_1-x_2})(1+e^{2x_2-x_3})\cdots(1+e^{2x_n-x_1}).$$

After taking logarithm on the both sides we obtain

$$\ln(1+e^{x_1})+\cdots+\ln(1+e^{x_n})\leq\ln(1+e^{2x_1-x_2})+\cdots+\ln(1+e^{2x_n-x_1}).$$

Let consider the sequences $a: 2x_1-x_2, 2x_2-x_3, \ldots, 2x_n-x_1$ and $b: x_1, x_2, \ldots, x_n$.

Since $f(x) = \ln(1 + e^x)$ is convex function on $\mathbb{R}$ by *Karamata's inequality* it suffices to prove that a (ordered in some way) majorizes the sequences b (ordered in some way), which can be done exactly as in Exercise 12.13, and therefore is left to the reader. ∎

294 Let a, b, c, d be positive real numbers such that $abcd = 1$. Prove the inequality

$$\frac{1}{(1+a)^2} + \frac{1}{(1+b)^2} + \frac{1}{(1+c)^2} + \frac{1}{(1+d)^2} \geq 1.$$

Solution 1 First we'll show that for all real numbers x and y the following inequality holds

$$\frac{1}{(1+x)^2} + \frac{1}{(1+y)^2} \geq \frac{1}{1+xy}.$$

We have

$$\begin{aligned} &\frac{1}{(1+x)^2} + \frac{1}{(1+y)^2} - \frac{1}{1+xy} \\ &\quad = \frac{xy(x^2+y^2) - x^2y^2 - 2xy + 1}{(1+x)^2(1+y)^2(1+xy)} = \frac{xy(x-y)^2 + (xy-1)^2}{(1+x)^2(1+y)^2(1+xy)} \geq 0. \end{aligned}$$

Now we obtain

$$\begin{aligned} &\frac{1}{(1+a)^2} + \frac{1}{(1+b)^2} + \frac{1}{(1+c)^2} + \frac{1}{(1+d)^2} \\ &\quad \geq \frac{1}{1+ab} + \frac{1}{1+cd} = \frac{1}{1+ab} + \frac{1}{1+1/ab} \\ &\quad = \frac{1}{1+ab} + \frac{ab}{1+ab} = 1. \end{aligned}$$

Equality holds iff $a = b = c = d = 1$. ∎

Solution 2 Let

$$f(a,b,c,d) = \frac{1}{(1+a)^2} + \frac{1}{(1+b)^2} + \frac{1}{(1+c)^2} + \frac{1}{(1+d)^2} \quad \text{and}$$
$$g(a,b,c,d) = abcd - 1.$$

Define

$$L = f - \lambda g = \frac{1}{(1+a)^2} + \frac{1}{(1+b)^2} + \frac{1}{(1+c)^2} + \frac{1}{(1+d)^2} - \lambda(abcd - 1).$$

For the first partial derivatives we have

$$\frac{\partial L}{\partial a}=\frac{-4}{(1+a)^2}-\frac{\lambda}{a}=0,\quad \text{i.e.}\quad \lambda=\frac{-4a}{(1+a)^2},$$
$$\frac{\partial L}{\partial b}=\frac{-4}{(1+b)^2}-\frac{\lambda}{b}=0,\quad \text{i.e.}\quad \lambda=\frac{-4b}{(1+b)^2},$$
$$\frac{\partial L}{\partial c}=\frac{-4}{(1+c)^2}-\frac{\lambda}{c}=0,\quad \text{i.e.}\quad \lambda=\frac{-4c}{(1+c)^2},$$
$$\frac{\partial L}{\partial d}=\frac{-4}{(1+d)^2}-\frac{\lambda}{d}=0,\quad \text{i.e.}\quad \lambda=\frac{-4d}{(1+d)^2}.$$

So we have $\frac{-4a}{(1+a)^2}=\frac{-4b}{(1+b)^2}=\frac{-4c}{(1+c)^2}=\frac{-4d}{(1+d)^2}=\lambda$, from which we get the following system of equations:

$$(a-b)(1-ab)=0,\qquad (a-c)(1-ac)=0,\qquad (a-d)(1-ad)=0,$$
$$(b-c)(1-bc)=0,\qquad (b-d)(1-bd)=0,\qquad (c-d)(1-cd)=0.$$

Solving this system we get that we must have $a=b=c=d$, and using $abcd=1$ it follows that $a=b=c=d=1$ and then we have

$$f(1,1,1,1)=\frac{1}{4}+\frac{1}{4}+\frac{1}{4}+\frac{1}{4}=1.$$

Since $f(1,1,1/2,2)=\frac{1}{4}+\frac{1}{4}+\frac{1}{9}+\frac{4}{9}=\frac{1}{2}+\frac{5}{9}>1$, by *Lagrange's theorem* we conclude that $f(a,b,c,d)\geq 1$, as required. ■

295 Let $a,b,c,d\geq 0$ be real numbers such that $a+b+c+d=4$. Prove the inequality

$$abc+bcd+cda+dab+(abc)^2+(bcd)^2+(cda)^2+(dab)^2\leq 8.$$

Solution Let us denote

$$f(a,b,c,d)=abc+bcd+cda+dab+(abc)^2+(bcd)^2+(cda)^2+(dab)^2.$$

Because of symmetry we may assume that $a\geq b\geq c\geq d$.

We have

$$\begin{aligned} &f\left(\frac{a+c}{2},b,\frac{a+c}{2},d\right)-f(a,b,c,d)\\ &\quad=\left(\frac{a-c}{2}\right)^2\left((b+d)+\left(\left(\frac{a+c}{2}\right)^2+ac\right)(b^2+d^2)-2b^2d^2\right)\\ &\quad\geq\left(\frac{a-c}{2}\right)^2(4abcd-2b^2d^2)\geq 0\quad (abcd\geq b^2d^2). \end{aligned}$$

So

$$f\left(\frac{a+c}{2}, b, \frac{a+c}{2}, d\right) \geq f(a,b,c,d).$$

According to the *SMV theorem* it suffices to show that

$$f(t,t,t,d) \leq 8$$

where $3t+d=4$ and clearly $0 \leq t \leq \frac{4}{3}$.

We have

$$f(t,t,t,d) \leq 8 \quad \Leftrightarrow \quad t^3 + 3t^2(3-3t) + 3t^4(4-3t)^2 + t^6 \leq 8$$
$$\Leftrightarrow \quad (t-1)^2(28t^4 - 16t^3 - 12t^2 - 8) \leq 0.$$

So it is enough to show that $28t^4 - 16t^3 - 12t^2 - 8 \leq 0$, which is easy to prove for $0 \leq t \leq \frac{4}{3}$.

Equality holds iff $a=b=c=d=1$. ∎

296 Let $a,b,c,d \geq 0$ such that $a+b+c+d=1$. Prove the inequality

$$a^4 + b^4 + c^4 + d^4 + \frac{148}{27}abcd \geq \frac{1}{27}.$$

Solution Denote $f(a,b,c,d) = a^4 + b^4 + c^4 + d^4 + \frac{148}{27}abcd - \frac{1}{27}$.

Since the given inequality is symmetric we may assume that $a \geq b \geq c \geq d$.

We have

$$f(a,b,c,d) - f\left(\frac{a+c}{2}, b, \frac{a+c}{2}, d\right) = \left(\frac{7}{8}(a-c)^2 + 3ac - \frac{37}{27}bd\right)(a-b)^2.$$

Since $ac \geq bd$ it follows that

$$f(a,b,c,d) - f\left(\frac{a+c}{2}, b, \frac{a+c}{2}, d\right) \geq 0,$$

i.e.

$$f(a,b,c,d) \geq f\left(\frac{a+c}{2}, b, \frac{a+c}{2}, d\right).$$

According to the *SMV theorem* it suffices to show that

$$f(t,t,t,d) \geq 0, \quad \text{where } t = \frac{1-d}{3}.$$

We have

$$f(t,t,t,d) = \frac{(1-d)^4}{27} + d^4 + \frac{148d(1-d)^3}{729} - \frac{1}{27} = \frac{2d(4d-1)^2(19d+20)}{729} \geq 0.$$

Equality occurs if and only if $a=b=c=d=1/4$ or $a=b=c=1/3, d=0$ (up to permutation). ■

297 Let a, b, c be positive real numbers such that $a^2+b^2+c^2=3$. Prove the inequality

$$a^2b^2+b^2c^2+c^2a^2 \le a+b+c.$$

Solution Without loss of generality we may assume that $a \le b \le c$. Then clearly $a \le 1$ and $b^2+c^2 \ge 2$, from which it follows that $b+c \ge \sqrt{2}$.

Let $f(a,b,c)=a+b+c-a^2b^2-b^2c^2-c^2a^2$. Then we have

$$f(a,b,c)-f\left(a,\sqrt{\frac{b^2+c^2}{2}},\sqrt{\frac{b^2+c^2}{2}}\right)$$

$$=(b-c)^2\left(\frac{(b+c)^2}{4}-\frac{1}{b+c+\sqrt{2(b^2+c^2)}}\right) \ge \left(\frac{2}{4}-\frac{1}{2+\sqrt{2}}\right)(b-c)^2 \ge 0.$$

Thus

$$f(a,b,c) \ge f\left(a,\sqrt{\frac{b^2+c^2}{2}},\sqrt{\frac{b^2+c^2}{2}}\right).$$

By the *SMV theorem* it suffices to prove that $f(a,t,t) \ge 0$, when $a^2+2t^2=3$.

We have

$$f(a,t,t) \ge 0$$

$$\Leftrightarrow \quad a+\sqrt{2(3-a^2)} \ge a^2(3-a^2)+\frac{1}{4}(3-a^2)^2$$

$$\Leftrightarrow \quad (a-1)^2\left(\frac{3}{4}(a+1)^2-\frac{3}{3-a+\sqrt{2(3-a^2)}}\right) \ge 0. \qquad (1)$$

Since $a \le 1$ it follows that

$$\frac{3}{3-a+\sqrt{2(3-a^2)}} \le \frac{3}{4} \le \frac{3}{4}(a+1)^2.$$

Therefore inequality (1) is true, and we are done.

Equality occurs iff $a=b=c=1$. ■

298 Let $a, b, c, d \ge 0$ be real numbers such that $a+b+c+d=4$. Prove the inequality

$$(1+a^2)(1+b^2)(1+c^2)(1+d^2) \ge (1+a)(1+b)(1+c)(1+d).$$

Solution Let

$$f(a,b,c,d)=(1+a^2)(1+b^2)(1+c^2)(1+d^2)-(1+a)(1+b)(1+c)(1+d),$$

and assume that $a\le b\le c\le d$ (symmetry).

We'll show that

$$f(a,b,c,d)\ge f\left(\frac{a+c}{2},b,\frac{a+c}{2},d\right).$$

Clearly

$$a+c\le 2, \tag{1}$$

so it follows that

$$\begin{aligned}
&f(a,b,c,d)-f\left(\frac{a+c}{2},b,\frac{a+c}{2},d\right)\\
&\quad=(1+b^2)(1+d^2)\left((1+a^2)(1+c^2)-\left(1+\left(\frac{a+c}{2}\right)^2\right)^2\right)\\
&\qquad+(1+b)(1+d)\left(\left(1+\frac{a+c}{2}\right)^2-(1+a)(1+c)\right).
\end{aligned}$$

Since

$$(1+a^2)(1+c^2)-\left(1+\left(\frac{a+c}{2}\right)^2\right)^2=(a-c)^2\left(\frac{1}{2}-\frac{(a+c)^2+4ac}{16}\right)\ge 0$$

(this inequality follows by (1)) and by $AM\ge GM$ it follows that

$$(1+a)(1+c)\le\left(1+\frac{a+c}{2}\right)^2.$$

So

$$f(a,b,c,d)-f\left(\frac{a+c}{2},b,\frac{a+c}{2},d\right)\ge 0,\quad \text{i.e.}$$

$$f(a,b,c,d)\ge f\left(\frac{a+c}{2},b,\frac{a+c}{2},d\right).$$

According to the *SMV theorem* it suffices to show that

$$f(t,t,t,d)\ge 0$$

where $3t+d=4$ i.e. $d=4-3t$.

We have

$$\begin{aligned}
f(t,t,t,d) &= (1+t^2)^3(1+(4-3t)^2) - (1+t)^3(5-3t) \\
&= 9t^8 - 24t^7 + 44t^6 - 72t^5 + 81t^4 - 68t^3 - 54t^2 - 36t + 12 \\
&= (t-1)^2(9t^6 - 6t^5 + 23t^4 - 20t^3 + 18t^2 - 12t + 12) \\
&= (t-1)^2(t^4(3t-1)^2 + 2t^4 + 5t^2(2t-1)^2 + 10t^2 + 3(t-2)^2) \geq 0.
\end{aligned}$$

Equality holds if and only if $a = b = c = d = 1$. ■

299 Let a, b, c be positive real numbers such that $abc = 1$. Prove the inequality

$$\frac{1}{a} + \frac{1}{b} + \frac{1}{c} + \frac{6}{a+b+c} \geq 5.$$

Solution Without loss of generality we may assume that $a \geq b \geq c$.

Let $f(a,b,c) = \frac{1}{a} + \frac{1}{b} + \frac{1}{c} + \frac{6}{a+b+c}$.

We'll prove that

$$f(a,b,c) \geq f(a, \sqrt{bc}, \sqrt{bc}).$$

We have

$$\begin{aligned}
& f(a,b,c) \geq f(a, \sqrt{bc}, \sqrt{bc}) \\
\Leftrightarrow \quad & \frac{1}{b} + \frac{1}{c} + \frac{6}{a+b+c} \geq \frac{2}{\sqrt{bc}} + \frac{6}{a+2\sqrt{bc}} \\
\Leftrightarrow \quad & c(a+b+c)(a+2\sqrt{bc}) + b(a+b+c)(a+2\sqrt{bc}) + 6bc(a+2\sqrt{bc}) \\
& \geq 2\sqrt{bc}(a+b+c)(a+2\sqrt{bc}) + 6bc(a+b+c) \\
\Leftrightarrow \quad & (\sqrt{b} - \sqrt{c})^2((a+b+c)(a+2\sqrt{bc}) - 6bc) \geq 0. \qquad (1)
\end{aligned}$$

Since $a \geq b \geq c$ we have $a \geq \frac{b+c}{2} \geq \sqrt{bc}$.

Thus

$$(a+b+c)(a+2\sqrt{bc}) \geq (\sqrt{bc} + 2\sqrt{bc})(\sqrt{bc} + 2\sqrt{bc}) = 9bc \geq 6bc.$$

So due to (1) and the last inequality we have

$$f(a,b,c) \geq f(a, \sqrt{bc}, \sqrt{bc}).$$

According to the *SMV theorem* we need to prove that $f(a,t,t) \geq 5$, with $at^2 = 1$.

We have

$$f(a,t,t) \geq 5 \quad \Leftrightarrow \quad \frac{1}{a} + \frac{2}{t} + \frac{6}{a+2t} \geq 5$$

which is equivalent to

$$(t-1)^2(2t^4+4t^3-4t^2-t+2)\ge 0,$$

which is true since $2t^4+4t^3-4t^2-t+2>0$ for $t>0$. ■

300 Let a,b,c be positive real numbers such that $a+b+c=3$. Prove the inequality

$$12\left(\frac{1}{a}+\frac{1}{b}+\frac{1}{c}\right)\ge 4(a^3+b^3+c^3)+21.$$

Solution Without loss of generality we may assume that $a\le b\le c$.

Let

$$f(a,b,c)=12\left(\frac{1}{a}+\frac{1}{b}+\frac{1}{c}\right)-4(a^3+b^3+c^3).$$

Then we have

$$\begin{aligned}
f(a,b,c)&-f\left(\frac{a+b}{2},\frac{a+b}{2},c\right)\\
&=12\left(\frac{1}{a}+\frac{1}{b}+\frac{1}{c}\right)-4(a^3+b^3+c^3)-12\left(\frac{4}{a+b}+\frac{1}{c}\right)+(a+b)^3+4c^3\\
&=12\left(\frac{1}{a}+\frac{1}{b}-\frac{4}{a+b}\right)+(a+b)^3-4(a^3+b^3)\\
&=3(a-b)^2\left(\frac{4}{ab(a+b)}-(a+b)\right). \qquad (1)
\end{aligned}$$

Since $a\le b\le c$ we must have $a+b\le 2$, and clearly $c\ge 1$.

By the $AM\ge GM$ we have

$$ab(a+b)^2\le\frac{(a+b)^4}{4}\le 4,\quad \text{i.e.}\quad \frac{4}{ab(a+b)}-(a+b)\ge 0.$$

Hence by (1) we deduce that

$$f(a,b,c)-f\left(\frac{a+b}{2},\frac{a+b}{2},c\right)\ge 0,\quad \text{i.e.}\quad f(a,b,c)\ge f\left(\frac{a+b}{2},\frac{a+b}{2},c\right).$$

So according to the *SMV theorem* it suffices to prove that $f(t,t,c)\ge 21$, when $2t+c=3$, $c\ge t$.

We have

$$\begin{aligned}
&f(t,t,c)\ge 21\\
\Leftrightarrow\quad &12\left(\frac{2}{t}+\frac{1}{c}\right)+(2t)^3-4c^3\ge 21
\end{aligned}$$

$$\Leftrightarrow \quad 12\left(\frac{4}{2t}+\frac{1}{c}\right)+(2t)^3-4c^3\geq 21$$

$$\Leftrightarrow \quad 12\left(\frac{4}{3-c}+\frac{1}{c}\right)+(3-c)^3-4c^3\geq 21$$

$$\Leftrightarrow \quad c^5-18c^3+48c^2-36c+12\geq 0$$

$$\Leftrightarrow \quad (c-2)^2(c-1)(c^2+3c-3)\geq 0$$

which is true since $c\geq 1$.

Equality occurs iff $(a,b,c)=(2,1/2,1/2)$. ■

301 Let a,b,c,d be non-negative real numbers such that $a+b+c+d+e=5$. Prove the inequality

$$4(a^2+b^2+c^2+d^2+e^2)+5abcd\geq 25.$$

Solution Without loss of generality we may assume that $a\geq b\geq c\geq d\geq e$.

Let us denote

$$f(a,b,c,d,e)=4(a^2+b^2+c^2+d^2+e^2)+5abcd.$$

Then we easily deduce that

$$f(a,b,c,d,e)-f\left(\frac{a+d}{2},b,c,\frac{a+d}{2},e\right)=\frac{(a-d)^2}{4}(8-5bce). \tag{1}$$

Since $a\geq b\geq c\geq d\geq e$, we have

$$3\sqrt[3]{bce}\leq b+c+e\leq\frac{3(a+b+c+d+e)}{5}=3.$$

Thus it follows that $bce\leq 1$.

Now, by (1) and the last inequality we get

$$f(a,b,c,d,e)-f\left(\frac{a+d}{2},b,c,\frac{a+d}{2},e\right)=\frac{(a-d)^2}{4}(8-5bce)$$
$$\geq\frac{(a-d)^2}{4}(8-5)\geq 0,$$

i.e.

$$f(a,b,c,d,e)\geq f\left(\frac{a+d}{2},b,c,\frac{a+d}{2},e\right).$$

According to the *SMV theorem* it remains to prove that $f(t,t,t,t,e)\geq 25$, under the condition $4t+e=5$.

Clearly $4t\leq 5$.

We have

$$\begin{aligned} & f(t,t,t,t,e) \geq 25 \\ \Leftrightarrow \quad & 4(t^2+e^2)+5t^4e \geq 25 \\ \Leftrightarrow \quad & 4t^2+4(5-4t)^2+5t^4(5-4t)-25 \geq 0 \\ \Leftrightarrow \quad & (5-4t)(t-1)^2(t^2+2t+3) \geq 0, \end{aligned}$$

which is true.

Equality occurs if and only if $a=b=c=d=e=1$ or $a=b=c=d=5/4$, $e=0$ (up to permutation). ■

302 Let a,b,c be positive real numbers such that $a+b+c=3$. Prove the inequality

$$\frac{1}{2+a^2+b^2}+\frac{1}{2+b^2+c^2}+\frac{1}{2+c^2+a^2} \leq \frac{3}{4}.$$

Solution Without loss of generality we may assume that $a \geq b \geq c$.

Let $f(a,b,c)=\frac{1}{2+a^2+b^2}+\frac{1}{2+b^2+c^2}+\frac{1}{2+c^2+a^2}$.

We have

$$\begin{aligned} & f\left(a,\frac{b+c}{2},\frac{b+c}{2}\right)-f(a,b,c) \\ & = \left(b^2+c^2-\frac{(b+c)^2}{2}\right) \\ & \quad \times \left(\frac{1}{(b^2+c^2+2)(2+\frac{(b+c)^2}{2})}-\frac{1}{(4+2a^2+b^2+c^2)(4+2a^2+\frac{(b+c)^2}{2})}\right). \end{aligned}$$

Since

$$b^2+c^2 \geq \frac{(b+c)^2}{2}, \qquad 4+2a^2+b^2+c^2 \geq b^2+c^2+2 \quad \text{and}$$

$$4+2a^2+\frac{(b+c)^2}{2} \geq 2+\frac{(b+c)^2}{2}$$

we have

$$f\left(a,\frac{b+c}{2},\frac{b+c}{2}\right)-f(a,b,c) \geq 0, \quad \text{i.e.}$$

$$f\left(a,\frac{b+c}{2},\frac{b+c}{2}\right) \geq f(a,b,c).$$

According to *SMV theorem* it suffices to prove that $f(a,t,t) \leq \frac{3}{4}$, when $a+2t=3$.

We have

$$\begin{aligned}
f(a,t,t) &\le \frac{3}{4} \\
\Leftrightarrow\quad & \frac{2}{2+a^2+t^2}+\frac{1}{2+2t^2}\le\frac{3}{4} \\
\Leftrightarrow\quad & \frac{8}{8+4a^2+(2t)^2}+\frac{2}{4+(2t)^2}\le\frac{3}{4} \\
\Leftrightarrow\quad & \frac{8}{8+4a^2+(3-a)^2}+\frac{2}{4+(3-a)^2}\le\frac{3}{4},
\end{aligned}$$

which can be easily transformed to $(a-1)^2(15a^2-78a+111)\ge 0$, and clearly holds.

Equality holds iff $a=b=c=1$. ■

303 Let a,b,c be positive real numbers such that $a^2+b^2+c^2=3$. Prove the inequality

$$ab+bc+ca\le abc+2.$$

Solution Without loss of generality we may assume that $a\ge b\ge c$.

Let $f(a,b,c)=ab+bc+ca-abc$.

We have

$$\begin{aligned}
& f(a,b,c)-f\left(\sqrt{\frac{a^2+b^2}{2}},\sqrt{\frac{a^2+b^2}{2}},c\right) \\
&= ab+bc+ca-abc-\frac{a^2+b^2}{2}-2c\sqrt{\frac{a^2+b^2}{2}}+c\frac{a^2+b^2}{2} \\
&= \left(ab-\frac{a^2+b^2}{2}\right)+c\left((a+b)-\sqrt{2(a^2+b^2)}\right)-c\left(ab-\frac{a^2+b^2}{2}\right) \\
&= \frac{-(a-b)^2}{2}-\frac{c(a-b)^2}{(a+b)+\sqrt{2(a^2+b^2)}}+\frac{c(a-b)^2}{2} \\
&= (a-b)^2\left(\frac{c}{2}-\frac{1}{2}-\frac{c}{(a+b)+\sqrt{2(a^2+b^2)}}\right). \qquad (1)
\end{aligned}$$

Notice that since $a\ge b\ge c$ we must have $c^2\le 1$, i.e. $c\le 1$ and $a^2+b^2\ge 2$.

By $AM\le QM$ we have

$$\begin{aligned}
\frac{c}{2}-\frac{1}{2}-\frac{c}{(a+b)+\sqrt{2(a^2+b^2)}} &\le \frac{c}{2}-\frac{1}{2}-\frac{c}{2\sqrt{2\cdot(a^2+b^2)}} \\
&\le \frac{c}{2}-\frac{1}{2}-\frac{c}{2\sqrt{2\cdot(a^2+b^2+c^2)}}
\end{aligned}$$

$$= \frac{c}{2} - \frac{1}{2} - \frac{c}{2\sqrt{6}} \leq \frac{1}{2} - \frac{1}{2} - \frac{c}{2\sqrt{6}}$$
$$\leq -\frac{c}{2\sqrt{6}} \leq 0.$$

Hence by (1) we get that

$$f(a,b,c) - f\left(\sqrt{\frac{a^2+b^2}{2}}, \sqrt{\frac{a^2+b^2}{2}}, c\right) \leq 0,$$

i.e.

$$f(a,b,c) \leq f\left(\sqrt{\frac{a^2+b^2}{2}}, \sqrt{\frac{a^2+b^2}{2}}, c\right).$$

According to the *SMV theorem* we need to prove that $f(t,t,c) \leq 2$, when $2t^2 + c^2 = 3$.

We have

$$f(t,t,c) \leq 2 \quad \Leftrightarrow \quad t^2 + 2ct - t^2c \leq 2$$
$$\Leftrightarrow \quad 2t^2 + 4ct \leq 2t^2c + 4 \quad \Leftrightarrow \quad 4ct \leq 2t^2c + 3 - 2t^2 + 1$$
$$\Leftrightarrow \quad 4ct \leq 2t^2c + c^2 + 1,$$

which is true due to $AM \geq GM$, i.e.

$$2t^2c + c^2 + 1 = t^2c + c^2 + 1 + t^2c \geq 4\sqrt[4]{t^2c \cdot c^2 \cdot t^2c} = 4ct.$$ ■

304 Let a, b, c be positive real numbers. Prove the inequality

$$\frac{a}{b} + \frac{b}{c} + \frac{c}{a} \geq \frac{a+b}{b+c} + \frac{b+c}{c+a} + \frac{a+c}{a+b}.$$

Solution Without loss of generality we may assume that $c = \min\{a,b,c\}$.

Notice that for $x, y, z > 0$ we have

$$\frac{x}{y} + \frac{y}{z} + \frac{z}{x} - 3 = \frac{1}{xy}(x-y)^2 + \frac{1}{xz}(x-z)(y-z).$$

Now we obtain

$$\frac{a}{b} + \frac{b}{c} + \frac{c}{a} - 3 \geq \frac{c+a}{c+b} + \frac{b+c}{b+a} + \frac{a+c}{a+b} - 3$$
$$\Leftrightarrow \quad \frac{1}{ab}(a-b)^2 + \frac{1}{ac}(a-c)(b-c)$$
$$\geq \frac{1}{(a+c)(b+c)}(a-b)^2 + \frac{1}{(a+c)(a+b)}(b-c)(a-c)$$

$$\Leftrightarrow \quad \left(\frac{1}{ab} - \frac{1}{(a+c)(b+c)}\right)(a-b)^2 + \left(\frac{1}{ac} - \frac{1}{(a+c)(a+b)}\right)$$
$$\times (a-c)(b-c) \geq 0.$$

The last inequality is true, since:

$$c = \min\{a, b, c\}, \qquad \frac{1}{ac} - \frac{1}{(a+c)(a+b)} > 0 \quad \text{and} \quad \frac{1}{ab} - \frac{1}{(a+c)(b+c)} > 0.$$

■

305 Let a, b, c be positive real numbers. Prove the inequality

$$\frac{a^2}{b^2+c^2} + \frac{b^2}{c^2+a^2} + \frac{c^2}{a^2+b^2} \geq \frac{a}{b+c} + \frac{b}{c+a} + \frac{c}{a+b}.$$

Solution We have

$$\frac{a^2}{b^2+c^2} - \frac{a}{b+c} = \frac{ab(a-b)+ac(a-c)}{(b^2+c^2)(b+c)},$$
$$\frac{b^2}{c^2+a^2} - \frac{b}{c+a} = \frac{bc(b-c)+ab(b-a)}{(c^2+a^2)(c+a)} \quad \text{and}$$
$$\frac{c^2}{a^2+b^2} - \frac{c}{a+b} = \frac{ac(c-a)+bc(c-b)}{(b^2+a^2)(b+a)}.$$

Now we obtain

$$\frac{a^2}{b^2+c^2} + \frac{b^2}{c^2+a^2} + \frac{c^2}{a^2+b^2} - \left(\frac{a}{b+c} + \frac{b}{c+a} + \frac{c}{a+b}\right)$$
$$= \frac{ab(a-b)+ac(a-c)}{(b^2+c^2)(b+c)} + \frac{bc(b-c)+ab(b-a)}{(c^2+a^2)(c+a)} + \frac{ac(c-a)+bc(c-b)}{(b^2+a^2)(b+a)}$$
$$= (a^2+b^2+c^2+ab+bc+ca) \cdot \sum \frac{ab(a-b)^2}{(b+c)(c+a)(b^2+c^2)(c^2+a^2)} \geq 0.$$

■

306 Let a, b, c be positive real numbers such that $a \geq b \geq c$. Prove the inequality

$$a^2b(a-b) + b^2c(b-c) + c^2a(c-a) \geq 0.$$

Solution We have

$$a^2b(a-b) + b^2c(b-c) + c^2a(c-a)$$
$$= a^2b(a-b) + b^2c(b-c) + c^2a(c-a) - ab^2(a-b) - ab^2(b-c)$$
$$- ab^2(c-a)$$

$$= (a^2b(a-b) - ab^2(a-b)) + (b^2c(b-c) - ab^2(b-c))$$
$$+ (c^2a(c-a) - ab^2(c-a))$$
$$= ab(a-b)^2 + (ab+ac-b^2)(a-c)(b-c).$$

So we need to show that

$$ab(a-b)^2 + (ab+ac-b^2)(a-c)(b-c) \geq 0,$$

which clearly holds since $a \geq b \geq c$. ■

307 Let a, b, c be the lengths of the sides of a triangle. Prove the inequality

$$\frac{(b+c)^2}{a^2+bc} + \frac{(c+a)^2}{b^2+ca} + \frac{(a+b)^2}{c^2+ab} \geq 6.$$

Solution We have

$$\frac{(b+c)^2}{a^2+bc} - 2 + \frac{(c+a)^2}{b^2+ca} - 2 + \frac{(a+b)^2}{c^2+ab} - 2 \geq 0$$
$$\Leftrightarrow \quad \frac{b^2+c^2-2a^2}{a^2+bc} + \frac{c^2+a^2-2b^2}{b^2+ca} + \frac{a^2+b^2-2c^2}{c^2+ab} \geq 0$$
$$\Leftrightarrow \quad \left(\frac{b^2-a^2}{a^2+bc} + \frac{a^2-b^2}{b^2+ca}\right) + \left(\frac{c^2-a^2}{a^2+bc} + \frac{a^2-c^2}{c^2+ab}\right)$$
$$+ \left(\frac{c^2-b^2}{b^2+ca} + \frac{b^2-c^2}{c^2+ab}\right) \geq 0$$
$$\Leftrightarrow \quad \frac{(b-a)^2(a+b)(a+b-c)}{(a^2+bc)(b^2+ca)} + \frac{(c-a)^2(c+a)(c+a-b)}{(a^2+bc)(c^2+ab)}$$
$$+ \frac{(b-c)^2(b+c)(b+c-a)}{(b^2+ca)(c^2+ab)} \geq 0,$$

which is clearly true. ■

308 Let a, b, c be positive real numbers. Prove the inequality

$$\frac{a+b}{b+c} + \frac{b+c}{c+a} + \frac{c+a}{a+b} + 3\frac{ab+bc+ca}{(a+b+c)^2} \geq 4.$$

Solution Without loss of generality we may assume that $c = \min\{a, b, c\}$.

Now we have

$$\frac{a+b}{b+c} + \frac{b+c}{c+a} + \frac{c+a}{a+b} - 3 = \frac{1}{(a+c)(b+c)}(a-b)^2$$
$$+ \frac{1}{(a+b)(b+c)}(a-c)(b-c)$$

and

$$3\frac{ab+bc+ca}{(a+b+c)^2}-1=-\frac{1}{(a+b+c)^2}(a-b)^2-\frac{1}{(a+b+c)^2}(a-c)(b-c).$$

The given inequality becomes

$$M(a-b)^2+N(a-c)(b-c)\geq 0, \qquad (1)$$

where $M=\frac{1}{(a+c)(b+c)}-\frac{1}{(a+b+c)^2}$ and $N=\frac{1}{(a+b)(b+c)}-\frac{1}{(a+b+c)^2}$.

We can easily prove that $M,N\geq 0$, and since $c=\min\{a,b,c\}$ we get inequality (1). ■

309 Let a,b,c be real numbers. Prove the inequality

$$3(a^2-ab+b^2)(b^2-bc+c^2)(c^2-ca+a^2)\geq a^3b^3+b^3c^3+c^3a^3.$$

Solution It is enough to consider the case when $a,b,c\geq 0$.

We have

$$(a^2-ab+b^2)(b^2-bc+c^2)(c^2-ca+a^2)=\sum_{\text{sym}}a^4b^2-\sum_{\text{cyc}}a^3b^3-\sum_{\text{cyc}}a^4bc+a^2b^2c^2.$$

The given inequality is equivalent to

$$3\sum_{\text{sym}}a^4b^2-4\sum_{\text{cyc}}a^3b^3-3\sum_{\text{cyc}}a^4bc+3a^2b^2c^2\geq 0,$$

which is equivalent to

$$\sum_{\text{cyc}}(2c^4+3a^2b^2-abc(a+b+c))(a-b)^2\geq 0. \qquad (1)$$

Assume $a\geq b\geq c$ and denote

$$S_a=2a^4+3b^2c^2-abc(a+b+c),$$

$$S_b=2b^4+3a^2c^2-abc(a+b+c)$$

and

$$S_c=2c^4+3a^2b^2-abc(a+b+c).$$

We have

$$S_a=2a^4+3b^2c^2-abc(a+b+c)\geq a^4+2a^2bc-abc(a+b+c)\geq 0,$$

$$S_c=2c^4+3a^2b^2-abc(a+b+c)\geq 3a^2b^2-abc(a+b+c)\geq 0,$$

$$\begin{aligned}S_a + 2S_b &= 2a^4 + 3b^2c^2 + 4b^4 + 6a^2c^2 - 3abc(a+b+c)\\ &\geq a^4 + 2a^2bc + 8b^2ca - 3abc(a+b+c) \geq 0\end{aligned}$$

and

$$\begin{aligned}S_c + 2S_b &= 2c^4 + 3a^2b^2 + 4b^4 + 6a^2c^2 - 3abc(a+b+c)\\ &\geq (3a^2b^2 + 3a^2c^2) + 3a^2c^2 - 3abc(a+b+c) \geq 0.\end{aligned}$$

(Since the given inequality is cyclic if we assume that $a \leq b \leq c$ similarly we can show that $S_a, S_c, S_a + 2S_b, S_c + 2S_b \geq 0$.)

According to the *SOS theorem* we obtain that inequality (1) holds, as required.

Equality holds iff $a = b = c$. ■

310 Let $a, b, c, d \in \mathbb{R}^+$ such that $a + b + c + d + abcd = 5$. Prove the inequality

$$\frac{1}{a} + \frac{1}{b} + \frac{1}{c} + \frac{1}{d} \geq 4.$$

Solution We'll use *Lagrange's theorem*.

Let

$$f(a,b,c,d) = \frac{1}{a} + \frac{1}{b} + \frac{1}{c} + \frac{1}{d} \quad \text{and} \quad g(a,b,c,d) = a+b+c+d+abcd-5 = 0.$$

We define

$$L = f - \lambda g = \frac{1}{a} + \frac{1}{b} + \frac{1}{c} + \frac{1}{d} - \lambda(a+b+c+d+abcd-5).$$

For the first partial derivatives we get

$$\begin{aligned}\frac{\partial L}{\partial a} &= -\frac{1}{a^2} - \lambda(1+bcd) = 0, & \frac{\partial L}{\partial b} &= -\frac{1}{b^2} - \lambda(1+acd) = 0,\\ \frac{\partial L}{\partial c} &= -\frac{1}{c^2} - \lambda(1+abd) = 0, & \frac{\partial L}{\partial d} &= -\frac{1}{d^2} - \lambda(1+abc) = 0.\end{aligned}$$

So

$$\lambda = -\frac{1}{a^2(1+bcd)} = -\frac{1}{b^2(1+acd)} = -\frac{1}{c^2(1+abd)} = -\frac{1}{d^2(1+abc)}.$$

From the first two equations we deduce

$$a^2(1+bcd) = b^2(1+acd), \quad \text{i.e.} \quad (a-b)(a+b+abcd) = 0.$$

Since $a + b + abcd > 0$ we must have $a = b$.

Analogously we deduce that $a = c = d$, i.e. $a = b = c = d$.
Using $a + b + c + d + abcd = 5$ we get

$$a^4 + 4a - 5 = 0 \quad \Leftrightarrow \quad (a - 1)(a^3 + a^2 + a + 5) = 0,$$

and it follows that we must have $a = 1$.
So $a = b = c = d = 1$.
Finally we have $f(1, 1, 1, 1) = 1 + 1 + 1 + 1 = 4$, and we are done. ■

Index of Problems

Chapter 1

1.8 BMO 2001
1.17 Russia MO 2002

Chapter 2

2.4 Viorel Vâjâitu, Alexandra Zaharescu, Gazeta Matematică
2.15 Ireland MO 2000

Chapter 3

3.6 IMO, shortlist 1969 (Romania)
3.8 India MO 2003

Chapter 4

4.1 France MO 1996
4.3 South Africa MO 1995
4.7 Crux Mathematicorum
4.8 Sefket Arslanagic
4.10 Art of problem solving
4.11 Art of problem solving
4.13 Andrei Ciupan, Romania 2007
4.15 Crux Mathematicorum
4.20 Pham Kim Hung
Corollary 4.5: Walther Janous

Chapter 5

5.3 IMO, shortlist 1974 (Finland)
5.10 Zdravko Cvetkovski
5.13 Zdravko Cvetkovski
5.14 Zdravko Cvetkovski
5.15 Zdravko Cvetkovski

Z. Cvetkovski, *Inequalities*,
DOI 10.1007/978-3-642-23792-8,

Chapter 6

Chapter 8

Chapter 9

Chapter 10

Chapter 12

Chapter 14

Chapter 15

Chapter 17

Chapter 18

18.3 Pham Kim Hung
18.5 Nguyen Minh Duc, IMO shortlist 1993

Chapter 20

6 Titu Andreescu, TST 2001 USA
9 Russia 2002
12 Czech and Slovak Republics 2005
14 Walther Janous, Crux Mathematicorum
16 Vasile Cîrtoaje, Gazeta Matematică
20 Titu Andreescu, Gabriel Dospinescu
22 Art of problem solving
23 Vasile Cîrtoaje
28 Art of problem solving
33 Art of problem solving
36 Belarus 1996
37 Art of problem solving
40 Zdravko Cvetkovski
41 Zdravko Cvetkovski
42 Zdravko Cvetkovski
45 Mircea Lascu, Gazeta Matematică
48 Art of problem solving
49 IMO 2000, Titu Andreescu
50 Bulgaria, 1997
56 Baltic Way, 2005
57 Gabriel Dospinescu, Marian Tetiva
58 Adrian Zahariuc
68 Vasile Cîrtoaje
71 MOSP 2001
72 Vasile Cîrtoaje, Mircea Lascu, Junior TST 2003 Romania
73 Marian Tetiva
75 Latvia 2002
81 Peru 2007
82 Romania 2003
84 Art of problem solving
86 Japan 2005
88 Art of problem solving
90 Canada 2008
91 Mathlinks contest
97 Kiran Kedlaya
98 Zdravko Cvetkovski, shortlist JBMO 2010
99 Vasile Cîrtoaje, Gazeta Matematică
105 JBMO 2002, shortlist
110 Pham Kim Hung
113 Pham Kim Hung

Abbreviations

APMO Asian-Pacific Mathematical Olympiad
BMO Balkan Mathematical Olympiad
IMO International Mathematical Olympiad
JBMO Junior Balkan Mathematical Olympiad
MYM Mathematics and Youth magazine, Vietnam
MO Mathematical Olympiad
MOSP Mathematical Olympiad Summer Program
USAMO United States of America Mathematical Olympiad

Z. Cvetkovski, *Inequalities*,
DOI 10.1007/978-3-642-23792-8, © Springer-Verlag Berlin Heidelberg 2012

References

1. Агаханов, Н.Х., и др.: Всероссийские Олимпиады Школьников по Математике 1993–2006. МЦНМО, Москва (2007) (Agakhanov, N.H., et al.: All-Russian Olympic in Mathematics 1993–2006, MCNMO, Moscow, 2007)
2. Алфутова, Б.Н., Устинов, А.В.: Алгебра и Теория Чисел для Математических Школ. МЦНМО, Москва (2003) (Alfutova, B.N., Ustinov, A.V.: Algebra and Number Theory for Mathematical Schools, MCNMO, Moscow, 2003)
3. Andreescu, T., Enescu, B.: Mathematical Olympiad Treasures. Birkhäuser, Basel (2003)
4. Andreescu, T., Cirtoaje, V., Dospinescu, G., Lascu, M.: Old and New Inequalities. Gil Publishing House, Zalau (2004)
5. Andreescu, T., Feng, Z., Lee, G.: Mathematical Olympiads 2000–2001; Problems and Solutions from Around the World. Math. Assoc. of America, Washington (2001)
6. Arslanagič, Š.: Matematika za Nadarene. Bosanska Riječ, Sarajevo (2005) (Mathematics for Gifted Students, Bosanska Riječ, Sarajevo, 2005)
7. Arslanagič, Š.: Metodička Zbirka Zadataka sa Osnovama Teorije iz Elementarne Matematike. Grafičar promet, Sarajevo (2006) (Methodical Collection of Problems with the Basic Theory of Elementary Mathematics, Grafičar promet, Sarajevo, 2006)
8. Art of Problem solving: http://www.artofproblrmsolving.com
9. Bottema, O., et al.: Geometric Inequalities. Wolters-Noordhoff Publishing, Groningen (1969)
10. Димовски, Д., и др.: Практикум по Елементарна Математика. Просветно дело, Скопје (1993). (Dimovski, D., et al.: Practicum for Elementary Mathematics. Prosvetno delo, Skopje (1993))
11. Djukić, D., Janković, V., Matić, I., Petrović, N.: The IMO Compendium, a Collection of Problems Suggested for the International Mathematical Olympiad: 1959–2004, 1st edn. Springer, Berlin (2006)
12. Engel, A.: Problem-Solving Strategies. Springer, New York (1997)
13. Hardy, G.H., Littlewood, J.E., Pólya, G.: Inequalities. Cambridge University Press, Cambridge (1967)
14. Ижболдин, О., Курляндчик, Л.: Неравенство Иенцена. Квант **4**, 7–10 (2000). (Izboldin, O., Kurlandcik, L.: Jensen's inequality. Quant **4**, 7–10 (2000))
15. Kedlaya, K.S.: $A < B$, http://www.unl.edu.com
16. Kim Hung, P.: The stronger mixing variables method. Math. Reflect. **6**, 1–8 (2006)
17. Kim Hung, P.: Secrets in Inequalities, vol. 1. Gil Publishing House, Zalau (2007)
18. Larson, C.L.: Problem Solving Through Problems. Springer, New York (1983)
19. Lee, H.: Topics in Inequalities—Theorems and Techniques. The IMO Compendium Group (2007)
20. Li, Y.K.: Math Problem Book I. Hong Kong Mathematical Society (2001)
21. Lozanski, E., Rousseau, C.: Wining Solutions. Springer, New York (1996)

Z. Cvetkovski, *Inequalities*,

DOI 10.1007/978-3-642-23792-8, © Springer-Verlag Berlin Heidelberg 2012

22. Малчески, Р., Малчески, А.: Избрани содржини од елементарна математика. СДМИ, Скопје (1994) (Malcheski, R., Malcheski, A.: Selected Content from Elementary Mathematics, SDMI, Skopje 1994)
23. MathLinks: http://www.mathlinks.ro
24. Matic, I.: Classical Inequalities. The IMO Compendium Group (2007)
25. Номировский, Д: Неравенство Караматы. Квант **4**, 43–45 (2000). (Nomirovski, D.: Karamata's inequality. Quant **4**, 43–45 (2000))
26. Прасолов, В.В.: Задачи по алгебре, арифметике и анализу (2005) (Prasolov, V.V.: Problems in Algebra, Arithmetic and Analysis, 2005)
27. Puong, T.: Diamonds in Mathematical Inequalities. HaNoi Publishing House (2007)
28. Седракян, Н.: О применении одного неравенства. Квант **2**, 42–44 (1997). (Sedrakian, N.: For the usage of recent inequality. Quant **2**, 42–44 (1997))
29. Shklarsky, D.O., Chentzov, N.N., Yaglom, I.M.: The USSR Olympiad Problem Book. Dover, New York (1994)
30. Slinko, A.: USSR Mathematical Olympiads 1989–1992. Australian Mathematical Trust, Belconnen (1997)
31. Verdiyan, V., Salas, D.C.: Simple trigonometric substitutions with broad result. Math. Reflect. **6**, 1–12 (2007)
32. Vo Quoc, B.: On a class of three-variable inequalities. Math. Reflect. **2**, 1–8 (2007)